The Running Athlete

Gian Luigi Canata • Henrique Jones
Werner Krutsch • Patricia Thoreux
Alberto Vascellari

Editors

The Running Athlete

A Comprehensive Overview of Running in Different Sports

ESSKA

Editors
Gian Luigi Canata
Centre of Sports Traumatology
Koelliker Hospital
Torino, Italy

Werner Krutsch
Sport Docs Franken
Nürnberg, Germany

Alberto Vascellari
Policlinico Città di Udine
Udine, Italy

Henrique Jones
Orthopedic and Sports Medicine Clinic
Montijo, Portugal

Patricia Thoreux
Centre d'Investigations en Médecine
du Sport
Hôpital Hôtel Dieu – APHP
Paris, France

ISBN 978-3-662-65066-0   ISBN 978-3-662-65064-6   (eBook)
https://doi.org/10.1007/978-3-662-65064-6

This Springer imprint is published by the registered company Springer-Verlag GmbH, DE part of Springer Nature.
The registered company address is: Heidelberger Platz 3, 14197 Berlin, Germany

Foreword

Since *Homo sapiens* has become bipedal, he developed a plantigrade support—as the only animal the bear to do so—and a powerful propulsion system involving a highly specialized foot for bipedalism and the Achilles tendon. This allows humans to walk rapidly and to increase their speed by "flying" over the ground. Running was certainly an early on mode of displacement for the hunters/gatherers, and since then allows the human being to escape dangers, play, compete, and have fun. Hence, running was from the beginning a natural way to move that permitted humans to develop themselves throughout history.

From the hunters/gatherers to the current ultra-trailers, running has always been a way to challenge themselves allowing men and women to compete in all modes and at all levels. Indeed, running is involved in almost all sports disciplines, alone or in combination with countless movements, as well as in training endurance and mobility. This is the originality of this book that not only includes the classical chapters devoted to running such as biomechanics, injury diagnosis and treatment, and prevention. This book further addresses all the specificity and peculiarities of running encountered in many different sports and activities, also looking at technique, mechanical and training load, as well as at the consequence of running misuse and overuse. Authors must be commended for this original and interesting initiative. It is a novel and magnificent book for all of us involved in sports medicine.

Jacques Menetrey
Orthopaedic Surgery Service
University Hospital of Geneva
Geneva, Switzerland

Acknowledgements

We wish to express our gratitude to:
    ESSKA who approved and made this project possible,
    the ESMA Board and the ESSKA team for their great support,
    Springer for the excellent editorial job,
    Catena Cottone who coordinated the various authors,
    All the renowned experts who accepted to write their chapter with invaluable cooperation.

Gian Luigi Canata, Torino, Italy
Henrique Jones, Montijo, Portugal
Werner Krutsch, Regensburg, Germany
Patricia Thoreux, Paris, France
Alberto Vascellari, Treviso, Italy

Contents

Part II Specific Aspects

Introduction

Our ancestors started running to escape from enemies or to hunt animals a long time ago. They found that running with an aerial phase without ground contact allowed a higher speed. Gradually, the running techniques improved allowing mankind to cover even long distances with an adequate training. This is the origin of the running athlete now engaged in a great variety of sports. Most sports activities require running abilities except for aquatic sports.

Every sport involving running has specific characteristics, technical peculiarities, and training differences. The aim of this publication is to give the reader a full insight into these surprisingly different running aspects. Even athletics, where running is the only reason to compete, presents a huge variety of disciplines covering different distances with different speeds, technical aspects, and training principles. This is coupled with different biomechanical and physiological requirements allowing athletes with inborn qualities to better compete at top level than others with different qualities so that selection at the beginning of training is an important means to address each young athlete to the most compliant activity.

In this book, the running aspects of each sport are deeply covered including technique, training, risk of injuries, and preventive measures. A proper running technique from foot strike to the swing phase with comfortable wear and shoes, an appropriate training with optimized workloads, and periodical screenings are the best ways to compete safely with cardiovascular, metabolic, and mental benefits.

Torino, Italy   Gian Luigi Canata

Part I

General Aspects

Biomechanics of Running

1

Karsten Hollander, Tim Hoenig,
and Pascal Edouard

1.1 Introduction

1.1.1 Definition of Biomechanics

Biomechanics is an interdisciplinary scientific field investigating the motion (mechanics) of living systems (bio). While there is no universal definition, biomechanics can be seen as the application of mechanical principles to biological systems, biological tissue, and medical problems [1]. Herewith, it is a scientific discipline working at the interface of scientific disciplines such as sports and exercise science, engineering, data science, physics, athletic training and many more as well as clinical disciplines such as orthopaedics, physiotherapy, sports and exercise medicine, physical medicine and rehabilitation, physiatry, osteopathy, or chiropractic. Biomechanics can be divided into two subcategories: kinematics as corresponding to the analysis of motion and dynamics (statics and kinetics) as corresponding to the analysis of the underlying causes of motion.

1.1.2 Musculoskeletal System: Passive and Active Structures

One of the aims of biomechanics is to better understand the musculoskeletal system and its implication in motion in order to develop methods to prevent or rehabilitate health problems such as injuries or to improve the athlete's performance. The musculoskeletal system can be divided into passive and active structures [1]. The passive musculoskeletal system includes bones, ligaments and joints, while the active musculoskeletal system consists of muscles and tendons. Bones and ligaments transmit forces and are responsible for the statics of the musculoskeletal system. Joints are special connections of bones and ligaments allowing movement in predefined directions and range of motion. Joints are moved by muscles that are able to contract and actively change their length to transmit forces by their corresponding tendons. Skeletal muscles are attached to a bone and typically span at least one joint. Muscles can

K. Hollander (✉)
Institute of Interdisciplinary Exercise Science and Sports Medicine, MSH Medical School Hamburg, Hamburg, Germany
e-mail: karsten.hollander@medicalschool-hamburg.de

T. Hoenig
Department of Trauma and Orthopaedic Surgery, University Medical Center Hamburg-Eppendorf, Hamburg, Germany
e-mail: t.hoenig@uke.de

P. Edouard
Inter-University Laboratory of Human Movement Science (LIBM EA 7424), University of Lyon, University Jean Monnet, Saint Etienne, France

Department of Clinical and Exercise Physiology, Sports Medicine Unit, University Hospital of Saint-Etienne, Faculty of Medicine, Saint-Etienne, France
e-mail: pascal.edouard@univ-st-etienne.fr

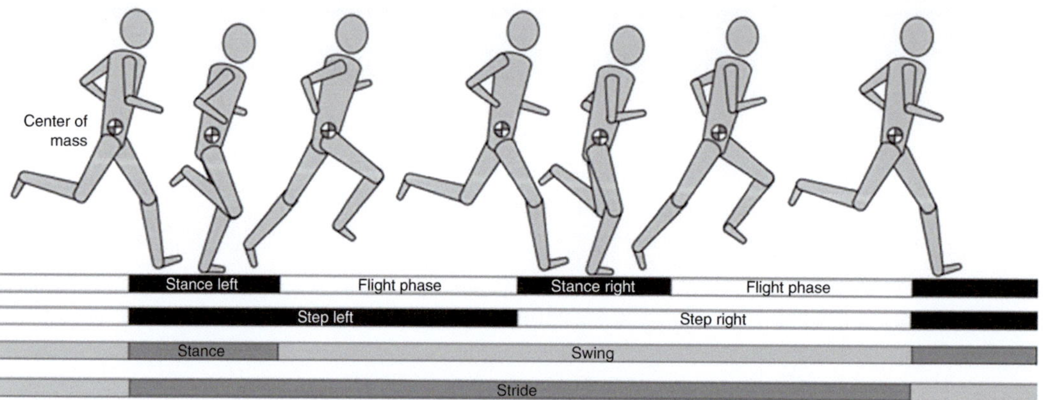

Fig. 1.1 Different phases of the running gait cycle. (Figure from Stöggl T., Wunsch T. (2016) Biomechanics of Marathon Running. In: Zinner C., Sperlich B. (eds) Marathon Running: Physiology, Psychology, Nutrition and Training Aspects. Springer, Cham. https://doi.org/10.1007/978-3-319-29728-6_2 [7])

actively reduce their length and have a specific function on a joint (agonist). An antagonist muscle with the opposite function on a joint (e.g. flexor vs. extensor) is needed in order to extent the length of an agonist muscle, and/or to counteract/control the action of the agonist muscle.

1.1.3 Duties of Biomechanics

There are different clinical or scientific questions that can be answered by the interdisciplinary field of biomechanics. Typically, biomechanical assessment of a healthy or injured athlete can be categorized into separate components such as motion and corresponding loads and forces [2]. These analyses can be used to detect potential injury mechanisms or to guide rehabilitation processes in the injured athlete [3]. Furthermore, biomechanics are used to optimize performance or equipment in a sports-related setting. For example, footwear has become a field of interest for clinical or sports science research with the motivation to reduce injury risk or to improve performance [4–6].

1.1.4 Running Biomechanical Analysis

Biomechanical analyses are frequently used in the evaluation of the individual running tech-

nique [2]. Running is a cyclic movement and differs from walking due to the presence of a flight phase in which no foot touches the ground. In a series of phases, a running cycle is defined from one foot ground contact to the next foot ground contact of the same foot. Herein, the running cycle can be differentiated into a stance and swing phase (Fig. 1.1). The stance phase consists of an initial foot strike, followed by a braking and a midstance phase and concluded by a take-off phase. The swing phase starts after the take-off phase and is divided into initial, mid and terminal swing phases. During these different phases, there are specific needs for the passive and active structures of the musculoskeletal system and a biomechanical analysis is determined to dissect different phases of the running cycle. Then, unnormal or changing biomechanical patterns can be detected and their relevance in the assessment of injury risk derived.

1.2 Brief History of Biomechanical Analysis

In order to investigate the musculoskeletal system and motion, various measurement methods can be used. Even though biomechanics is a relatively young scientific discipline, already Aristotle is said to have observed the movement patterns of Olympic athletes with his eyes in order to determine superordinate patterns in 350 B.C.E. The Italian polymath Leonardo da Vinci also wondered about biomechanics by observation which, for example, led to his drawing of the Vitruvian Man (~1490). The development of biomechanics was further influenced by the English natural philosopher Isaac Newton and his books *"Mathematical Principles of Natural Philosophy"* and *"Principia"* in which the Newtonian laws of motion and gravitation were formulated [1]. However, it lasted until the 1870s until the first biomechanical recording was carried out. The French physician Etienne-Jules Marrey, along with photographers Ottomar Anschütz, Albert Londe and Eadweard Muybridge, developed the single image photography for the reconstruction of motion sequences. This technical innovation enabled a serial recording and elements of the movement that were not visible to the human eye could be made visible [8]. However, the term biomechanics did not develop as an independent scientific field until the 1960s. In August 1967, the first international scientific conference was held in Switzerland and in 1973 the International Society of Biomechanics was founded. Since then, several technical advances enabled increasingly accurate measurements of biomechanics.

Nowadays, motion capture systems can be used to measure human movement (kinematics); and instrumented treadmills force plates and insoles are used to measure pressures and forces (kinetics). Motion capture systems use several optical cameras and can track retro-reflective markers which then transfer human movements via triangulation into 3D models generated in the computer (Fig. 1.2). Pressure and force plates are frequently used in running biomechanics analysis and are either embedded into the ground or into an instrumented treadmill. Pressure and force data can then be captured and ground reaction forces calculated. These ground reaction forces can be used to determine joint load when combined with motion data. Furthermore, muscle activity can be measured via electromyography (EMG). In general, surface electrodes are used to measure electrical muscle activity which then can be graphically displayed in the corresponding gait cycle.

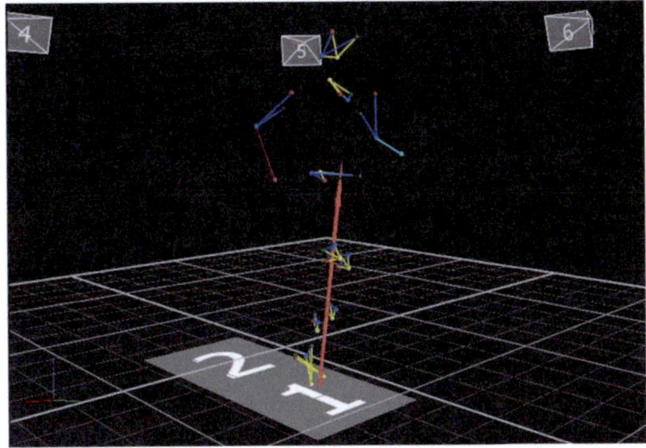

Fig. 1.2 Example of a 3D motion capture system

While these investigations are typically performed in a highly specialized laboratory setting, only recently, portable sensors ("wearables") have improved measurement accuracy and thus gained in popularity to obtain information on individual biomechanics outside a biomechanics laboratory. The latest technical innovation in the field of artificial intelligence is a promising field for markerless motion tracking and offers many practical application possibilities [9].

All these methods can be used to analyse motion (kinematics) or the occurrence of forces (kinetics) which are further described in the next chapter.

1.3 Inside the Lab: Three-Dimensional Biomechanical Analysis of Running

1.3.1 Kinematics

In kinematics, the movement of points is analysed in a three-dimensional space. These points are evaluated in terms of their position in space, speed and acceleration. When viewed together the data can for example be used to make statements about joint angles. While elaborate systems with infrared cameras and special reflective markers are usually used for this purpose, normal high-speed cameras coupled with machine learning systems can also make good approximations of these values [9–11]. Kinematic variables are measured without the causes of their motion, i.e.

the forces that occur and are based on the relationship of displacement, velocity and acceleration vectors. Joint angles can be determined at specific time points of the running gait cycle (e.g. sagittal ankle angle at foot strike) or maxima or minima can be calculated (e.g. maximum sagittal knee angle during stance phase) (Fig. 1.3). While sagittal plane kinematics are most frequently investigated [13], also frontal and transverse (horizontal) plane angles can be determined.

A frequently investigated variable in running biomechanics is the foot strike pattern [14, 15]. The foot strike can be divided into a forefoot, midfoot or rearfoot strike pattern and can be actively changed or influenced by footwear or gait retraining [16–19]. It is generally thought that the forefoot pattern may be more efficient and allow for better running performance. Reasons for this may be that sprinting is seen as the model for fast running. Furthermore, when looking at elite track runners, an analysis from the men's 10,000 m final race during the World Athletics Championships in London 2017 showed that at no point during the race a rearfoot strike was used [20]. For longer distances and outside of competitions, the rearfoot strike is the most prevalent [21, 22]. Regarding injuries, a recent study from a biomechanics laboratory showed an association of Achilles tendon and posterior calf injuries with particular foot strike patterns (midfoot strike and Achilles tendon; forefoot strike and posterior calf injuries) [23]. Therefore, the implication of foot strike pattern is widely debated regarding performance [20] and injury occurrence [24] without any consensus on whether one-foot strike pattern is better than another [25, 26].

When looking at the foot movement, in addition to assessing sagittal foot strike behaviour, supination and pronation must also be considered. Supination is when the outer edge of the foot is lowered ("bending outward") while pronation causes the inner edge of the foot to be lowered ("bending inward"). For many years, these movements have been suspected of being associated with running-related injuries, and numerous manufacturers of running footwear have developed shoe models to counteract this movement. However, no such association of supination and pronation with running-related injuries has been proven yet. Interestingly, the opposite may actually be the case. In a study of over 1800 runners,

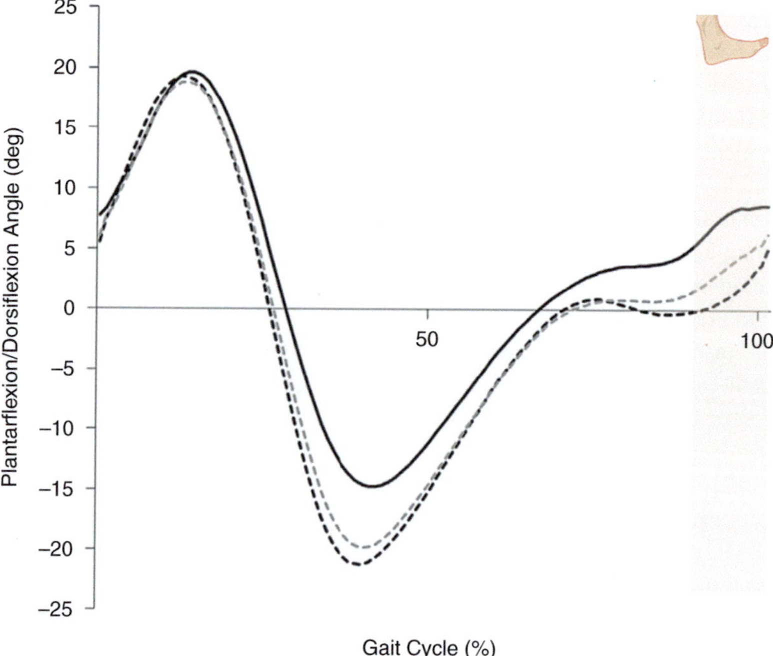

Fig. 1.3 Example graphs for different sagittal ankle angles (adapted from [12])

it was shown that those with a pronated foot position even had lower injury rates [27]. Only very extreme movements in the ankle joint (overpronation or oversupination) may be associated with certain injuries, such as bone stress injuries or Achilles tendinopathies [28, 29].

1.3.2 Kinetics

For the measurement of forces (kinetics), special force plates are used which are typically either embedded in the floor or installed under a treadmill. These force plates record the occurring forces with a high sampling frequency (typically at least 1000 Hz) and can be used to calculate the forces and mass, but also torques and impulses. Together with the kinematic variables, kinetic data can be used to calculate joint moments to assess an approximation of the loading on individual structures.

Ground reaction forces are the forces generated by the body colliding with the ground. Especially, *vertical* ground reaction forces are frequently investigated in running biomechanics. Here, the morphology (one or two peaks) or the maximum value can be investigated. A variable of increasing interest is the vertical load rate, reflected as the increment of the first part of the ground reaction force curve (Fig. 1.4) [30]. The load rate is measured as force/body weights per

time and a high value is thought to be stress for the body [30]. One randomized controlled study has shown that a gait retraining aiming to reduce vertical load rates has the potential to reduce occurrence of injuries [31]. While some studies have found associations between high load rates and certain injury pattern [32], no conclusive evidence is available today [3, 33]. However, vertical load rates can be altered by cadence, footwear and gait retraining and therefore are of interest for the development of injury prevention strategies [34, 35].

For running at maximal or close to maximal velocity (i.e. acceleration or maximal sprint), the *horizontal* ground reaction forces are frequently investigated. Indeed, from a sports performance point of view, sprint acceleration performance has been shown to be associated with the ability to produce and apply high levels of force in the horizontal direction over the entire acceleration [36]. And from a health/injury point of view, a clear decrease in horizontal force production at low velocities has been reported at the time of return to football after hamstring injury [37, 38].

1.3.3 Temporospatial

In addition, temporospatial parameters such as stride length, cadence (stride frequency) and ground contact times can be determined in a

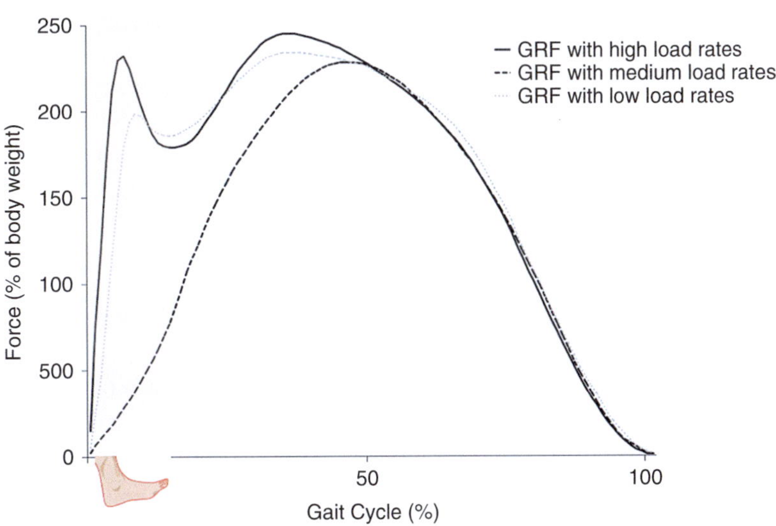

Fig. 1.4 Examples of vertical ground reaction force (GRF) curves with high, medium and low load rates

biomechanical analysis. The running performance (velocity) is determined from the product of the space-time parameters stride length and cadence. Stride length is defined as the distance from one ground contact to the next ground contact of the other foot. Cadence is the number of steps in a given time, typically per minute. Both are dependent on the forces that occur and on the ground contact and flight times [39].

For running biomechanics, cadence is a temporospatial variable which is in the focus of research. For example, a study of U.S. high school runners showed that a higher stride rate (above 174 to 178 strides per minute) was associated with a reduced number of shin injuries [40]. Furthermore, a very recent prospective study has shown that low cadence is an important predictor for injuries, more specifically bone stress injuries in collegiate runners [40]. With this in consideration, the potential of increasing cadence to reduce running-related injuries is frequently discussed. Cadence can be modified by gait retraining and is a possible target for reducing injury rates.

1.4 Outside the Lab: Wearables and New Technologies

1.4.1 Inertial Measurement Units (IMUs)

Three-dimensional optoelectronic biomechanical analysis was (and is) considered the gold standard to capture an athlete's running pattern [41]. More recently, however, scholars and clinicians have focused on wearable technology in the evaluation of running biomechanics [9, 42, 43]. Not only does it provide an accurate, effective and less expensive alternative to camera-based systems, but it also allows for assessing human movement in a real-world "out-of-the-lab" setting [44]. In particular, inertial measurement units (IMUs) are regarded as a promising tool to study the biomechanics of running. IMUs are typically devices which combine accelerometers, gyroscopes and magnetometers. Although IMUs do not provide direct in-vivo measurements, they may indirectly represent internal stress distribution and tissue strain [45, 46]. Thus, special attention was given to the accelerometry system as causal links between accelerometry-based loading and overuse injuries were assumed [47]. During foot strike, ground reaction forces are generated with consecutive deceleration and acceleration of the body. A variety of mechanisms modify the impact shock, and measuring of accelerations at the tibia or other anatomical sites could offer opportunities for performance enhancement, injury prevention and gait retraining [45, 48, 49]. Due to advances in technology, IMUs nowadays allow for much more than acceleration measures. Among others, IMUs have been successfully used for the assessment of spatiotemporal parameters and 3D motion capture.

1.4.2 Pressure Insoles

As mentioned above, assessment of musculoskeletal loading during foot strike has been mainly based on force plates and instrumented treadmills. To allow for out-of-the lab measurements, pressure distributions of the foot can be determined by wearable shoe insoles. For this purpose, multiple sensors located on varying positions over the entire sole measure and estimate biomechanical parameters including ground reaction forces and centre of pressure motion. Although much effort has been spent on investigating validity, reliability and usefulness of these devices, many research questions are yet to be answered [50]. Nevertheless, clinical application is making enormous progress and insole technology is increasingly used in sports science and rehabilitation.

1.4.3 Smartphone Apps

Information on running biomechanics collected in laboratory settings are of great interest from a performance or injury risk reduction perspective.

However, such measurements are not accessible for all athletes. To this aim, some researchers developed simple methods for field applications in order to measure or estimate several aspects of running biomechanics, such as lower limb stiffness, foot strike patterns, force-velocity profiles and joint angle estimations [51]. Such measurements or estimations are now possible by only using smartphone applications [52–54], which makes evaluation of some aspects of the running biomechanics easily accessible for the end-users.

1.5 New Analysis Methods and Future Areas of Research

1.5.1 Finite Element Analysis

One of the main disadvantages of the assessment methods outlined above is the lack of direct information on internal tissue loading (stress) and tissue deformation (strain). Finite element analysis is a powerful opportunity to evaluate mechanical characteristics that cannot be measured in vivo [55]. In sports science, it is used as a computational method for simulating the behaviour of tissue under biomechanical loading. At first, a three-dimensional model of the human body or parts of it are modelled by using data from magnetic resonance imaging or computed tomography. Generally speaking, the anatomical structures are then partitioned into "finite elements". The tissue components are then combined in a computational model and data on key tissue properties (e.g. elasticity) are added. Once finished, external forces such as human running can be applied to the computational model and the resulting stresses and strains can be studied. Recently, finite element analysis has been successfully used to study stress and strain of anatomical structures that are related to running-related injuries (e.g. metatarsals, plantar fascia) [56–58]. As "overuse" is a key risk factor for most running-related injuries, finite element analysis can be seen as a promising approach for better injury prevention [42].

1.5.2 Machine Learning

Machine Learning is a subfield of artificial intelligence, and its general idea is to develop a prediction algorithm (or model) based on learning from a labelled database [59]. Such machine learning algorithms are able to detect features using classification methods or a score using regression methods as an output from an input data, based on the learning from the labelled database. In addition, it is possible to determine the important features corresponding to the weight of each parameter in the algorithm, as well as the uncertainty of this prediction (i.e. performance of the algorithm). Such an analytical approach allows a complementary option in addition to classical statistics to help in the understanding of running biomechanics and its relationships with performance or injury [9, 46, 60, 61]. Furthermore, machine learning has successfully been implemented into kinematic analysis [10]. However, despite ongoing developments, machine learning cannot *yet* substitute the conventional full biomechanical analysis including kinetics [11].

1.6 Conclusions

The interdisciplinary field of biomechanics offers a scientific examination of motion in living systems. In running biomechanics, precise analyses of kinematics, kinetics and temporospatial variables offer a foundation to improve performance and reduce injuries by detecting potentially injurious mechanics. As biomechanical variables do interrelate and are influenced manifoldly (speed, footwear, running surface, etc.), biomechanics cannot simply be divided into "good" and "bad" but need a holistic assessment approach. While the gold standard is an assessment in a controlled laboratory setting, recent developments have increased the options to measure relevant biomechanics in the field. Wearables and new computer algorithms using artificial intelligence have made the biomechanical analysis more feasible and relevant than ever in optimizing performance and injury prevention in different sports.

References

1. Richard HA, Kullmer G. Biomechanik. 2013.
2. Nicola TL, Jewison DJ. The anatomy and biomechanics of running. Clin Sports Med. 2012;31(2):187–201.
3. Ceyssens L, et al. Biomechanical risk factors associated with running-related injuries: a systematic review. Sports Med. 2019;49(7):1095–115.
4. Hoogkamer W, et al. A comparison of the energetic cost of running in Marathon racing shoes. Sports Med. 2018;48(4):1009–19.
5. Hoogkamer W, Kipp S, Kram R. The biomechanics of competitive male runners in three Marathon racing shoes: a randomized crossover study. Sports Med. 2019;49(1):133–43.
6. Lindlein K, et al. Improving running economy by transitioning to minimalist footwear: a randomised controlled trial. J Sci Med Sport. 2018;21(12):1298–303.
7. Stöggl T, Wunsch T. Biomechanics of Marathon running. In: Zinner C, Sperlich B, editors. Marathon running: physiology, psychology, nutrition and training aspects. Cham: Springer International Publishing; 2016. p. 13–45.
8. Gao L, et al. Single-shot compressed ultrafast photography at one hundred billion frames per second. Nature. 2014;516(7529):74–7.
9. Hollander K. Biomechanik des Laufens—Implikationen für laufbedingte Verletzungen und zukünftige Forschungsfelder. Deutsche Zeitschrift für Sportmedizin. 2020;71(3):53–4.
10. Cronin NJ, et al. Markerless 2D kinematic analysis of underwater running: a deep learning approach. J Biomech. 2019;87:75–82.
11. Cronin NJ. Using deep neural networks for kinematic analysis: challenges and opportunities. J Biomech. 2021;123:110460.
12. Hollander K, et al. Effects of footwear on treadmill running biomechanics in preadolescent children. Gait Posture. 2014;40(3):381–5.
13. Roberts M, Mongeon D, Prince F. Biomechanical parameters for gait analysis: a systematic review of healthy human gait. Phys Ther Rehab. 2017;4(1):6.
14. Almeida MO, Davis IS, Lopes AD. Biomechanical differences of foot-strike patterns during running: a systematic review with meta-analysis. J Orthop Sports Phys Ther. 2015;45(10):738–55.
15. Hoenig T, Rolvien T, Hollander K. Footstrike patterns in runners: concepts, classifications, techniques, and implications for running-related injuries. Deutsche Zeitschrift für Sportmedizin. 2020;71(3):55–61.
16. Davis IS, Rice HM, Wearing SC. Why forefoot striking in minimal shoes might positively change the course of running injuries. J Sport Health Sci. 2017;6(2):154–61.
17. Squadrone R, et al. Acute effect of different minimalist shoes on foot strike pattern and kinematics in rearfoot strikers during running. J Sports Sci. 2015;33(11):1196–204.
18. Hollander K, et al. Comparison of minimalist footwear strategies for simulating barefoot running: a randomized crossover study. PLoS One. 2015;10(5):e0125880.
19. Hollander K, et al. Long-term effects of habitual barefoot running and walking: a systematic review. Med Sci Sports Exerc. 2017;49(4):752–62.
20. Hanley B, et al. Footstrike patterns and race performance in the 2017 IAAF world championship men's 10,000 m final. Sports Biomech. 2021:1–10.
21. de Almeida MO, et al. Is the rearfoot pattern the most frequently foot strike pattern among recreational shod distance runners? Phys Ther Sport. 2015;16(1):29–33.
22. Hasegawa H, Yamauchi T, Kraemer WJ. Foot strike patterns of runners at the 15-km point during an elite-level half marathon. J Strength Cond Res. 2007;21(3):888–93.
23. Hollander K, et al. Multifactorial determinants of running injury locations in 550 injured recreational runners. Med Sci Sports Exerc. 2021;53(1):102–7.
24. Yong JR, et al. Acute changes in foot strike pattern and cadence affect running parameters associated with tibial stress fractures. J Biomech. 2018;76:1–7.
25. Hamill J, Gruber AH. Is changing footstrike pattern beneficial to runners? J Sport Health Sci. 2017;6(2):146–53.
26. Anderson LM, et al. What are the benefits and risks associated with changing foot strike pattern during running? A systematic review and meta-analysis of injury, running economy, and biomechanics. Sports Med. 2020;50(5):885–917.
27. Nigg BM, et al. Running shoes and running injuries: mythbusting and a proposal for two new paradigms: 'preferred movement path' and 'comfort filter'. Br J Sports Med. 2015;49(20):1290–4.
28. Millar NL, et al. Tendinopathy. Nature Reviews Disease Primers. 2021;7:1. https://doi.org/10.1038/s41572-020-00234-1
29. Hoenig T, et al. Bone stress injury. Nature Reviews Disease Primers. 2022. Epub, ahead of print.
30. Futrell EE, et al. Relationships between habitual cadence, footstrike, and vertical load rates in runners. Med Sci Sports Exerc. 2018;50(9):1837–41.
31. Chan ZYS, et al. Gait retraining for the reduction of injury occurrence in novice distance runners: 1-year follow-up of a randomized controlled trial. Am J Sports Med. 2018;46(2):388–95.
32. Davis IS, Bowser BJ, Mullineaux DR. Greater vertical impact loading in female runners with medically diagnosed injuries: a prospective investigation. Br J Sports Med. 2016;50(14):887–92.
33. van der Worp H, Vrielink JW, Bredeweg SW. Do runners who suffer injuries have higher vertical ground reaction forces than those who remain injury-free? A systematic review and meta-analysis. Br J Sports Med. 2016;50(8):450–7.

34. Rice HM, Jamison ST, Davis IS. Footwear matters: influence of footwear and foot strike on load rates during running. Med Sci Sports Exerc. 2016;48(12):2462–8.

35. Hollander K, et al. Adaptation of running biomechanics to repeated barefoot running: a randomized controlled study. Am J Sports Med. 2019;47(8):1975–83.

36. Morin JB, et al. Mechanical determinants of 100-m sprint running performance. Eur J Appl Physiol. 2012;112(11):3921–30.

37. Mendiguchia J, et al. Progression of mechanical properties during on-field sprint running after returning to sports from a hamstring muscle injury in soccer players. Int J Sports Med. 2014;35(8):690–5.

38. Mendiguchia J, et al. Field monitoring of sprinting power-force-velocity profile before, during and after hamstring injury: two case reports. J Sports Sci. 2016;34(6):535–41.

39. van Oeveren BT, et al. The biomechanics of running and running styles: a synthesis. Sports Biomech. 2021:1–39.

40. Luedke LE, et al. Influence of step rate on shin injury and anterior knee pain in high school runners. Med Sci Sports Exerc. 2016;48(7):1244–50.

41. Cuesta-Vargas AI, Galan-Mercant A, Williams JM. The use of inertial sensors system for human motion analysis. Phys Ther Rev. 2010;15(6):462–73.

42. Reenalda J, et al. Continuous three dimensional analysis of running mechanics during a marathon by means of inertial magnetic measurement units to objectify changes in running mechanics. J Biomech. 2016;49(14):3362–7.

43. Hoenig T, et al. Analysis of running stability during 5000 m running. Eur J Sport Sci. 2019;19(4):413–21.

44. Renggli D, et al. Wearable inertial measurement units for assessing gait in real-world environments. Front Physiol. 2020;11:90.

45. Tenforde AS, et al. Tibial acceleration measured from wearable sensors is associated with loading rates in injured runners. PM R. 2020;12(7):679–84.

46. Rahlf AL, et al. A machine learning approach to identify risk factors for running-related injuries: study protocol for a prospective longitudinal cohort trial. BMC Sports Science, Medicine and Rehabilitation. 2022. Epub, ahead of print.

47. Crowell HP, Davis IS. Gait retraining to reduce lower extremity loading in runners. Clin Biomech (Bristol, Avon). 2011;26(1):78–83.

48. Johnson CD, et al. Comparison of tibial shock during treadmill and real-world running. Med Sci Sports Exerc. 2020;52(7):1557–62.

49. Sheerin KR, Reid D, Besier TF. The measurement of tibial acceleration in runners—a review of the factors that can affect tibial acceleration during running and evidence-based guidelines for its use. Gait Posture. 2019;67:12–24.

50. Blazey P, Michie TV, Napier C. A narrative review of running wearable measurement system accuracy and reliability: can we make running shoe prescription objective? Footwear Sci. 2021;13(2):117–31.

51. Morin JB, Samozino P. Biomechanics of training and testing. 2018.

52. Balsalobre-Fernandez C, Agopyan H, Morin JB. The validity and reliability of an iPhone app for measuring running mechanics. J Appl Biomech. 2017;33(3):222–6.

53. Romero-Franco N, et al. Sprint performance and mechanical outputs computed with an iPhone app: comparison with existing reference methods. Eur J Sport Sci. 2017;17(4):386–92.

54. Rodríguez SM. Runmatic for running mechanics assessment (mobile app user guide). Br J Sports Med. 2018;52(2):139–40.

55. Zienkiewicz OC, et al. The finite element method, vol. 3. London: McGraw-Hill; 1977.

56. Gu YD, et al. Computer simulation of stress distribution in the metatarsals at different inversion landing angles using the finite element method. Int Orthop. 2010;34(5):669–76.

57. Chen TL, et al. Prediction on the plantar fascia strain offload upon fascia taping and low-dye taping during running. J Orthop Transl. 2020;20:113–21.

58. Chen TL-W, et al. Plantar fascia loading at different running speed: a dynamic finite element model prediction. HKIE Trans. 2021;28(1):14–21.

59. Edouard P, Verhagen E, Navarro L. Machine learning analyses can be of interest to estimate the risk of injury in sports injury and rehabilitation. Ann Phys Rehabil Med. 2020:101431.

60. Claudino JG, et al. Current approaches to the use of artificial intelligence for injury risk assessment and performance prediction in team sports: a systematic review. Sports Med Open. 2019;5(1):28.

61. Casals M, Finch CF. Sports biostatistician: a critical member of all sports science and medicine teams for injury prevention. Inj Prev. 2017;23(6):423–7.

Physiology of Running: Sprint and Short Distances

2

Pascal Edouard, Jean-Benoît Morin, Karsten Hollander, and François Lhuissier

2.1 Introduction

Sprint running consists in covering a given distance in the shortest time possible (or the largest distance in a given time of action). This action is one of the key performance determinants in many sports (e.g., track and field events or team sports). The determinants of sprint running performance have been widely investigated in scientific studies with practical implications for improving them [1]. Most of these studies explored the biomechanical aspects of sprint running performance (e.g., kinematics, kinetics, and temporo-spatial parameters) (see Chap. 1), but very few explored the physiological component related to energy expenditure [1]. This could be in part due to experimental and methodological challenges [1, 2], but also due to the fact that effort duration is so short that it has been hypothesized that performance limitation factors were likely not energetics but rather biomechanical. Since energetics always play a role in energy bases for muscle action and in turn human motion and sports, there is a need to better understand this aspect specifically in the sprint running setting.

2.2 What Are the Main Energy Source of Sprint Running?

In team sports such as football, rugby, basket, or handball, sprint running is performed on a distance from 0 to ~60–80 m and lasts less than 15 s. In track and field, sprints are considered the distances from 50 to 400 m (https://www.iaaf.org/

P. Edouard (✉)
Inter-University Laboratory of Human Movement Biology, University Lyon,
UJM-Saint-Etienne, Saint Etienne, France

Department of Clinical and Exercise Physiology,
Sports Medicine Unit, University Hospital of
Saint-Etienne, Faculty of Medicine,
Saint-Etienne, France
e-mail: pascal.edouard@univ-st-etienne.fr

J.-B. Morin
Inter-University Laboratory of Human Movement Biology, University Lyon, UJM-Saint-Etienne,
Saint Etienne, France
e-mail: jean.benoit.morin@univ-st-etienne.fr

K. Hollander
Institute of Interdisciplinary Exercise Science and Sports Medicine, MSH Medical School Hamburg,
Hamburg, Germany
e-mail: karsten.hollander@medicalschool-hamburg.de

F. Lhuissier
Medecine de L'Exercice et du Sport, Hopital Jean Verdier, AP-HP, Bondy, France

UMR INSERM U1272 'Hypoxie & Poumon',
Université Sorbonne Paris Nord, Bobigny, France
e-mail: francois.lhuissier@aphp.fr

disciplines/sprints/100-metres) and last from about 6 to 60 s. Given these efforts duration, the main energy source involved in sprint running is anaerobic metabolism.

For shorter durations (<5–8 s), since immediate energy is needed, the energy is mainly provided by intramuscular high-energy phosphates [3]. Energy is released from the intramuscular energy-rich phosphates: adenosine triphosphate (ATP) and phosphocreatine (PCr) [3, 4]. This allows a higher rate of energy production through ATP (i.e., number of ATP per unit of time) than with the aerobic system, allowing attaining high power outputs [3, 4]. However, the limited storage of ATP and PCr leads to limitation in the duration of such very intense (maximal) efforts.

For longer durations (up to 2–3 min depending on individual capabilities), other sources of energy are needed to maintain high levels of power output. An alternative is the lactic acid system via rapid anaerobic glycolysis that allows rapid regeneration of ATP [3]. This provides a higher rate of energy through ATP than the aerobic system, but a lower rate of energy than with high-energy phosphates and with the concurrent production of pyruvate [3]. In anaerobic glycolysis, pyruvate is converted to lactate, which inhibits an important enzyme of the glycolysis (phosphofructokinase) [3]. Consequently, with increased effort duration, it is not energetically possible to maintain the same level of power output. And such metabolic approach cannot be maintained or repeated because of a limiting factor such as acidosis.

2.3 What Is the Part of Energy in the Determinants of Sprint Running Performance?

In long-duration efforts, such as long-distance running or endurance exercises, the capability of the organism to provide energy and the higher rate of energy throughout the efforts is one of the main determinants of the performance [5]. The

question of the importance of energy supply for sprint running performance has been addressed by Bundle and Weyand in a narrative review concluding that "for burst-type sprints that last only a few seconds, a wealth of data spanning multiple levels of biological organization is fully consistent in indicating that the availability of metabolic power neither determines nor directly limits performance" and performance "predominantly reflect musculoskeletal function" [5].

With increased effort duration, the metabolic part in the performance production increases. The decrease in available energy per unit of time induced by the change from high-energy phosphates to anaerobic glycolysis leads to impairment in the whole-body musculoskeletal performance, and consequently sprint running performance [5]. This has been represented by Bundle and Weyand [5] in a performance-duration curve describing the relative contribution of more mechanical and more metabolical components. They also provided "three basic conclusions regarding sprint running performance:

1. The view that brief all-out exercise performance is directly limited by rates of chemical energy provision to the contractile machinery in skeletal muscle is no longer supportable.
2. The metabolic energy released during sprinting is demand-driven and not supply-limited.
3. Sprint exercise performance is determined by the application of musculoskeletal forces with a duration dependency dictated by how rapidly these forces are compromised by rates of fatigue in vivo" [5].

For sprint running performance, metabolic energy is thus not a major limiting factor. Contrary to endurance exercises in which metabolism predominantly determines performance, for sprint running performance Bundle and Weyand [5] concluded that "the intensity of the mechanical activity that the musculoskeletal system can transiently achieve determines the quantities of metabolic energy released and the level of performance attained".

2.4 What Is the Energy Cost of Sprint Running?

Although energy supply is not a major limiting factor of sprint running performance [5], it is of interest to better understand the influencing factors of the energy cost of sprint running. However, measurements or estimations are challenging due to experimental limitations [1, 2]. Therefore, di Prampero et al. [2] developed an innovative approach based on acceleration biomechanics, by considering sprint running acceleration as mechanically and energetically equivalent to constant-speed uphill running. This modelling energetic approach allowed them to calculate energy cost and metabolic power during sprint running acceleration based on speed, ground speed, and equivalent body mass and equivalent slope (calculated from body mass, gravity, acceleration, and distance) [1, 2]. Using this innovative method [2], di Prampero et al. [1, 6] calculated the mean energy cost of a 100-m sprint in medium-level sprinters 10.7 ± 0.6 J/kg/m and during the 100-m World Record 18.4 J/kg/m, as well as the metabolic power of a 100-m sprint in medium-level sprinters 61.0 ± 4.66 J/kg/m and during the 100-m World Record 105.0 J/kg/m. They also were able to calculate the mean and peak energy costs and metabolic power during sprints running distances in Track and field from 100 to 400 m as well as during sprint running in football [1, 6] They concluded that such measurement could be an additional approach to better understand sprint running performance with practical implications in sports.

2.5 Conclusions

Sprint running consists in covering a given distance in the shortest time possible (or the largest distance in a given time of action) and is one of the key performance determinants in many sports. From a metabolic standpoint, sprint running performance determines the quantities of metabolic energy needed to achieve the corresponding time-displacement, yet those are not a major limiting factor of performance. Sprint energy cost and metabolic power could be calculated based on measurable biomechanical data, and help to better understand sprint running performance with practical implications in sports.

References

1. Morin J, Samozino S. Biomechanics of training and testing: innovative concepts and simple field methods. Springer International Publishing; 2018.
2. di Prampero PE, Fusi S, Sepulcri L, Morin J-B, Belli A, Antonutto G. Sprint running: a new energetic approach. J Exp Biol. 2005;208:2809–16.
3. McArdle WD, Katch FI, Katch VL. Exercise physiology: nutrition, energy, and human performance. 7th ed. Philadelphia: Lippincott Williams & Wilkins; 2010.
4. Spriet LL. Anaerobic metabolism in human skeletal muscle during short-term, intense activity. Can J Physiol Pharmacol. 1992;70:157–65.
5. Bundle MW, Weyand PG. Sprint exercise performance: does metabolic power matter? Exerc Sport Sci Rev. 2012;40:174–82.
6. di Prampero PE, Botter A, Osgnach C. The energy cost of sprint running and the role of metabolic power in setting top performances. Eur J Appl Physiol. 2015;115:451–69.

Training Principles

3

Renato Canova and Claudio Gaudino

3.1 Training Concept

Training can be defined as any type of repetitive activity underlying a learning concept.

Therefore, we will have a training in everyday life, referring to the technical and cognitive learning of absolutely natural actions in each individual (i.e., walking, drinking, eating, communicating through articulated words and concepts) which can be referred to as *education*, and a training more directed to the development of personal qualities, so as to lead to physiological improvements. In this case, the term *Sports Training* indicates the physiological and psychological adaptation process with the ultimate goal of improving sports performance [1–3].

In order to build an effective workout for achieving the best possible results, we must take into account the following fundamental principles:

1. The identification and classification of training means and methods, and their correlation with the race performance.
2. Personalization of the training. It must be subjective and linked to individual peculiar char-

acteristics. In fact, multiple factors differentiate athletes from each other, such as:

- Hereditary factors
- Chronological and biological age
- Age-related development
- Level of preparation depending on previous training
- Training periodization: from a general phase toward the highly specific final prematch period
- Modulation of the training load, and particularly the relationship between volume and intensity
- Variation of the stimuli
- Individual athlete involvement in the training process

The path to personal growth can be pursued only by learning the reasons underlying the programmatic choices of the coach, who, for his part, must not "order", but "explain" and "convince" the athlete.

3.2 Changes Induced by Training

Training is responsible for physiological changes so as to improve the physical qualities of the subject and maximize the athletic performance.

Besides the improvement of the running technique, the development of organic qualities plays an essential role in endurance sports training. It

R. Canova (✉)
International Running Coach, Milan, Italy

C. Gaudino
Speed and Hurdles Track and Field Coach and
Football Fitness Coach, Milan, Italy

G. L. Canata et al. (eds.), *The Running Athlete*, https://doi.org/10.1007/978-3-662-65064-6_3

aims to improve the central aerobic components (that is the ability of the heart to contract very quickly, working close to its maximum potential), as well as to enhance the use of oxygen by the muscle fibers, through an increased concentration of mitochondria and aerobic enzymes, which in practical terms is induced by a concentration of lactic acid slightly higher than the anaerobic threshold.

Finally, a better use of lactate by muscle fibers is of fundamental importance to increase the levels of lactacid power, capacity, and resistance, as a result of improving the levels of the anaerobic threshold.

The lactacid power is the capacity to produce the maximum amount of energy, resulting from the maximum level of adenosine triphosphate (TPA) by means of glycolysis in the unit of time. On a practical level, the maximum values can be obtained through a high-intensity performance, ranging between 30″ and 1′30″, depending on the characteristics of the athletes and their specialty.

The lactacid capacity consists in the capacity of producing lactic acid for a prolonged time, using it as an energy source, and it is closely linked to the power of the lactacid mechanism, since it is achieved during a close to the maximum intensity workout (not below 90%), through tests lasting between 30″ and 1′30″, interspersed with long recovery times (from 8′ to 12′).

The lactacid resistance is the ability to increase the tolerance level of lactate in the muscle fibers, more linked to the development of aerobic power than to lactacid capacity.

3.3 Training Methods and Physiological Effects Related

3.3.1 Aerobic Endurance Training

Aerobic endurance training is the first real step of the workout for middle-distance runners and cross-country skiers. Different methods are adopted:

(a) Long continuous run at uniform speed, with a variable duration ranging from 40′ to 2 h 30′
(b) Long continuous run in progression
(c) Continuous running with rhythmic variations, with an overall tempo between 40′ and 1 h 30 min, with a progressive increase in heart rate due to the higher production of lactate during the variations

3.3.2 Aerobic Power Training

Aerobic power training constitutes the critical point of the entire strategy, being the key factor leading to strong chronometric progress in all distances. Methods of execution may be:

(a) Continuous medium or short running distance carried out at a speed slightly higher than that corresponding to the lactacid threshold.

   The overall duration of the exercise varies from 20′ for athletes of medium/long distances, to 1 h for marathon runners.
(b) Continuous run with long variations, with the insertion of sections lasting from 3′ to 6′ performed at a pace higher than the threshold speed. The total duration of the work varies from 30′ to 1 h 15′, depending on the athlete's specialty.
(c) Continuous run with average variations, in which sections lasting between 1′30′ and 3′ are inserted on the basic speed, and a total duration varying from 30′ to 1 h.
(d) Continuous running with short variations, lasting between 30″ and 60″, performed at a high pace while recovery run speed is decidedly more limited.
(e) Long repeated tests, which are usually carried out on the track. The length of each single test, its speed, and the overall volume depend on the race to be prepared.
(f) Mixed repeated tests, performed on different distances, always according to the race distance.
(g) Repeated short tests, to increase the speed, and especially to improve the resistance to the specific speed.

3.3.2.1 Lactacid Resistance Training

Ways of improving lactacid endurance are similar to those used to train the aerobic power; however, a slight reduction of the kilometers performed must be applied, while slightly increasing the intensity.

The aim is to enhance the tolerance to the accumulation of lactate in the muscle fibers. This is the essential difference with both aerobic power (whose purpose is to make the elimination of lactate from the muscle faster, with a consequent raising of the threshold anaerobic), and from the lactacid capacity, whose purpose is to produce a greater quantity of lactate in the same amount of time.

3.3.2.2 Lactacid Capacity Training

Aerobic power and lactacid resistance share a wide part of training programs, whereas lactacid capacity is more connected with lactacid power, and from a methodological point of view it does not show particular analogies with lactacid resistance.

On a practical level, the difference between a power, or capacity, or lactacid endurance training session emerges from the following example, using the same distance:

- Lactacid power: 1×600 m in $1'17''$ (personal record)
- Lactate capacity: 3×600 m at 95% (equal to $1'21''$) with a $8'/10'$ recovery
- Lactacid resistance: 8×600 m at 87% (equal to $1'28''$) with a $3'/4'$ recovery
- Aerobic power: 15×600 m at 82% (equal to $1'34''$) with a $1'30''$ recovery

3.3.2.3 Lactacid Power Training

Lactacid power training consists in a single running distance between 250 and 600 m to the best capacities, in order not to accumulate large quantities of lactate.

3.3.3 Speed Training

Speed training is particularly important at a youth level, as it is connected with the improvement of nervous qualities and the ability to use fast fibers. In the adult athlete, speed training is essential for the preparation of short distances (800 m in particular), but it must also be performed by long distance runners in order to improve the characteristics enabling the final sprint.

Training the absolute speed means to focus on a limited number of repetitions, running very short distances at the maximum effort: 40 m to get the ability to accelerate, 60–80 m to train the phase launched, grouped in "series" with large recoveries.

3.3.4 Resistance Training Using Uphill Running

The aim of uphill running in training protocols is the improvement of resistance, or the ability to use a high percentage of the maximum force possessed for a prolonged time.

Multiple variables must be combined in this type of work: test duration, speed, slope, and progressive loss of the elastic reactivity.

(a) Long ascents are used, during the main period, by long-distance athletes. In this case, the slope must not be too pronounced (usually between 3 and 6%), the length between 6 and 12 km, the level of effort required is sub-maximal.

(b) Medium length ascents, with gradients ranging between 5 and 10% which last between 1 and 3 min.

   The execution takes place at high intensity, and the result is a rather high production of lactate, which requires long recovery times $(4'–6')$.

(c) Short climbs are typically used in all training periods. There are characterized by a short running distance (50–100 m as maximum) at the maximum personal speed. The slope may reach even 15%.

   The main purpose of uphill sprints is to encourage the recruitment of a high percentage of fast fibers, thus constituting an essential stimulus for the neuro-muscular system.

3.3.5 Hill Running as a Training Strategy

Running on hilly routes is characterized by uphill and downhill sections, with considerable variations in the slopes, and consequently in running speed.

This work, by enabling a continuous change in the percentages of muscle fibers utilized, helps to improve muscle strength and running in its entirety, thus reducing the energy cost required.

Since an important part of the hill running is constituted by downhill sections during which the eccentric contraction system is used, it also contributes to developing the athlete's proprioceptive skills, and as such, it plays a primary role in the preparation of the 3000-meter steeplechase and cross-country running.

References

1. Magness S. The science of running: how to find your limit and train to maximize your performance. Origin Press; 2014. 332 p.
2. Napier C. Science of running: analyse your technique, prevent injury, revolutionize your training. Dorling Kindersley Ltd; 2020. 224 p.
3. Higdon H. Marathon, revised and updated 5th edition: the ultimate training guide: advice, plans, and programs for half and full marathons. Harmony/Rodale; 2020. 370 p.

Nutrition for Track Running and Ultra-Running: Practical Recommendations

Corinne Fernandez

4.1 Nutrition Requirements in Track Running

Track running has special dietary requirements, as this sport is characterized by short and repetitive efforts. Optimizing athletic performance is important to distribute energy and hydration intake over the day, taking into account running times, repetition of effort over a day or over several days. To repair muscle fibre and eliminate fatigue metabolites (including lactates), it is necessary to replenish energy stocks after exercise. The diet plan of the athlete should be adapted to the sports events taking place on the track during the whole day.

It is recommended to adopt a nutritional strategy that will allow to wait for his selective sets without wasting energy [1, 2].

The nutritional goals will be to maintain constant blood sugar levels and protein tone throughout the day, especially during running activities.

If the competition takes place in the morning, the usual breakfast must be planned 3 h before the competition. The athlete should take into account the warm-up period before the start of the heats. If necessary, a ration of an energy drink can be considered, diluted according to the manufacturer's recommendations to facilitate the absorption of CHOs.

If the competition takes place in the morning and/or afternoon, the athlete should have a light meal such as water, hot sweet drink, gingerbread and a compote or banana. Regular hydration (water and/or energy drink) between sets is recommended. After the sets or at the end of the morning, it is advisable a light and digestible meal for lunch with low-fat proteins such as some chicken or fish or tofu (100–130 g), a portion of vegetables (salad, carrots or beets), little fat (1–2 tablespoons of vegetable oil for cooking or vinaigrette) and 1 portion of carbohydrates (1/2–1 plate) of pasta or rice or semolina or quinoa, and for dessert: a fruit compote. This meal optimizes digestion and recharges water, mineral salts, vitamins, proteins and energy.

In the afternoon, between sets, the stability of blood sugar levels must be ensured with a ration of rapid assimilation sugars (fruit juice, fruit paste, ripe banana or energy drink and water).

When the athlete has finished his day of trials, it will be important to recover with a balanced ration, re-hydrating, and predefined in qualitative and quantitative energy standards adapted to his well-being, his health and his athletic performance. The recovery dinner may include a soup or a starter of raw vegetables, meat/fish or an omelette, or tofu accompanied by vegetables and a plate of starchy foods, bread in addition, and a dairy product and a fruit for dessert.

C. Fernandez (✉)
Sports Nutrition, UPMC Sports Medicine, Paris,
France

4.2 Nutrition in Ultra-Running

Ultra-running concerns running events longer than marathon (>42.2 km). The prolonged duration of the ultra-race led to changes and decrease of most or all physiological parameters, including energy expenditure [3]. The ultra-running suffers considerable energy deficits. The success on this race results from a combination of physiological, nutritional and psychological factors. Concerning the nutrition, there are three essential phases to remember: The Preparation, The Race, and The Recovery.

Race strategies nutritional include CHO-rich eating the days before the race to store sufficient glycogen for event fuel needs [3, 4]. The topics of "train low", ketogenic diet, training with few carbohydrates and so a low store of glycogen, aren't discussed here.

4.3 The Preparation

4.3.1 Seven Days Before the Race: Saturating Muscle Glycogen Stores

Day D-7: meals with proteins, low fat, slightly high carbohydrates. Ration of 2000–3000 kcal by athlete with proteins intakes $\approx$1.6–2.5 g·kg·day; fat = 20–25%; carbohydrates = 60% + hydration.

During the final week, it is important anticipating the energy needs of the race and maintain the metabolism in top form. That means the athlete should structure his daily dietary intakes in three to four meals, without excess, without snacking all day. Hydration: 1.5–2 L per day.

Breakfast will change during this week. It will include an additional protein such as a slice of ham or one egg or some white cheese (low fat) or one soya yoghurt.

Day D-3: High carbohydrate diet for the endurance athlete. Ration of 2400–3500 kcal with proteins = 1.6–2.5 g·kg·day; fat = 15–20%; carbohydrates = 65–70% with complex carbohydrates, cereals and bread at every meal + hydration.

The nutritional goal is to increase and even double the muscle glycogen, with one complete plate of pasta or rice and some bread at every meal (e.g. cereals at breakfast) according to stomach capacities.

Some foods must be avoided before the race for the gut comfort: spicy foods, fatty meats, cooked fats, milk, cheese rich in fat, including hard cheeses such as Parmesan or Cheddar. The vegetables with hard fibres such as salsifies, peppers, onions, garlic, artichokes, and cabbage are not recommended.

For complex carbohydrates, it will be better to choose low to medium glycaemic index cereals: pasta, basmati rice or brown rice, or quinoa or sweet potatoes. For a digest meal, better favorate eating low fat meats or low fat fishes or veggie proteins such as tofu, seïtan, tempeh, natto....

Cooking by steaming and non-grilled, better in broth, in a bain-marie, in the oven, or in foil is a good habit. Eating cooked non-irritant fibrous vegetables such as: leeks, asparagus tips, carrots, beets, endives, courgettes (seeded), salad leaves excluding the centre which contains too irritant fibres and green beans.

For dessert, choose among apples or slightly ripe bananas, mangoes, stewed fruits, low-fat yoghurts or soya yoghurts, semolina of milk or rice of milk, or dessert with vegetable beverage such as almond milk or chestnut milk if more tolerable for digestive system.

Day D-1: Non-alcohol beverages and basic comportment with food.

The tips are:

- Do not skip meals.
- Opting for cereal products and easily digestible foods at each meal.
- Regular hydration throughout the day with a low-mineralized water.
- Do not gorge at the pasta party, or prior to the race. Too many complex and simple carbohydrates during this meal may interfere with the digestion, create gastric reflux, or promote intestinal fermentation during the night and sabotage the last night of sleep.

D-day, race day: The nutritional rules are to have one breakfast with high proteins and high carbohydrates, to eat quantities according to the stomach capacities and to respect time to digest. Count at least 3 h of digestion time before the race for a complete classic meal (breakfast, lunch and dinner). Or 1 h if a sport cake is preferred for breakfast.

- Example, for breakfast: 100–200 g of low-fat white cheese or soya yoghurt, toasts with a little butter and defatted ham or two boiled or scrambled eggs or buckwheat flakes, oats flakes or muesli with milk without lactose or vegetable beverages, or cereals porridges or banana bread, honey or agave syrup or chestnut cream, fruit puree or stewed fruits or sport foods like energy cakes. Breakfast drinks: water, coffee or tea according to habits. Staying careful before the race with caffeine and his potential laxative effects (risk of Diarrhea, risk of intestinal cramps and pain).

D-day, race day: The meal prior competition. Basically, the profile of this meal is the same as breakfast. Firstly, an easily digestible meal to obtain a few hours later gastric emptiness.

- It will consist of complex carbohydrates (cereals, bread), low-fat proteins, uncooked fat, a little salt, a few simple carbohydrates for dessert and water for hydration. Specifically, this meal can be divided into 4–5 components if there is a tall hungry: cooked vegetables as a starter, cereals (basmati rice, spaghetti, quinoa …) with meat or fish or vegetable proteins (tofu, seïtan, tempeh), white cheese or yoghurt and sweetened dessert: fruit pie, rice or semolina cake, or a stewed fruit and gingerbread. Drink water moderately during the meal and according to gastric tolerance. No alcohol.
- In a restaurant, for lunch or dinner, it is better to opt for a low-fat menu and avoid dishes in sauce, too heavy to digest. It is recommended to eat simple meals, without excess quantities or excess fat, in order to facilitate digestion sometimes disturbed by the stress of the upcoming competition.

If for any reason the formula meal prior to the race is a sandwich, it is better to make it yourself at home for protecting hygiene and for the rule of common sense "at least, we know what is inside".

The ham butter classic sandwich (two or three slices of ham) or tuna sandwich can be garnished with lettuce leaves and grated carrots. One or two veggie wraps can be also added. This practical and digest meal can be complemented by stewed fruit, rice or semolina cake, or an apple—one of the easiest fruit to digest—and two to three slices of gingerbread if hunger persists. For drinking: water or apple juice or coconut water. Alcohol is of course not recommended.

D-day, race day: 2 h 30 min before departure and every 60 min only if long distances.

Moderate caloric intakes only if hungry will maintain blood glucose levels and protect glycogen: 20–30 g in a single intake cereal bars or biscuits (fat < 10 g per 100 g) + water. The hydration (water) before the race should be regular and moderate.

D-day, race day: 30–15 min of departure

Stress hormones and catecholamines (adrenaline, cortisol) can induce a slight hyperglycaemia before the race. To prevent the decline of glycogen, it is possible to prepare a drink, with mainly fructose and by measuring the drink carefully: 20–30 g maximum fructose per litre. The use of fructose should be carefully taken into account, especially if the runner has a sensitive gut. This sugar may be a cause of flatulence and diarrhoeas which is common in the case of oral ingestion, when the absorbed doses are greater than 50 g per litre.

Grape juice diluted with water (1 volume of grape juice for 2 volumes of water as tolerated) can also be adapted: to be tested beforehand in training or during a minor race.

In the case of digestive disorders, hydrate regularly with small amounts of water and do not eat. Remain calm, because all will be well, of course, everything will be all right!

4.4 The Race

4.4.1 Two Goals: (1) Facilitate Athletic Performance and (2) Prevent or Limiting Gut Problems During the Race

The athletic performance needs energy and caloric foods during the race. These are meal, energy intakes, and nutrient timing guidelines which are consumed on race. The outcomes of researches indicate that runners should consume on the race amount 150–400 kcal/h (carbohydrates 30–50 g, proteins 5–10 g/h [5]).

The nutritional objectives of the race are:

- Manage the depletion of muscle glycogen and the liver (by the official supplies and regular spontaneous food intake).

- Protect the muscle fibres (by Food and Hydration).
- Mobilize fat (for intensity sub-maximal effort—aerobic spinneret).

No innovations during the race but many rehearsals beforehand during the training, to complete the nutritional preparation and be in good form and perform during the race.

1. **Food during the race can be composed mainly carbohydrates brought by sport beverages or gels or cereals bars or may be some personal choice but must be remain caloric**

   For the long-distance runners in ultramarathon or ultra-trail, the usual energy intake of complex carbohydrates (bread, cereal bars, dry cookies and so on) can be completed with homemade preparation (sandwiches, cakes, energy balls…) and by the official refuelling during the race. For them, it is interesting to take a real small meal every 2–3 h or more, depending upon their digestive capacity.

2. **Thinking of a combination of salty and sweet foods**

   With carbohydrate intakes ranging from 15 to 20 g per hour for the fragile digestive systems, up to 50–60 g per hour and for big eaters which have a very hunger among 120 g per hour [6], with a variety of calories-dense foods (gels or sports drinks with carbohydrate intake for effort, cereal bars, and salty foods such as cashew nuts or macadamia, crackers, stewed fruit in gourd, dried dates, chestnut cream …) and enjoying official refuelling, eating both in salty and sweet foods among the food palatability, the individual tolerance and the preference for tasty foods in longer races (soup and pasta if adapted at the stomach, bread, cheese, cake, gingerbread, dried meat, ham, cereal bars, chocolate etc.…). Be careful with solids (high-fibre foods), gels and sport beverages which cause, in some cases, gastrointestinal discomfort [7, 8].

3. **Protecting the digestive system to prevent or limit a bad digestion which can stop the race prematurely**

Nutrition. The act of chewing is essential for a runner to facilitate the absorption of food. If the size of food particles is greater than 2 mm, there may be partial or total inhibition of gastric emptying. Concerning the quantities, portion size should be tested during training. It is strongly recommended that you test and train the digestive capacity on the long efforts, making longer races with a full stomach, or by reducing energy intake to very small quantities (but beware of hypoglycaemia). Or taking small size portions but frequently.

Basically, all nutritional strategies must be tested at the training or before the race.

Hydration. Water is an essential nutrient for life and responsible for maintaining physiological functions vital on competitions. The dehydration during exercise increases muscle glycogen utilization and affects lipid metabolism, due to reductions in the uptake of free fatty acids [7]. Hypohydration may also alter cognitive performance [9–11].

Despite the importance of hydration, this topic remains one of the major causes of digestive disorders or gastrointestinal discomfort. Dehydration and a deep temperature body increase due to the effort and the weather will slow gastric emptying. Also the stress, the fluid volumes absorbed, the intensity of the race, the steepness of the slope, whether it is a very steep descent or not, all things play their role. One of the reasons is that the portal vein blood flow (the access to the liver) is reduced by 50–80% according to individual runners. Drink isotonic sports drinks containing carbohydrates and electrolytes is therefore highly recommended, and only if this type of drink is well tolerated. The usual hydration recommendations are 150–250 mL every 20 min (450–750 mL/h),

to be tested beforehand during a long training of running.

4. **The evaluation of the factors affecting gastric emptying**

The athlete must be paying attention to the two following parameters: the volume and the flux of liquids and solids ingested.

- To properly dose the drink is very important. A too higher concentration (too much level of carbohydrates) may interfere indeed with gastric emptying. It is better before buying a sport drink to check its osmolality and pH which are normally shown on the packaging, or to ask the manufacturer. The osmolality of a sport drink is approximately 290 mOsmol. The hypertonic drinks (osmolality >290 mosmol) cause the transfer of water from the cells to the middle of the intestines, which cause the risk of osmotic diarrhoea and can increase dehydration.

- Planning small recovery breaks after long and hard downhill stretches is a good idea because the shocks of the stride can severely affect the digestive system.

- Avoiding to be dehydrated before drinking and conversely, you should not drink in excess to catch up with dehydration. On average, it is recommended to drink one-two sips every 15 min (hydration should be personalized to each runner).

4.4.2 Why to Adopt a Hydration Strategy?

- To limit the sense of exhaustion (due to the rise of body temperature).
- To limit dehydration (protection of the digestive system, maintain plasma volume ...)
- To limit the loss of minerals through sweat (by a moderate and regular consumption of effort's drinks).

4.4.3 How to Measure Out the Exercise Drink's Powder?

Use energetic powder by dosing it 30–60 g per litre of water, to be adapted depending on the outside temperature.

A simple rule to effect a good dosage of the sportive beverage is *the highest the outside temperature gets, the less carbohydrates must be put in the drink. The colder it gets, the more carbohydrates the beverage could contain.*

In any case, be careful not to overdose the beverage: 120 g carbohydrates per litre must not be exceeded.

Nevertheless drinking only water for hydrating may induce the risk of hyponatremia (dilution of sodium in the blood). The electrolytes must be absolutely present in the drink, and mostly sodium.

Therefore, the runner can prepare his own sport drink with tea, for example, sugar AND salt, even on the official refuelling. Six to 12 lumps of white sugar added to a salts of about 1.2–1.5 g per litre (400–575 mg sodium) in hot tea can provide an adapted sport drink. The sodium thus will facilitate the absorption of carbohydrates in the cells.

4.5 The Recovery

4.5.1 Two Goals: (1) Countering the Effects of Dehydration. (2) Fill the Glycogen Stores Emptied by the Effort of the Race

To facilitate recovery [10], glycogen levels have to be restored in a relatively short space of time (countdown) immediately after the arrival. The body is particularly sensitive to this metabolic window, especially because the new synthesis of the glycogen also consumes energy.

After the competition, it is absolutely necessary to:

- Hydrate to restore the water reserves of the body and remove metabolic waste

- Fill, as fast as possible, glycogen stores emptied by the duration and intensity of the race
- Soothe the inflammation of the muscles consecutive to efforts eccentric of the running race
- Fight against free radicals produced during the race, by an increased intake of antioxidants
- Eat and hydrate consciously to eradicate tissue acidity
- Soothe the digestive system weakened by the effort

Ideally, the athlete should firstly rehydrate with a recovery drink enriched with amino acids branched (AAB), alternately with mineral water enriched in bicarbonates for an anti-acid effect.

The recovery meals can be consisted of solid foods and take into account the state of the digestive system on arrival: a tasty real breakfast with eggs, fruits, or biscuits and cakes, bread and some cheese, with fruit juice or mineral water with bicarbonates.

Basically, it is recommended to provide 50 g of carbohydrates per hour every 2 h until the main meal:

- either with a recovery energy drink, drinking it very regularly
- either with solid foods and a mineral water-enriched with bicarbonates: fruits like apples or pears or ripe bananas or apple puree to soothe the digestive system, dried fruits (apricots or prunes) and nuts (almonds or hazelnuts), ice cream, soya yoghurt and cake or cookies, also some dietetic energy biscuits.

At the main meal which follow the race, runners have to eat according to their hunger but without exceeding and opting for very digestible foods. Vegetables and fruits are essential to restore the antioxidants status [12]:

- Fish, poultry or soya burger, cooked vegetables, as for example soups or vegetable flan, potatoes steamed or boiled for their alkalizing property, pasta or rice are also welcome.
- Using colza oil or nuts for seasonings (for omega3 intakes).

- Cooked fruits and oilseeds (i.e. almonds, hazelnuts…) during or after food.
- 1 yoghurt (manufactured with goat milk, cow milk or soya milk) with sugar (honey, marple, jam).
- Always keep to hydrate oneself with bicarbonates water and/or the recovery energy drink.

This ration provides the minerals and trace elements that will facilitate the recovery such as potassium, magnesium, antioxidants, alkaline foods and essential fatty acids.

On arrival, quickly restoration and well hydration must be respected: not abuse sodas (acidifying drink) or beers (diuretic effects), avoid dark beer (toxins), but a light beer with friends upon arrival to celebrate the passage under the banner is not ruled out!

References

1. Slater GJ, Sygo J, Jorgensen M. Sprinting…dietary approaches to optimize training adaptation and performance. Int J Sport Nutr Exerc Metab. 2019;29(2):85–94. PMID: 30943814. https://doi.org/10.1123/ijsnem.2018-0273.
2. Stellingwerth T, Bovim IM, Whitfield J. Contemporary nutrition interventions to optimize performance in middle-distance runners. Int J Sport Nutr Exerc Metab. 2019;29(2):106–16. https://doi.org/10.1123/ijsnem.2018-0241. Epub 2019 Feb 12.
3. Berger N, Cooley D, Graham M, Harrison C, Best R. Physiological responses and nutritional intake during a 7-day treadmill running world record. Int J Environ Res Public Health. 2020;17(16):5962. PMID: 32824531. PMCID: PMC7459626. https://doi.org/10.3390/ijerph17165962.
4. Tiller NB, et al. International Society of Sports nutrition position stand: nutritional considerations for single-stage ultra-marathon training and racing. J Int Soc Sports Nutr. 2019;16(1):50. PMID: 31699159. PMCID: PMC6839090. https://doi.org/10.1186/s12970-019-0312-9.
5. Vitale K, Getzin A. Nutrition and supplement update for the endurance athlete: review and recommendations. Nutrients. 2019;11(6):1289. PMID: 31181616. PMCID: PMC6628334. https://doi.org/10.3390/nu11061289.
6. Urdempilleta A, Arribalzaga S, Viribay A, Castaneda-Barbarro A, Seco-Calvo J, Mielgo-Ayuso J. Effects of 120 vs. 60 and 90 g/h carbohydrate intake during a trail marathon on neuromuscular function and high intensity run capacity recovery. Nutrients. 2020;12(7):2094. PMID: 32679728. PMCID: PMC7400827. https://doi.org/10.3390/nu12072094.
7. Guillochon M, Rowlands DS. Solid, gel, and liquid carbohydrate format effects on gut comfort and performance. Int J Sport Nutr Exerc Metab. 2017;27(3):247–54. PMID: 27997257. https://doi.org/10.1123/ijsnem.2016-0211.
8. Erdman KA, Jones KW, Madden RF, Gammak N, Parnell JA. Dietary patterns in runners with gastrointestinal disorders. Nutrients. 2021;13(2):448. PMID: 33572891. PMCID: PMC7912258. https://doi.org/10.3390/nu13020448.
9. Burke LM, Jeukendrup AE, Jones AM, Mooses M. Contemporary nutrition strategies to optimize performance in distance runners and race walkers. Int J Sport Nutr Exerc Metab. 2019;29(2):117–29. PMID: 30747558. https://doi.org/10.1123/ijsnem.2019-0004.
10. Gonzalez-Alonso J, Calbet JAL, Nielsen B. Metabolic and thermodynamic responses to dehydration-induced reductions in muscle blood flow in exercising humans. J Physiol. 1999;520:577–89. PMID: 10523424. PMCID: PMC2269598. https://doi.org/10.1111/j.1469-7793.1999.00577.x.
11. Gopinathan PM, Pichan G, Sharma VM. Role of dehydration in heat stress-induced variations in mental performance. Arch Environ Health 1988;43:15–7 PMID: 3355239. https://doi.org/10.1080/00039896.1988.9934367.
12. Neubauer O, Reichhold S, Nics L, Hoelzl C, Valentini J, Stadlmayr B, Knassmüller S, Wagner K-H. Antioxidant responses to an acute ultra-endurance exercise: impact on DNA stability and indications for an increased need for nutritive antioxidants in the early recovery phase. Brit J Nut. 2010;104(8):1129–38. https://doi.org/10.1017/S0007114510001856. Epub 2010 Jul 19

Age and Running: Children and Adolescents, Elder People

Sergio Rocha Piedade, Larissa Oliveira Viana, and Bruno Paula Leite Arruda

5.1 Introduction

In sports practice, running features its straight connection to movement, psychological motivation, and high-intensity physical activity. Running may represent our first high-demand activity, starting when a child stands up, starts walking, and then follows the innate desire to go as fast as possible—our lifetime's first run [1].

There is no doubt that running could be an exciting physical activity or sport, but more than that, it may be a lifestyle, especially for people searching for health benefits to their cardiovascular, respiratory, immunological, neurological, and mental health [2, 3].

Nowadays, running is, by far, one of the most popular and practiced sport on earth [4, 5], and a good reason for that is that it demands no more than a pair of running shoes to start it, making it a world passion.

Like any sports, running performance is affected by biological development, which accounts for the body process of maturity and degeneration. Each phase of our lives, childhood, adolescence, and old age plays an essential role in this physiological development—lifetime effect.

The growing number of sexagenarian runners, men and women, participating in long-distance races has stimulated researchers to address the metabolic changes, gender, performance, and injury rate in this population [6–8].

Sometimes the passion for running may become a nightmare under some circumstances, mainly if the professional or recreational runner, regardless of age range (child, adolescent, or aged), adopts extreme physical behavior to achieve their desired goals. At his point, the body could become more vulnerable to the deleterious effect of running and intense physical efforts. Consequently, the body systems may be affected separately or together—the "dark side" of running and other sports practice [9–11].

This chapter explores the biological and physiological aspects, aerobic metabolism, running performance at different age range (children, adolescents, and elderly people), deleterious impact of inadequate running training program on their psychological status and performance, injuries and prevention.

S. R. Piedade (✉)
Exercise and Sports Medicine, Department of Orthopedics, Rheumatology and Traumatology, School of Medical Sciences, University of Campinas —UNICAMP, Campinas, SP, Brazil
e-mail: piedade@unicamp.br

L. O. Viana · B. P. L. Arruda
School of Medical Sciences, University of Campinas—UNICAMP, Campinas, SP, Brazil

© The Author(s), under exclusive license to Springer-Verlag GmbH, DE, part of Springer Nature 2022
G. L. Canata et al. (eds.), *The Running Athlete*, https://doi.org/10.1007/978-3-662-65064-6_5

5.2 The Role of Aerobic Metabolism in Running

Although short and high-intensity running training works as anaerobic exercise, running is, in essence, an aerobic physical activity; therefore, aerobic metabolism is the prevalent method of energy production [12]. In sports, a physical training program designed and focused on running helps build up cardiac and pulmonary efficiency to supply the muscle with energy from oxygen consumption and improve running performance.

The evaluation of the peak of oxygen consumption (VO_2 max) and anaerobic threshold offers valuable data of a runner or athlete's cardiorespiratory fitness—a functional index of the pulmonary, cardiovascular, and hematological components of oxygen supply to the muscular oxidative mechanisms in exercise [13]. It is essential to state that high cardiorespiratory fitness in childhood and adolescence will probably promote a healthy cardiovascular condition in adulthood and old age [14].

5.3 Running Performance at the Age Range

Running undergoes age-related changes, a natural and physiological process of human body development and muscle strengthening, affected by chronological age, growth, biological maturation, and also by the runner's anthropometric characteristics [15]. A marked and continuous improvement in running performance will happen from childhood to adolescence until reaching the peak at the age of 30–35 years old [13]. From 40 to 50 years old, a runner's performance tends to decline progressively and continues in the next decades [16, 17]. However, around 70 years of age, this decline is more pronounced in women than men, reflecting the expected reduction in cardiorespiratory fitness (CRF) or aerobic fitness due to the effect of aging according to gender [18].

5.4 Children and Adolescents

As a function of time, the human body growth is related to quantitative changes in its dimensions, while its development involves quantitative and qualitative changes. The individual's biological and behavioral settings are included in this arranged and sequential process. VO_2 max is an excellent indicator of aerobic fitness in young people [19]. However, it is necessary to consider that methodological and ethical restrictions may limit child and adolescent aerobic fitness analysis [20].

The peak of oxygen increases with age, growth, and maturation, up to 150% in boys and 80% in girls aged 8–16 years [21, 22]. Moreover, the oxygen peak rises during adolescence, with higher values for men than women, confirming gender differences. It could be explained by male greater muscle mass that is more evident at puberty and hemoglobin concentration [21, 23].

Armstrong et al. performed a multilevel regression analysis of the influence of gender, growth and maturation on the peak of expressed oxygen consumption concerning body mass in the population of healthy and untrained adolescents aged 11–13 years. The authors conclude that values are more consistent in boys over the development period, while girls tend to stabilize at around 14 years of age [21, 24].

The maturation speed may vary for each individual, and therefore, it can be classified as normal, early (fast), or late (slow). Moreover, genetic and environmental factors may reflect physiological, morphological, and psychological differences in the maturity among individuals of the same sex.

Some child and adolescent runners or athletes may present a better sports performance due to their fast maturation rate compared to normal or slower maturation ones—a "dopping" resulted from a faster individual maturation process. Therefore, when evaluating development and sports skills, the physician should systematically assess the neurological, cognitive, somatic, and physiological functions before classifying a runner or athlete as sports talent without considering the individual's maturation phase.

5.5 Elderly Runner

Maximum aerobic capacity is a fundamental metabolic characteristic, responsible for 40–70% of the variation in performance [25]. With aging, the VO_2 max falls due to the decline of maximum cardiac output, maximum heart rate, maximum stroke volume, and the artery-venous capacity of O_2. Consequently, the running performance progressively reduces [26]. Studies indicate that the O_2 absorption capacity decreases 10% per decade [27], with a 1–2% loss beginning after 25–35 years [28].

The endocrine system presents a decrease in growth hormone (GH), insulin, IGF-1 (insulin-like growth factor-1), glucagon with aging, and lower concentration of the sex hormones estrogen and testosterone [29, 30]. Regarding the musculoskeletal system, loss of muscle mass, sarcopenia, and lower ability to generate muscle strength will appear with aging [31]. Also, the thickness of quadriceps and plantar flexor muscles will decline more than 5% each decade. Type II muscle fibers by area reduce about 7% per decade. The ratio of the occupied area of type I fibers to type II decreases by 4.5% per decade. The type II/I fiber ratio is 1.2–20 years old, reaching 0.7–80 years old [32].

Nevertheless, the causes for the loss of strength in both sexes are not known deeply. Some authors suggest menopause and andropause as possible primary causes of sarcopenia [33, 34]. However, the onset and progression of this hormonal reduction and their impact on the neuromuscular system are still unknown [35].

5.6 Biomechanical Changes of Aging in Running Performance

From a clinical point of view, aging causes mechanical changes in running performance secondary to joint stiffness [36], reduced function of the sural triceps and quadriceps muscle, increasing the vulnerability of the runner to musculoskeletal injuries. Biomechanically, elderly runners decrease stride length and running speed, lower front and vertical ground reaction and propulsion forces [37]. With aging, the runner increases his flight time, decreases the propulsive peak and displacement of the center of mass, diminishes the knee and ankle (particularly the ankle eversion mechanism) [26].

5.7 "Forrest Gump Runner's Behavior" Syndrome [1]

5.7.1 An Irrational Impulse for Body Performance or Aesthetics

Running could be a turning point for many sedentary, overweight, stressed, and low self-esteem people to change these unhealthy conditions and achieve a healthy lifestyle. There is no doubt that running is an exciting activity, but we must be aware of how prepared we should be before defining the level of running or training. For most recreational runners, the passion for running is born after overcoming their own physical limits for the first time. However, some of them will go beyond, dreaming of new physical challenges and consequently developing a trigger to adopt a non-stop behavior of faster-increasing training and running to reach their self-imposed goals.

In my clinical practice, I refer to this as the **"Forrest Gump" runner's behavior syndrome:** a clinical condition triggered by an uncontrolled runners or athletes' obsession for achieving the maximum outcome with no awareness of their body limits to physical activity manifested by an irrational impulse for body performance or aesthetics. It may affect children, adolescents, and elderly runners indistinctly. It is essential to state that the process involves a psychological vulnerability portraited by a sequence of **passion, dreaming, and "nightmare"**—the dark side of running and other sports practice. Over time, runners will become more physically vulnerable and predisposed to injuries such as knee and shin pain, muscle injuries, bad sleeping, insomnia, resulting in the gradual decline of their sports performance (Fig. 5.1).

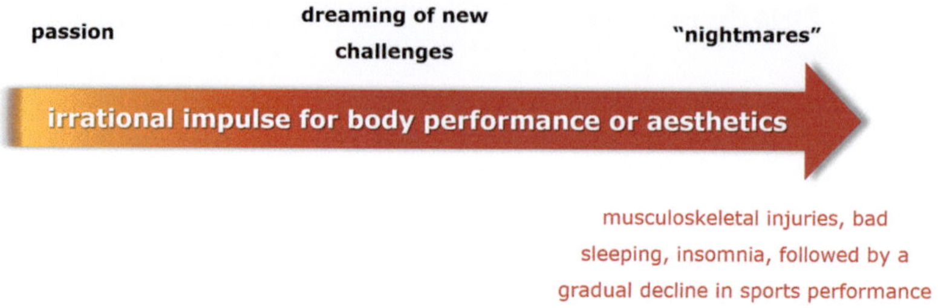

Fig. 5.1 Dark side of running and other sports—"Forrest Gump" runner's behavior syndrome

5.7.2 Injury and Prevention

Inadequate running program regarding the volume, intensity, frequency of training, and muscle imbalances may create a "perfect" scenario for injuries where the lower limb is the most prevalent site—knee (25%), leg (20%), foot (16%), and ankle (15%) [38]. The common injuries reported are located in hamstring muscles, calf, plantar fasciitis, ankle sprain, shin pain, and stress fractures. Another point to be emphasized is that previous injuries and the duration, speed, and interval of training play an essential role [39].

Master runners tend to have a higher prevalence of injuries to the calf, Achilles tendon, and hamstrings [40, 41]. However, Achilles tendinopathy in master runners is the most common pathology. The strength decrease in the plantar flexors as well as stiffness of ankle and leg tendons may be involved in the genesis of the injury pathophysiology [26, 42]. Hes Espanhol et al. [39] 2016 found no relationship between the static alignment of the lower limbs and lower limb injuries.

As we know, aging is followed by degenerative joint changes, especially in lower limbs— osteoarthritis (OA) that may impact the running performance negatively. However, it is important to state that recreational runners have a minor association with hip and knee OA than non-runners or sedentary individuals and elite runners—a protective effect of continuous and balanced physical exercises.

Injuries prevention should identify the pitfalls and adjust the training program (volume, intensity, and frequency) [43]. Moreover, concentric and eccentric strength training for the lower limbs focused on quadriceps, sural triceps, hip musculature [44], warm-up and cool-down, improvement on running technique and recovery time may reduce the injury rate [45]. Although flexibility training is a usual strategy, its effectiveness in protecting from injuries has not been confirmed in the literature.

Therefore, considering that running injuries may have different scenarios, the sports medicine physician must also be aware of the runner's biological body development, physical and psychological stress related to running or other sports practice, and its mental balance status.

References

1. Piedade SR, Fereira DM, Magro DA, Colombo CSSS. Marathon. In: Rocha Piedade S, Neyret P, Espregueira-Mendes J, Cohen M, Hutchinson MR, editors. Specifc sports-related injuires. Cham: Springer International Publishing; 2021. https://doi.org/10.1007/978-3-030-66321-6.
2. Dc L, Brellenthina AG, Thompsonb PD, Suic X, Leed IM, Lavie CJ. Running as a key lifestyle medicine for longevity. Prog Cardiovasc Dis. 2017;60(1):45–55. https://doi.org/10.1016/j.pcad.2017.03.005.
3. Lee DC, Pate RR, Lavie CJ, Sui X, Church TS, Blair SN. Leisure-time running reduces all-cause and cardiovascular mortality risk. J Am Coll Cardiol. 2014;64(5):472–81. https://doi.org/10.1016/j.jacc.2014.04.058. PMID: 25082581.
4. Hulteen RM, Smith JJ, Morgan PJ, Barnett LM, Hallal PC, Colyvas K, Lubans DR. Global participation in sport and leisure-time physical activities: a systematic review and meta-analysis. Prev Med. 2017;95:14–25. https://doi.org/10.1016/j.ypmed.2016.11.027.
5. van Mechelen W. Running injuries. A review of the epidemiological literature. Sports Med. 1992 Nov;14(5):320–35. https://doi.org/10.2165/00007256-199214050-00004.
6. Piedade SR, et al. Physical activity at adulthood and old age. In: Rocha Piedade S, Imhoff A, Clatworthy M, Cohen M, Espregueira-Mendes J, editors. The sports medicine physician. Cham: Springer; 2019. https://doi.org/10.1007/978-3-030-10433-7_6.

7. Cuk I, Nikolaidis PT, Markovic S, Knechtle B. Age differences in pacing in endurance running: comparison between Marathon and half-Marathon men and women. Medicina (Kaunas). 2019;55(8):479. https://doi.org/10.3390/medicina55080479.

8. Clark BC, Manini TM. What is dynapenia? Nutrition. 2012;28(5):495–503. [PubMed: 22469110].

9. Arnold MJ, Moody AL. Common running injuries: evaluation and management. Am Fam Physician. 2018;97(8):510–6.

10. Francis P, Whatman C, Sheerin K, Hume P, Johnson MI. The proportion of lower limb running injuries by gender, anatomical location and specific pathology: a systematic review. J Sports Sci Med. 2019;18(1):21–31.

11. Ramskov D, Rasmussen S, Sørensen H, Parner ET, Lind M, Nielsen R. Progression in running intensity or running volume and the development of specific injuries in recreational runners: run clever, a randomized trial using competing risks. J Orthop Sports Phys Ther. 2018;48(10):740–8.

12. Patel H, Alkhawam H, Madanieh R, Shah N, Kosmas CE, Vittorio JT. Aerobic vs anaerobic exercise training effects on the cardiovascular system. World J Cardiol. 2017;9(2):134–8.

13. Tanaka H, Seals DR. Physiology of aging. Invited review: dynamic exercise performance in master athletes: insight the effectos of primary human aging on physiological functional capacity. J Appl Physiol. 2003;95:2152–62. https://doi.org/10.1152/japplphysiol.00320.2003.

14. Andersen LB, Hasselstrøm H, Grønfeldt V, Hansen SE, Karsten F. The relationship between physical fitness and clustered risk and tracking of clustered risk from adolescence to young adulthood: eight years follow-up in the Danish youth and sport study. Int J Behav Nutr Phys Act. 2004;1(1):6. https://doi.org/10.1186/1479-5868-1-6.

15. Ortega F, Ruiz J, Castillo M, Sjöström M. Physical fitness in childhood and adolescence: a powerful marker of health. Int J Obes. 2008;32:1–11.

16. Fair RC. Estimated age effects in athletic events and chess. Exp Aging Res. 2007;33(1):37–57.

17. Wilson TM, Tanaka H. Meta-analysis of the age-associated decline in maximal aerobic capacity in men: relation to training status. Am J Physiol Heart Circ Physiol. 2000;278(3):H829–34.

18. Reaburn P, Dascombe B. Endurance performance in masters athletes. Eur Rev Aging Phys Act. 2008;5(1):31–42.

19. Armstrong N. Aerobic fitness of children and adolescents. J Pediatr. 2006;82:406–8.

20. Castro-Piñeiro J, Ortega FB, Keating XD, González-Montesinos JL, Sjöstrom M, Ruiz JR. Percentile values for aerobic performance running/walking field tests in children aged 6 to 17 years: influence of weight status. Nutr Hosp. 2011;26(3):572–8. https://doi.org/10.1590/S0212-16112011000300021.

21. Armstrong N, Welsman JR. Assessment and interpretation of aerobic fitness in children and adolescents. Exerc Sport Sci Rev. 1994;22:435–76.

22. Krahenbuhl GS, Skinner JS, Kohrt WM. Developmental aspects of maximal aerobic power in children. Exerc Sport Sci Rev. 1985;13:503–38.

23. Eisenmann JC, Pivarnik JM, Malina RM. Scaling peak VO2 to body mass in young male and female distance runners. J Appl Physiol. 2001;90(6):2172–80. https://doi.org/10.1152/jappl.2001.90.6.2172. PMID: 11356780.

24. Armstrong N, Welsman JR, Nevill AM, Kirby BJ. Modeling growth and maturation changes in peak oxygen uptake in 11-13 yr olds. J Appl Physiol. 1999;87(6):2230–6. https://doi.org/10.1152/jappl.1999.87.6.2230. PMID: 10601172.

25. Loftin M, Sothern M, Koss C, Tuuri G, Vanvrancken C, Kontos A, et al. Energy expenditure and influence of physiologic factors during marathon running. J Strength Cond Res. 2007;21(4):1188–91. https://doi.org/10.1519/R-22666.1.

26. Willy RW, Paquette MR. The physiology and biomechanics of the master runner. Sports Med Arthrosc Rev. 2019;27:15–21.

27. Evans WJ. Effects of exercise on body composition and functional capacity of the elderly. J Gerontol A Biol Sci Med Sci. 1995;50:147–50.

28. Trappe WS, Costill DL, Vukovich MD, Jones J, Melham T. Aging among elite distance runners: a 22-yr longitudinal study. The American Physiological Society; 1996.

29. Sipilä S, Narici M, Kjaer M, Pöllänen E, Atkinson RA, Hansen M, et al. Biogerontology. 2013;14:231–45. https://doi.org/10.1007/s10522-013-9425-8.

30. Sellami M, Bragazzi NL, Slimani M, Hayes L, Jabbour G, de Giogio A, et al. The effect of exercise on glucoregulatory hormones: a countermeasure to human aging: insights form a comprehensive review of the literature. Int J Environ Res Public Health. 2019;16:1709. https://doi.org/10.3390/ijerph16101709.

31. Manini TM, Clark BC. Dynapenia an aging: an update. J Gerontol A Biol Sci Med Sci. 2012;67A(1):28–40. https://doi.org/10.1093/gerona/glr010.

32. Korohen MT, Mero AA, Alén M, Sipilä S, Häkkinen K, Liikavainio T, et al. Biomechanis and skeletal muscle determinants of maximum running speed with aging. Med Sci Sports Exerc. 2009;41:884–56. https://doi.org/10.1249/MSS.0b013e3181998366.

33. Collins BC, Laakkonen DA, Lowe DA. Aging of the musculoskeletal system: how the loss of estrogen impacts muscle strength. Bone. 2019;123:137–44. https://doi.org/10.1016/j.bone.2019.03.033.

34. Diamanti-Kandarakis E, Dattilo M, Macut D, Duntas L, Gonos ES, Goulis DG, Gantenbein CK, Kapetanou M, Koukkou E, Lambrinoudaki I, Michalaki M, Eftekhari-Nader S, Pasquali R, Peppa M, Tzanela M, Vassilatou E, Vryonidou A. Aging and anti-aging: a combo-endocrinology overview. Eur J Endocrinol.

2017;176(6):R283–308. https://doi.org/10.1530/EJE-16-1061.

35. Haynes EMK, Neubauer NA, Cornett KMD, Jones GR, Jakobi JM. Age and sex-related decline of muscle strength across the adult lifespan: a scoping review of aggregated data. Appl Physiol Nutr Metab. 2020;45(11):1185–96.

36. Butler RJ, Crowell HP, Davis IM. Lower extremity stiffness: implications for performance and injury. Clin Biomech. 2003;18:511–7.

37. Fukuchi RK, Duarte M. Comparison of three-dimensional lower extremity running kinematics of young adult and elderly runners. J Sports Sci. 2008;26(13):1447–54. https://doi.org/10.1080/02640410802209018.

38. Fields KB. Running injuries—changing trends and demographics. Curr Sports Med Rep. 2011;10(5):299–303. https://doi.org/10.1249/jsr.0b013e31822d403f.

39. Hespanhol LC Jr, Costa LOP, Lopes AD. Previous injuries and some training characteristics predict running-related injuries in recreational runners: a prospective studuy. J Physiother. 2013;59:263–9.

40. Fields KB, Rigby MD. Muscular calf injuries in runners. Curr Sports Med Rep. 2016;15(5):320–4. https://doi.org/10.1249/jsr.0000000000000292.

41. Knobloch K, Yoon U, Vogt PM. Overuse injuries correlated to hours of training in master running athletes. Foot Ankle Int. 2008;29:671. https://doi.org/10.3113/FAI.2008.0671.

42. Devita P, Fellin RE, Seay JF, Edward IP, Stavro N, Messier SP. The relationship between age and running biomechanics. Med Sci Sports Exerc. 2016;48(1):98–106. https://doi.org/10.1249/MSS.0000000000000744.

43. van der Worp MP, ten Haaf DSM, van Cingel R, de Wijer A, Nijhuis-van der Sanden MWG, Staal JB. Injuries in runners: a systematic review on risk factors and sex differences. PLoS One. 2015;10(2):e0114937. https://doi.org/10.1371/journal.pone.0114937.

44. Lauersen JB, Bertelsen DM, Andersen LB. The effectiveness of exercise interventions to prevent sports injuries: a systematic review and meta-analysis of randomised controlled trials. Br J Sports Med. 2014;48:871–7.

45. Loudon JK. The master female triathlete. Phys Ther Sport. 2016;22:123–8. https://doi.org/10.1016/j.ptsp.2016.07.010.

Sex Differences Between Women and Men in Running

6

Beat Knechtle and Pantelis T. Nikolaidis

6.1 Sex Difference in Participation in Running Races

There are differences in the participation trends in running races between women and men [1–6]. It seems that women prefer to compete in shorter running races. In the USA, 55% of the finishers in road races are women [5]. In shorter running races such as the 10-km distance, increasingly more women compete compared to men [2, 4]. An analysis of 10 of the largest 10-km running races in the USA showed significant annual decrease in the ratio of men to women finishers [4]. In 10 km, half-marathon and marathon races held between 2008 and 2018 in Oslo with 115,725 finishers, relatively more women finished a 10 km and less a half-marathon and a marathon [?].

The trend is opposite in longer running distances such as ultra-marathon distances. In ultra-marathon running, the participation of women is considerably lower compared to men [1, 7, 8]. In time-limited ultra-marathon from 6 h to 10 days, the participation of women and men varied by age, indicating a relatively low participation of women in the older age groups [1].

6.2 Performance Difference Between Women and Men

The sex difference in running performance has been investigated for different running distances [5, 6, 9] such as 3000 m track running, 5000 m track running, 10,000 m track running [10], 10 km road running, half-marathon, marathon [2, 11–14] and ultra-marathon running of different distances and/or durations for time-limited races [1, 3, 8, 15–18].

Generally, men are running faster than women for all distances [1, 3, 5, 6, 8–15, 17–21]. The sex difference has been well investigated in marathon running. In marathon running, the sex difference in running time remained stable at ~18.7 ± 3.1% from 18 to 57 years of age [14]. An analysis of elite marathoners competing in the World Marathon Majors Series showed that men were 11.6 ± 1.8% faster than women [20].

In the world records of marathon running, the sex difference increased non-linearly from 5 to ~20 years, remained unchanged at ~20 min from ~20 to ~50 years and increased thereafter. The sex difference was lowest (7.5%, 10.5 min) at the age

B. Knechtle (✉)
Medbase St. Gallen Am Vadianplatz,
Gallen, Switzerland

Institute of Primary Care, University of Zurich,
Zurich, Switzerland
e-mail: beat.knechtle@hispeed.ch

P. T. Nikolaidis
Exercise Physiology Laboratory, Nikaia, Greece

School of Health and Caring Sciences, University of
West Attica, Athens, Greece
e-mail: pnikolaidis@uniwa.gr

© The Author(s), under exclusive license to Springer-Verlag GmbH, DE, part of Springer Nature 2022
G. L. Canata et al. (eds.), *The Running Athlete*, https://doi.org/10.1007/978-3-662-65064-6_6

of 49 years [13]. In ultra-marathon running, the sex difference seems to be higher than in marathon running. In the Badwater ultra-marathon, men were 19.8 ± 4.8% faster than women, and in the Spartathlon, men were 19.6 ± 2.5% faster [19].

The trend in sex difference has also been investigated over calendar years [1, 3, 5, 8, 11, 17–19, 22]. Generally, women were able to reduce the gap to men over time [1, 3, 5, 8, 11, 17–19, 22]. In the Boston Marathon from 1972 (first women ever running a marathon) to 2017, average performance worsened over the years, but the differences between the sexes decreased [11]. More analyses were performed for ultra-marathon running. In the "Comrades Marathon" from 1994 to 2017, the sex difference decreased across years [22]. In 100-mile ultra-marathon running races held in North America between 1977 and 2008, race times of women improved relative to men through the 1980s, but were then stable over the past two decades with the fastest women running about 20% slower than the fastest men [8]. An analysis of 50-mile, 100-mile, 200-mile, 1000-mile, and 3100-mile events held worldwide between 1971 and 2012 showed a linear decrease in sex difference for 50-mile and 100-mile events suggesting that women were reducing the sex gap for these distances [17]. An analysis of 50 km, 100 km, 200 km, and 1000 km events held worldwide from 1969 to 2012 showed that sex differences in running speeds decreased non-linearly in 50 km and 100 km but remained unchanged in 200 km and 1000 km. The sex differences in running speeds showed no change with increasing length of the race distance [18].

In earlier years, it has been assumed that women would be able to beat men in ultra-marathon running [23, 24]. It has been shown that the rate of improvement for women has been extraordinary and was larger for longer distance events [25]. However, in ultra-marathon running, the fastest men are faster than the fastest women [3, 7, 15–18]. This has been confirmed in 50-mile races [16], in 100-km races [3], in 100-mile races [16], in the UTMB (Ultra-trail du Mont Blanc) [15], in races from 50 to 1000 km [18], and in races from 50 miles to 3100 miles where the fastest men were ~17–20% faster than the fastest

women for all distances [17]. Obviously, women are not able to reduce the gap to men in very long ultra-marathons. In the two ultra-marathons "Badwater" and "Spartathlon", the sex difference remained unchanged at ~20% between 2000 and 2012 [19].

Women can, however, reduce the gap to men with increasing age [3, 9, 16]. In 50-mile and 100-mile races, the sex differences decreased with older age [16]. An analysis of "American Master Road Running Records" for 5 km, 8 km, 10 km, 10 miles, 20 km, half-marathon, 25 km, 30 km, marathon, 50 km, 50 miles, 100 km, 100 miles, 12 h, 24 h, 48 h and 144 h, for athletes in age groups ranging from 40 to 99 years old showed that the sex gap decreased with increasing age [9]. A reduction in the sex difference has also been shown for time-limited ultra-marathons from 6 h to 10 days where women were able to reduce the gap to men for most of timed ultra-marathons and for those age groups (i.e., older age groups) where they had relatively high participation [1]. In 100-km ultra-marathon races, the performance gap disappeared in athletes older than 60 years. The performance gap between the sexes was not significant in the oldest age groups (>90 years) among recreational athletes and among top-three athletes older than 70 years [3].

In some instances, women can reduce the gap to men with increasing race distance [9, 16]. In 50-mile and 100-mile races, the sex difference was smaller in 100-mile (4.41%) than in 50-mile races (9.13%) [16]. However, an analysis of "American Master Road Running Records" for 5 km, 8 km, 10 km, 10 miles, 20 km, half-marathon, 25 km, 30 km, marathon, 50 km, 50 miles, 100 km, 100 miles, 12 h, 24 h, 48 h and 144 h, for athletes in age groups ranging from 40 to 99 years old showed that the sex gap did not decrease with increasing event length [9].

The sex difference in ultra-marathon running is depending upon the participation of women. The sex difference in running speed was largest when there were fewer women than men finishers in a race and lower participation rates of women than men in the lower-distance ultramarathons and less depth among lower-placed women runners inflate the sex difference in ultra-marathon performance [7]. It has also been shown that elite women who perform similarly to men in 50 km ultra-marathon running achieve a similar performance to men in 80 km and 100 miles ultra-marathon running [26].

Women exhibit numerous phenotypes that would be expected to confer an advantage in ultra-marathon running (e.g., greater fatigue resistance, greater substrate efficiency, and lower energetic demands), they also exhibit several characteristics that unequivocally impinge on performance (e.g., lower O_2-carrying capacity, increased prevalence of gastrointestinal distress, and sex-hormone effects on cellular function/injury risk) [27]. It is unrealistic to believe that women will once outrun men. Current evidence suggests that women will not run as fast as men, which speaks against eliminating sex segregation in running. The plateau in the performance gap—although women reduced the gap to men—suggests a persistent dominance of biological influences (e.g., longer limb levers, greater muscle mass, greater aerobic capacity, and lower fat mass) on performance [28].

6.3 Age of Peak Running Performance

The age of peak running performance has been investigated for different running distances such as 10 km [21], half-marathon [21, 29], marathon [13, 14, 20, 21, 30], and ultra-marathon running [8, 15, 21]. There seem to be differences regarding the different race distances whether women or men were older. However, the age of peak performance seemed to be higher in the longer race distances [21].

In the world records of half-marathon running, the age of the fastest race time does not differ between men and women, with 26.62 years in women and 26.80 years in men [29]. Running times of the top ten men and women at 1-year intervals (from 18 to 75 years) were analyzed for the 2010 and 2011 races in the "New York City marathon" where the lowest race time was obtained at 27 years (149 ± 14 min) in men and at 29 years (169 ± 17 min) in women [14]. An analy-

sis of elite runners competing at world-class level (i.e., World Marathon Majors Series, World Championships, and Olympic Games) showed that women were older than men, but for only two of the seven marathons, the Chicago and the London Marathons. There was no sex difference in age for the Berlin, Boston, New York City, World Championship, and Olympic Marathons [20].

In ultra-marathon running, the age of the best performance is considerably higher compared to shorter running distances [8]. There are differences in the age of the best performances [8, 15]. In 100-mile ultra-marathon running races held in North America between 1977 and 2008, the fastest times were produced by the 30–39 year age group among men and the 40–49 year age group for women [8]. In the UTMB (Ultra-trail du Mont Blanc), the fastest women were older than the fastest men [15].

6.4 Pacing in Running

The aspect of pacing has been investigated for both women and men for different running distances such as 5 km [31], 10 km [32], half-marathon [33, 34], marathon [33–35], and ultra-marathon running [15]. In 5-km running races, women slowed significantly more than men. It was assumed that the sex difference in pacing partly reflected a sex difference in some aspects of decision making, such as overconfidence, risk perception, or willingness to tolerate discomfort [31]. In contrast, in 10-km running races, men slowed significantly more than women where it was assumed that the sex difference in pacing partly reflected a sex difference in decision making [32].

Differences in pacing between the sexes have been reported in half-marathon and marathon running. In both half-marathon and marathon running, women showed a more even pacing than men [33]. However, the sex difference in pacing was smaller in half-marathon than in marathon running [33]. In half-marathon running, women show a similar pacing as men [34]. In marathon running, there is a sex difference regarding pacing for all age and performance groups and may reflect sex differences in physiology, decision making, or both [35]. Also, in ultra-marathon running, a sex difference in pacing has been reported. In the UTMB (Ultra-trail du Mont Blanc), women had a higher pace variation than men [15].

6.5 Specific Physiological Differences

The better performance in male runners compared to female runners can be explained by physiological differences [36–41]. Male runners have a 14% higher maximum oxygen uptake compared to female runners [42] and women have a reduced O_2 carrying capacity [40]. Women are different compared to men regarding body core temperature and sweating during running. In a 15-km running race under cool conditions, absolute heat production was higher in men than in women, men demonstrated a greater increase in core body temperature, and whole-body sweat rate was larger in men [43]. The maximal accumulated oxygen deficit is a measure of anaerobic capacity. It has been shown that this deficit is 32% higher in men than in women. It has been shown that during a 1-km run, the metabolic cost was higher in women than in men [44]. Also, peak post-exercise blood lactate as a reflection of glycolytic contribution is 23% higher in men compared to women [42]. Men also seemed to have a better running economy than women [45]. The higher body fat in women might induce larger utilization of O_2 per unit of fat-free mass during submaximal exercise and lower VO_2max [46]. Sex differences in 12-min run performance were attributed mostly to body fat (74%), followed by VO_2max (20%) and running economy (2%) [41].

6.6 Mental Fatigue

The aspect of mental fatigue has also been investigated in female and male runners where female athletes were suffering less from mental fatigue compared to male athletes [47].

6.7 Nutritional Aspects

Sex differences in nutrition have been investigated for female and male runners competing in different running distances [48]. Female runners choose more often food to obtain specific phytochemicals compared to men [48].

6.8 Medical Aspects

There are differences in female and male runners regarding diseases and medication [48]. Female runners show a higher prevalence of hypothyroidism and female runners have a higher use in thyroid medication and intake of hormones and supplements compared to men [48].

6.9 The Aspect of Nationality

It is well-known that East African runners from Ethiopia and Kenya are dominating the marathon distance [21]. An interesting aspect is fact of differences between Ethiopian women and men. In the "World Marathon Majors" (Boston, Berlin, Chicago, and New York) and the "Stockholm Marathon" between 2000 and 2014, Ethiopian men improved marathon race times, but not Ethiopian women. Age increased in Ethiopian men, but not in Ethiopian women [30].

6.10 Motivational Aspects

Although men have physiological advantages regarding running performance, the sex difference can also be explained by differences in motivation. Deaner reviewed different running studies regarding sex difference and concluded that this sex difference in relative performance can be attributed, at least in part, to men's greater training motivation, and that this pattern has been stable for several decades [49].

In summary, although some studies suggested that women might be able to outrun men, the sex difference remains stable in running at ~10–11% (marathon) up to 20% in ultra-marathon running and women will never outrun men in the future [50–53]. Women's times have now reached a plateau similar to that observed for men [54]. For longer distances, the sex difference increases where this might be confounded by the reduced number of women competing in longer events [55]. The sex difference in performance will remain fairly constant because of biological differences between men and women that give men an advantage in distance running [51]. The remaining sex gaps in performance appear biological in origin [56]. Success in distance running and sprinting is determined largely by aerobic capacity and muscular strength, respectively [56]. Because men have a larger aerobic capacity and greater muscular strength, the gap in running performances between men and women is unlikely to narrow naturally [54]. Current sex differences in performance may now reasonably reflect the true physiological differences between women and men [53] (Tables 6.1 and 6.2).

Table 6.1 Age of peak performance for female and male runners regarding distance sorted by length

Distance	Women (years)	Men (years)	Reference
10 km	32.0 ± 6.0	25.3 ± 4.3	[21]
Half-marathon (21.1 km)	27.5 ± 4.7	25.9 ± 4.1	[21]
Half-marathon	26.62	26.80	[29]
Marathon (42.2 km)	29.8 ± 4.2	28.9 + 3.8	[20]
Marathon	29.5 ± 5.5	29.1 ± 4.3	[21]
Marathon	29	27	[14]
50 miles (80.46 km)	33	33	[16]
Comrades (90 km)	33.9 ± 4.6	36.3 ± 5.9	[22]
100 km	36.6 ± 6.1	35.9 ± 5.5	[21]
100 km	34.9 ± 3.2	34.5 ± 2.5	[57]
100 miles (160.93 km)	33		[16]
Badwater (135 miles, 217.26 km)	42.3 ± 3.8	39.8 ± 5.7	[19]
Spartathlon (246 km)	44.6 ± 3.2	39.7 ± 2.4	[19]

Table 6.2 Sex difference in running performance for different distances sorted by length

Distance	Sex difference	References
Marathon (42.2 km)	10.7%	[11]
Marathon	18.7 ± 3.1%	[14]
Marathon	11.6 ± 1.8%	[20]
50 km	19.3 ± 5.8%	[7]
50 km	15.4%	[18]
50 miles (80.46 km)	9.13%	[16]
50 miles	17.1 ± 1.9%	[17]
100 km	14.9 ± 4.2%	[7]
100 km	14.0 ± 1.2%	[57]
100 km	10.0 ± 3.0%	[18]
100 miles (160.93 km)	4.41%	[16]
100 miles	19.2 ± 1.5%	[17]
200 km	27.3 ± 5.7%	[18]
Badwater (135 miles, 217.26 km)	19.8 ± 4.8%	[19]
Spartathlon (246 km)	19.6 ± 2.5%	[19]
1000 miles (1609.34 km)	16.7 ± 1.6%	[17]

References

1. Knechtle B, Valeri F, Nikolaidis PT, Zingg MA, Rosemann T, Rüst CA. Do women reduce the gap to men in ultra-marathon running? Springerplus. 2016;5(1):672.
2. Nikolaidis PT, Cuk I, Clemente-Suárez VJ, Villiger E, Knechtle B. Number of finishers and performance of age group women and men in long-distance running: comparison among 10km, half-marathon and marathon races in Oslo. Res Sports Med. 2021;29(1):56–66.
3. Stöhr A, Nikolaidi PT, Villiger E, Sousa CV, Scheer V, Hill L, et al. An analysis of participation and performance of 2067 100-km ultra-marathons worldwide. Int J Environ Res Public Health. 2021;18(2):1–12.
4. Cushman DM, Markert M, Rho M. Performance trends in large 10-km road running races in the United States. J Strength Cond Res. 2014;28(4):892–901.
5. Deaner RO, Addona V, Mead MP. U.S. masters track participation reveals a stable sex difference in competitiveness. Evol Psychol. 2014;12(5):848–77.
6. Deaner RO, Mitchell D. More men run relatively fast in U.S. road races, 1981-2006: a stable sex difference in non-elite runners. Evol Psychol. 2011;9(4):600–21.
7. Senefeld J, Smith C, Hunter SK. Sex differences in participation, performance, and age of ultra-marathon runners. Int J Sports Physiol Perform. 2016;11(5):635–42.
8. Hoffman MD. Performance trends in 161-km ultra-marathons. Int J Sports Med. 2010;31(1):31–7.

9. Sousa CV, Da Silva Aguiar S, Rosemann T, Nikolaidis PT, Knechtle B. American masters road running records—the performance gap between female and male age group runners from 5 km to 6 days running. Int J Environ Res Public Health. 2019;16(13):2310.

10. Coquart JB, Mercier D, Tabben M, Bosquet L. Influence of sex and specialty on the prediction of middle-distance running performances using the Mercier et al.'s nomogram. J Sports Sci. 2015;33(11):1124–31.

11. Knechtle B, Di Gangi S, Rüst CA, Nikolaidis PT. Performance differences between the sexes in the Boston Marathon from 1972 to 2017. J Strength Cond Res. 2020;34(2):566–76.

12. Senefeld J, Joyner MJ, Stevens A, Hunter SK. Sex differences in elite swimming with advanced age are less than marathon running. Scand J Med Sci Sports. 2016;26(1):17–28.

13. Knechtle B, Assadi H, Lepers R, Rosemann T, Rüst CA. Relationship between age and elite marathon race time in world single age records from 5 to 93 years. BMC Sports Sci Med Rehabil. 2015;6(1):31.

14. Lara B, Salinero JJ, Del Coso J. The relationship between age and running time in elite marathoners is U-shaped. Age. 2014;36(2):1003–8.

15. Suter D, Sousa CV, Hill L, Scheer V, Nikolaidis PT, Knechtle B. Even pacing is associated with faster finishing times in ultramarathon distance trail running—the "ultra-trail du Mont Blanc" 2008–2019. Int J Environ Res Public Health. 2020;17(19):1–11.

16. Waldvogel KJ, Nikolaidis PT, Di Gangi S, Rosemann T, Knechtle B. Women reduce the performance difference to men with increasing age in ultra-marathon running. Int J Environ Res Public Health. 2019;16(13):2377.

17. Zingg MA, Knechtle B, Rosemann T, Rüst CA. Performance differences between sexes in 50-mile to 3,100-mile ultramarathons. Open Access J Sports Med. 2015;6:7–21.

18. Zingg MA, Karner-Rezek K, Rosemann T, Knechtle B, Lepers R, Rüst CA. Will women outrun men in ultra-marathon road races from 50 km to 1,000 km? Springerplus. 2014;3(1):97.

19. da Fonseca-Engelhardt K, Knechtle B, Rüst CA, Knechtle P, Lepers R, Rosemann T. Participation and performance trends in ultra-endurance running races under extreme conditions—'Spartathlon' versus 'Badwater'. Extreme Physiol Med. 2013;2(1):15.

20. Hunter SK, Stevens AA, Magennis K, Skelton KW, Fauth M. Is there a sex difference in the age of elite marathon runners? Med Sci Sports Exerc. 2011;43(4):656–64.

21. Nikolaidis PT, Onywera VO, Knechtle B. Running performance, nationality, sex, and age in the 10-km, half-marathon marathon, and the 100-km ultra-marathon IAAF 1999-2015. J Strength Cond Res. 2017;31(8):2189–207.

22. Nikolaidis PT, Knechtle B. Russians are the fastest and the youngest in the "comrades Marathon". J Sports Sci. 2019;37(12):1387–92.

23. Bam J, Noakes TD, Juritz J, Dennis SC. Could women outrun men in ultramarathon races? Med Sci Sports Exerc. 1997;29(2):244–7.

24. Whipp BJ, Ward SA. Will women soon outrun men? Nature. 1992;355(6355):25.

25. Chatterjee S, Laudato M. Gender and performance in athletics. Soc Biol. 1995;42(1–2):124–32.

26. Hoffman MD. Ultramarathon trail running comparison of performance-matched men and women. Med Sci Sports Exerc. 2008;40(9):1681–6.

27. Tiller NB, Elliott-Sale KJ, Knechtle B, Wilson PB, Roberts JD, Millet GY. Do sex differences in physiology confer a female advantage in ultra-endurance sport? Sports Med. 2021;51(5):895–915.

28. Millard-Stafford M, Swanson AE, Wittbrodt MT. Nature versus nurture: have performance gaps between men and women reached an asymptote? Int J Sports Physiol Perform. 2018;13(4):530–5.

29. Nikolaidis PT, Di Gangi S, Knechtle B. World records in half-marathon running by sex and age. J Aging Phys Act. 2018;26(4):629–36.

30. Knechtle B, Aschmann A, Onywera V, Nikolaidis PT, Rosemann T, Rüst CA. Performance and age of African and non-African runners in world Marathon majors races 2000–2014. J Sports Sci. 2017;35(10):1012–24.

31. Deaner RO, Lowen A. Males and females pace differently in high school cross-country races. J Strength Cond Res. 2016;30(11):2991–7.

32. Deaner RO, Addona V, Carter RE, Joyner MJ, Hunter SK. Fast men slow more than fast women in a 10 kilometer road race. PeerJ. 2016;4:e2235.

33. Cuk I, Nikolaidis PT, Knechtle B. Sex differences in pacing during half-marathon and marathon race. Res Sports Med. 2020;28(1):111–20.

34. Nikolaidis PT, Ćuk I, Knechtle B. Pacing of women and men in half-marathon and marathon races. Medicina (Lithuania). 2019;55(1):14.

35. Deaner RO, Carter RE, Joyner MJ, Hunter SK. Men are more likely than women to slow in the marathon. Med Sci Sports Exerc. 2014;47(3):607–16.

36. Padilla S, Bourdin M, Barthélémy JC, Lacour JR. Physiological correlates of middle-distance running performance—a comparative study between men and women. Eur J Appl Physiol Occup Physiol. 1992;65(6):561–6.

37. Helgerud J, Ingjer F, Strømme SB. Sex differences in performance-matched marathon runners. Eur J Appl Physiol Occup Physiol. 1990;61(5–6):433–9.

38. Støa EM, Helgerud J, Rønnestad BR, Hansen J, Ellefsen S, Støren Ø. Factors influencing running velocity at lactate threshold in male and female runners at different levels of performance. Front Physiol. 2020;11:585267.

39. Bunc V, Heller J. Energy cost of running in similarly trained men and women. Eur J Appl Physiol Occup Physiol. 1989;59(3):178–83.

40. Lewis DA, Kamon E, Hodgson JL. Physiological differences between genders implications for sports conditioning. Sports Med. 1986;3(5):357–69.

41. Sparling PB, Cureton KJ. Biological determinants of the sex difference in 12-min run performance. Med Sci Sports Exerc. 1983;15(3):218–23.

42. Hill DW, Vingren JL. Effects of exercise mode and participant sex on measures of anaerobic capacity. J Sports Med Phys Fitness. 2014;54(3):255–63.

43. Bongers CCWG, ten Haaf DSM, Ravanelli N, Eijsvogels TMH, Hopman MTE. Core temperature and sweating in men and women during a 15-km race in cool conditions. Int J Sports Physiol Perform. 2020;15(8):1132–7.

44. Bhambhani Y, Singh M. Metabolic and cinematographic analysis of walking and running in men and women. Med Sci Sports Exerc. 1985;17(1):131–7.

45. Daniels J, Daniels N. Running economy of elite male and elite female runners. Med Sci Sports Exerc. 1992;24(4):483–9.

46. Cureton KJ, Sparling PB. Distance running performance and metabolic responses to running in men and women with excess weight experimentally equated. Med Sci Sports Exerc. 1980;12(4):288–94.

47. Lopes TR, Oliveira DM, Simurro PB, Akiba HT, Nakamura FY, Okano AH, et al. No sex difference in mental fatigue effect on high-level runners' aerobic performance. Med Sci Sports Exerc. 2020;52(10):2207–16.

48. Boldt P, Knechtle B, Nikolaidis P, Lechleitner C, Wirnitzer G, Leitzmann C, et al. Sex differences in the health status of endurance runners: results from the NURMI study (step 2). J Strength Cond Res. 2019;33(7):1929–40.

49. Deaner RO. Distance running as an ideal domain for showing a sex difference in competitiveness. Arch Sex Behav. 2013;42(3):413–28.

50. Thibault V, Guillaume M, Berthelot G, Helou NE, Schaal K, Quinquis L, et al. Women and men in sport performance: the gender gap has not evolved since 1983. J Sports Sci Med. 2010;9(2):214–23.

51. Sparling PB, O'Donnell EM, Snow TK. The gender difference in distance running performance has plateaued: an analysis of world rankings from 1980 to 1996. Med Sci Sports Exerc. 1998;30(12):1725–9.

52. Tucker R, Santos-Concejero J. The unlikeliness of an imminent sub-2-hour marathon: historical trends of the gender gap in running events. Int J Sports Physiol Perform. 2017;12(8):1017–22.

53. Seiler S, De Koning JJ, Foster C. The fall and rise of the gender difference in elite anaerobic performance 1952-2006. Med Sci Sports Exerc. 2007;39(3):534–40.

54. Cheuvront SN, Carter Iii R, Deruisseau KC, Moffatt RJ. Running performance differences between men and women: an update. Sports Med. 2005;35(12):1017–24.

55. Coast JR, Blevins JS, Wilson BA. Do gender differences in running performance disappear with distance? Can J Appl Physiol. 2004;29(2):139–45.

56. Joyner MJ. Physiological limits to endurance exercise performance: influence of sex. J Physiol. 2017;595(9):2949–54.

57. Cejka N, Knechtle B, Rüst CA, Rosemann T, Lepers R. Performance and age of the fastest female and male 100-km ultramarathoners worldwide from 1960 to 2012. J Strength Cond Res. 2015;29(5):1180–90.

José Gomes Pereira

7.1 Introduction

Walking is a form of natural locomotion for humans. As long as one has neuro-muscular capabilities to practice running, it can also be considered basically a form of natural locomotion. The act of running is the form of displacement used in a multiplicity of sports. In this context, we can consider running on two levels, running as a sport, such as athletics in its different variants and specialties—speed and endurance events, and running as a means of specific mobility in other sports, such as football, rugby, handball, basketball, tennis and many others. It is obvious that the running technique varies depending on the type of sport considered.

Another way of considering running is as a leisure activity or prophylactic procedure for many diseases where cardio-metabolic dysfunctions occupy a suitable place. In fact, running is a form of cardiovascular exercise that is easily accessible and has a great aerobic impact. Indeed, by improving aerobic fitness, running is a great way to help improve cardiopulmonary and metabolic health. In addition, it burns calories and can increase strength. It is also essential to consider the psychological benefits. There is no doubt that running has benefits for physical and mental health. However, as with all physical and sporting activities, it is not risk-free. Therefore, the prescription of running exercises should consider the individual's physical condition and a correct training methodology.

When it is adapted to the individual's specific condition and if there are no musculoskeletal or disease-induced contraindications, running is a safe activity with health benefits. Within the scope of health benefits and in the context of health promotion, running is a form of physical exertion and a very effective method to improve the cardiometabolic condition. However, a cardiovascular system well adapted to effort and a trained and healthy musculoskeletal system is not the only adaptations caused by exercise. Therefore, it is necessary to consider a set of metabolic changes that result from exercise and are present in regular joggers [1]. In addition to cardiovascular benefits, physical exercise can have an anti-inflammatory effect, which implies a beneficial effect of regular exercise in chronic metabolic diseases. In this context, it is essential to consider the anti-inflammatory and immunomodulatory action associated with exercise because the immune system and inflammation play a crucial role in developing several chronic metabolic diseases.

Because it is not possible to fully address the various and numerous metabolic diseases in the writing of this chapter, which is necessarily summarized, we sought to emphasize the effect of

J. G. Pereira (✉)
School of Human Kinetics, University of Lisbon, Lisbon, Portugal
e-mail: jgpereira@fmh.ulisboa.pt

exercise, particularly running, on the individual's metabolic health, expressed in diseases and metabolic disorders that clearly benefit from exercise in general and running in particular. In this case, we considered type 2 diabetes, dyslipidaemia, obesity and metabolic syndrome.

7.2 Concept of Metabolic Disease

7.2.1 What Is a Metabolic Disorder or Disease?

There are hundreds of metabolic diseases, each with its specificities. For this reason, it is not easy to establish a generalist definition. However, in general, and simplistic terms, we can see that a metabolic disease or metabolic disorder is a pathological condition in which a dysfunctional metabolic process causes an organic malfunction. The metabolic diseases can be classified as congenital dictated by heredity or acquired due to a disease of an endocrine organ or another metabolically important organ. In addition, it can be defined as any disorder that disrupts normal metabolism. A metabolic disease occurs when an important metabolic organ becomes dysfunctional, such as the liver or pancreas, to mention just two examples. Type 2 diabetes is the most common metabolic disease. However, there are many others with different causes and characteristics. In general terms, we can say that a metabolic disease occurs when the body's normal metabolic processes are disturbed. In fact, metabolic disorders incorporate a range of diseases that affect the chemical processes in the body. The management of these disorders is disease-specific and usually incorporate dietary restrictions, medications and exercise in some cases, or a combination between them.

Metabolism is a set of biochemical processes that occur in our body. The metabolism comprises two phases, anabolism when related to the generation phase and catabolism when associated with destruction. A metabolic dysfunction occurs when the metabolic process alters essential substances necessary for maintaining health. The human organism is susceptible to errors in metabolism, and a metabolic disease encompasses all conditions where these processes are functioning incorrectly.

As mentioned before, metabolic dysfunctions can be classified as hereditary or acquired. Hereditary metabolic disorders are due to innate errors in metabolism. They are heritable or genetic disorders. There are hundreds of hereditary metabolic diseases, for example, phenylketonuria, Hurler syndrome, Tay-Sachs disease and Fabry disease, among many others.

The acquired metabolic disorders are associated with external factors. Such as sedentary lifestyle, dietary errors with excessive caloric intake and unhealthy lifestyles are considered. For Eckel et al. [2], the human lifestyle is associated with an inherited epigenetic pattern, which leads to the development of metabolic disorders. The well-known metabolic syndrome corresponds to a set of factors that we will address later, strongly associated with obesity and insulin resistance, which can be considered, at present, as a global epidemic. In fact, the concept of metabolic syndrome has a predisposition for the onset of cardiovascular and metabolic disorders that are well expressed in the concept of cardiometabolic risk. In this context, strategies include prevention through exercise, where running is an option, in addition to nutrition and healthy lifestyles.

As we saw before, there are numerous metabolic diseases, but we will address those that are most related to the effect that running can have on these diseases for the purposes of the present chapter. In this context, we will consider the effect of exercise on the regulation of metabolism and its influence on cardiometabolic risk, metabolic syndrome, obesity, dyslipidaemia and type 2 diabetes.

7.3 Health Benefits of Running. General Aspects

Conceptually, running is a type of terrestrial locomotion in which human beings move quickly. Running is very different from walking both biomechanically and physiologically. As it is known, running is characterized by a typical aerial phase in which the two feet for a moment do not touch the ground, which is very different from walking where one of the two feet is always on the ground, progressing in a cyclical and repeated way. Ancestrally, man developed the ability to move around running, often long distances, to satisfy his survival needs. The use of running for sporting purposes came later and for therapeutic purposes and health promotion much later. Nowadays, there is a generalization of running for leisure and well-being and not just as sports. In fact, derived from its biomechanical and physiological conditions, running has more impact than walking and can be considered the most accessible sport in the world. It is precisely based on its biomechanical and physiological characteristics that its impact on health can be analysed.

Concerning jogging and slow running, we can say that there are differences. These differences are substantially reflected in the intensity of the exercise. Basically, both are forms of aerobic exercise. As is known, aerobic exercise provides cardiorespiratory and musculoskeletal adaptations. In this sense, it is important to consider the effects of running on pulmonary function, cardiovascular system and skeletal muscle—muscle strength and endurance.

Regarding the VO_2max, it is a physiological parameter considered the best indicator of cardiorespiratory fitness. It can be defined as the ability to capture oxygen (pulmonary ventilation), fixation (alveolar-capillary exchanges), transport (cardiovascular system) and peripheral use (musculoskeletal tissue) in an effort of the whole body and maximum intensity. A high VO_2max value, common in runners, indicates better aerobic fitness and a better cardiometabolic condition. In this context, it can be classified as a health indicator. The VO_2 max is influenced by several factors. Among these factors, we highlight heredity, systematic aerobic training, age and gender. A high VO_2max allows more energy to be produced (aerobic capacity). The performance of medium and long-distance running constitute an effective mode of aerobic training. In fact, VO_2max is the "gold standard"

measure of general fitness and preventive health for cardiovascular and some metabolic diseases.

Considering the response and hemodynamic adaptations to exercise expressed by blood pressure, an important cardiovascular indicator, it can be said that running provides adaptive processes with a clear hypotensive effect. However, according to Igarashi and Nogami [3], there are pros and cons for the relationship between running regularly and changes in resting blood pressure. Therefore, for these authors, running regularly at moderate intensity and a restrained volume is recommended to lower resting blood pressure in subjects with hypertension.

The implementation of regular running programs of moderate-intensity, not exceeding 10 METs and adapted to the physical condition of each one has a clear and safe positive effect in reducing mortality from cardiovascular causes and mortality from all causes. This benefit can be in the range of 20 to 30% [4]. Based in Warburton et al. [5] and Hu et al. [6], an increase in energy expenditure related to the exercise of 1000 kcal per week, or an increase in the physical condition of 1 MET, results in a 20% benefit in mortality. For the same authors (5 ,6), middle-aged women who do less than 1 hour of exercise per week experience a 52% increase in all-cause mortality, compared to physically active women, main related to cardiovascular diseases—a doubling related mortality and cancer-related mortality (29%). A clear indicator in favour of physical exercise is that mortality from inactivity presents similar values to several modifiable risk factors, such as hypertension and obesity.

Regular physical activity reduces the risk of death from cardiovascular disease. Mortality from all causes and mortality from cardiovascular disease is estimated to be reduced by 20–35% [4]. Because exercise promotes energy expenditure, the additional energy expenditure of 1600 kcal/week can slow the progression of heart disease with a 20% mortality benefit [6]. More than 2200 kcal/week energy expenditure can reduce atheromatous plaque [7]. In randomized clinical trials, exercise contributes to the secondary prevention of cardiovascular disease. Contrary to what was thought, recent data from current investigations prove that exercise attenuates or reverses the risk of cardiovascular disease [8].

Running also has a musculoskeletal impact. Running has repercussions for gains in muscle mass and strength, especially in the lower limbs. This effect of running on the muscular function of the lower limbs and musculature, in general, adds benefits to the effects of running on health. In fact, running is a complete exercise and an option for whole-body training. Although running is a full-body workout, it has the particularity of requesting the stabilizing muscles of the core and pelvis. This is important because these muscles are responsible for the stability of the skeleton.

Exercise can also be considered primary prevention for osteoporosis. Peak bone mass is reached between 25 and 35 years old, and regular exercise minimizes age-related loss. Exercise may be helpful for the prevention of osteoporosis and for the increase of bone density. Exercise can also be helpful as a therapeutic modality for patients with osteoporosis already present. However, in this case, the major benefit of exercise can be related to better muscle strength and coordination, important aspects for fall prevention. It is important to point out that hight impact exercises as running are contraindicated in people with osteoporosis already installed because the compressive forces in the spine and lower extremities can cause fractures in the osteoporotic bones. However, such exercises exhibit efficacy for maintaining bone mineral density in non-osteoporotic persons and can be used to prevent or delay osteoporosis. To benefit osteoporosis through a running exercise program, it is important to know that the gains made in bone density will only be maintained as long as the exercise is continued. The changes in bone mass required approximately 1 year to be significant. If significant bone density gains are not achieved, several benefits from an exercise program like muscle strength remain, like balance and motor coordination. In this case, resistance exercises seem to be more advisable compared to low impact exercises. Conventionally oriented exercise is estimated to prevent 1% of

bone loss per year. This loss is greater for postmenopausal women, with exercise playing an essential role in attenuating this loss of bone mass and reducing fracture risk [9]. A 12-year follow-up study of more than 60,000 postmenopausal women proves that the risk of hip fracture was reduced by 6% for each increase in 3 MET-hours per week of activity [10]. For the same authors, active women had a 55% lower risk of hip fracture than sedentary women. Walking at least 4 h a week was associated with a 41% lower risk of hip fracture [10].

Exercise can be considered the main component of a healthy lifestyle. It also has a therapeutic effect. However, like any therapy, it is essential to use the correct intensity and duration. Several studies demonstrate the beneficial effect of running on longevity due to the control of risk factors and cardiorespiratory and metabolic physical condition improvement.

The regular practice of physical activity, namely walking, jogging or running, positively influences the functioning of most organs and tissues, with well-established relationships between physical activity and a variety of biomarkers.

Walking and running at low intensities and low impact is a natural way of exercising, requiring any special preparation or equipment (except the shoes and suitable clothes adapted to the environment climate).

Running regularly can have a favourable impact on health status. It is believed that the benefits of running can appear with only 5–10 min of moderate-intensity with a positive influence on mortality, particularly in specific diseases, namely cardiovascular and stroke [11]. It is also believed to influence oncological and neurological diseases, namely Alzheimer's and Parkinson's diseases. It is curious to note that these benefits can be obtained with a low running volume.

In fact, the benefits of exercise in oncologic diseases are well established. Exercise is believed to have a primary preventive effect. In this context, the most available studies focus on two types of cancer, breast and colon [12]. The reduction is estimated to be 20% for breast cancer in women and 30% for colon cancer in men. In order to have benefits from these diseases, exercise does not need to be very intense. The use of low intensities, in the order of 5 MET, which corresponds to a fast walk, already seems to affect [13]. In fact, exercise improves the quality of life and general health in cancer patients. These benefits may be related to adipose tissue metabolism, improved immune function, and reduced oxidative stress [5]. Adapting the exercise prescription for cancer patients is very difficult because the ability to respond to different types of exercise differs between individuals and depends on the type of cancer and factors related to the individual, such as age, general health status, physical conditioning and presence of comorbidities. The proposed guidelines for exercise prescription include moderate to vigorous exercises performed 3–5 times a week for 30–60 min each session [14].

Exercise can reduce the risk of developing Alzheimer's disease by up to 50%. In the context of this disease, an exercise program can also delay deterioration in people who have already started to develop cognitive problems. This statement is supported by the Alzheimer's Research and Prevention Foundation [15]. In a literature review on the effectiveness of physical exercise, there was an improvement in cognitive function and neurogenesis. Most of these studies indicate that physical inactivity is one of the most common preventable risk factors for the development of Alzheimer's disease, and a higher level of physical activity is associated with reduced risk and the symptoms associated with the disease, namely the cognitive function, clearly improve with exercise [16]. For this purpose, the exercises that have proven to be most effective are aerobic, such as brisk walking or even running, for those who are physically fit for that. The recommendation of 30 min of moderately vigorous aerobic exercise, 3–4 days per week, has consensus. The muscle-strengthening activities also proved to be useful and recommendable.

Considering the influence of running on longevity, it is estimated that the reduction in premature mortality is between 25 and 40%. In this context, the life expectancy in runners is about 3 years longer when compared to non-runners [17]. Thus, the impact of running has repercussions on the different indicators of health and premature mortality. Therefore, it is essential to consider the mechanisms that have connections between running and preventing chronic diseases and longevity. However, it is doubtful that high doses of running will have an added benefit. According to Peter Schnohr et al. [18], running 2.5 h a week, or 30 min 5 days a week, is sufficient to produce benefits in longevity.

It is curious to emphasize that physical inactivity is a real risk factor [19]. The advantage is that it is a modifiable risk factor. It is important to consider that it is possible to reverse this situation at any point in our life and start a practice of physical activity, where running is a possible option. Because physical inactivity predisposes to various diseases, such as cardiovascular diseases, obesity, type 2 diabetes, cancer, hypertension, osteoporosis, sarcopenia, depression, and anxiety, it is essential to eliminate this risk factor—physical inactivity. Following what was previously mentioned, it is also proven that physical activity reduces the risk of premature death and promotes longevity. In this context, it can be considered as a treatment modality included in the scope of primary prevention.

The effects of exercise on ageing can also be considered in the context of general benefits. These benefits are also seen for people with risk factors, such as hypertension, dyslipidaemia, diabetes, chronic obstructive pulmonary disease, smoking, etc. The studies by Garatachea et al. [20] clarify the effects of exercise on physiological changes associated with ageing. This positive influence has repercussions at the cellular level, with clear implications for the inflammatory process and intercellular signalling. These concepts extend to the concept of exercise seen as a therapy with a high potential for applicability. However, in this perspective of exercise seen as a medicine, the amount of physical activity necessary to optimize health has not yet been adequately established. Miller et al. [19] postulate that the specific dose of exercise necessary to obtain benefits concerning a certain disease process is difficult to determine.

7.4   Cardiometabolic Risk

Cardiometabolic diseases are the number one cause of death and are mainly caused by an unhealthy lifestyle. Exercise reduce their level of risk. Cardiometabolic disorders represent a cluster of interrelated risk factors that can also be called cardiometabolic risk biomarkers, namely: primarily hypertension, elevated fasting blood sugar, dyslipidaemia, abdominal obesity and elevated triglycerides. Cardiometabolic risk is significantly associated with central obesity and visceral adipose tissue [21]. Visceral fat is the result of an imbalance between intake and energy expenditure. Considering its metabolically active function, adipose tissue produces various pro-inflammatory and pro-thrombotic cytokines [22]. The cardiometabolic syndrome is considered a disease by The World Health Organization [21]. People with cardiometabolic syndrome increase two to three times the mortality from coronary heart disease or stroke [23]. Programs for the prevention and treatment of cardiometabolic syndrome use strategies that combine diet and exercise. They mainly affect behavioural changes in order to reduce cardiometabolic risk factors. Of all the strategies, aerobic physical activity seems to be the most consensual, with running being a good example.

Cardiometabolic risk is identified with an increased risk and a greater probability of vascular events, development of diabetes and other metabolic diseases. Hypertension and dyslipidaemia are cardiometabolic risk factors. Two of the main factors that influence cardiometabolic risk are abdominal adiposity and insulin resistance. Family history, hypertension, dysglycaemia, dyslipidaemia, smoking, abdominal obesity, insulin resistance, inflammation, diet, sedentary lifestyle and psychosocial stress are also recognized as risk factors.

Running is basically an aerobic exercise and has a favourable influence on cardiometabolic risk factors. Therefore, it is crucial to identify the types of training based on running that are most effective in preventing cardiometabolic risk. Unfortunately, the literature on this specific subject is limited. For this reason, the previous finding is based on logical reasoning in the field of cardiorespiratory physiology. Based on the work of Kemmler et al. [24], both high-intensity interval training and low-impact continuous aerobic training were comparatively effective, in untrained middle-aged men, for the development of cardiometabolic indices and cardiorespiratory condition.

Running training and the belief that the most efficient way to improve cardiometabolic risk factors is aerobic exercise has undoubtedly not yet been confirmed. However, Kemmler et al. [25] determined the effect of high-intensity interval training, 80–100% of maximum heart rate compared to continuous running training of moderate-intensity, 65–77.5% of maximum heart rate cardiometabolic risk factors, in selected subjects. Both protocols were comparable for energy consumption. However, the changes in VO_2max differed significantly, and the high-intensity interval training showed a more significant impact on cardiometabolic health.

Tuttor et al. [26] also studied the possible positive effects of high-intensity interval training on cardiometabolic, cardiac and morphometric indicators closely related to cardiometabolic health in overweight men 30–50 years old. The results obtained point out that the effects of high-intensity interval training were, on average, more pronounced and positively affected the cardiometabolic risk.

7.5   Myokines and Exercise: Role of Myokines in the Regulation of Metabolism and Metabolic Health

Physical exercise, which includes running, improves the general physical condition and allows combating or delaying the onset of chronic diseases, including metabolic diseases. It is believed that this effect of metabolic regulation through muscular activity may be related to the secretory activity of the muscle that produces substances called cytokines that can function as hormones and the muscle as an endocrine organ

[27]. In addition, cytokines will exert their effect on other several target organs such as the liver, adipose tissue and bone, among others, contributing to their metabolic regulation. Therefore, another designation for cytokines produced by the skeletal muscle tissue is myokines. What are myokines? The term "myokine" was introduced by Pederson et al. [28]. Myokines are a particular type of cytokines synthesized and released by myocytes during muscular activity. They are implicated in regulating metabolism in the muscle as well in other tissues and target organs through their receptors [29]. Muscle exercise as a stimulus for synthesizing myokines already identified in studies both in vivo and in vitro [30].

Running has a beneficial effect on metabolic diseases. The mechanisms by which exercise produces these beneficial effects are not yet fully explained. Running exercises seem to promote a favourable response to the action of myokines. Conversely, a sedentary lifestyle harms it. An explanation for the association between physical inactivity and chronic diseases may lie here.

For Leal Luana et al. [31], myokines function as an interface of an immunometabolic factor through the release of humoral factors that interact with other tissues, especially adipose tissue. On the other hand, physical inactivity would be associated with inflammation and the production of pro-inflammatory cytokines (ex. adipokines).

Muscular activity has a favourable effect on metabolic disorders. The interest in the effects of exercise on muscle metabolism has given rise to recent research [32]. Some of these studies focus on the effect of muscle exercise on myokine secretion, establishing a relationship between the contractile activity of skeletal muscle and humoral changes. Some of the metabolic changes induced by exercise have repercussions on other organs, such as the liver and adipose tissue [33]. Several myokines have already been identified, with particular interest for some of them, namely interleukin-6, irisin, myostatin, interleukin-15 and a neurotrophic factor derived from the brain (BDNF), among several others. Brain-derived neurotrophic factor (BDNF) is essential for plasticity, is a member of the neurotrophy family of growth factors and increases following physical exertion. Skeletal muscle expresses several myokines. The contractile activity plays a key role in regulating the expression of cytokines in skeletal muscle. Anti-inflammatory, immunoregulatory and metabolic functions have been attributed to it [34].

The comprehension of the myokines functioning activity allowed us to clarify the regulatory effect of muscle activity on metabolism. It is not only the regulation of local muscle metabolism but also systemic metabolism. This regulation can be carried out in an autocrine, paracrine and endocrine manner [32]. Knowing how myokines work allows us to understand the role of exercise in metabolism regulation, with evident effects in preventing and treating metabolic diseases [35]. In this context, it has been shown that a 10 km run induces an increase in plasma levels of some myokines [36].

How muscle activity regulates metabolic and physiological effects with beneficial health effects is not yet fully elucidated. The discovery that muscle contraction induces the synthesis and release of cytokines with an effect on target organs creates a new perspective: the endocrine functioning of skeletal muscle. Myokines produced in muscle interfere with the metabolism of tissues and organs. This opens a horizon for the understanding of metabolic dysfunctions and their relationship with the effects of exercise.

Severinsen and Pedersen [32], in a recent paper, postulate that exercise stimulates the muscle to produce myokines, which allows communication between the muscle and other organs – "muscle-organ crosstalk". It is also mentioned that myokines can be useful biomarkers to monitor exercise prescription for people with pathological conditions, including metabolic diseases.

7.6 Metabolic Syndrome

The metabolic syndrome is closely related to overweight and obesity. It is also associated with a lack of activity—a sedentary lifestyle. Metabolic syndrome is a set of risk factors, essentially lipid and cardiovascular, which are based on abdominal obesity. Individuals with this syndrome are at risk of developing type 2 diabetes and cardiovascular diseases. It is preferable and more correct to consider metabolic syndrome as a pre-pathologic situation typically related to modern civilization. The human being is not genetically adapted for high caloric intake and low energy expenditure (sedentary lifestyle). The immune system and inflammation play a crucial role in the development of several chronic metabolic diseases. In this case, the energy is accumulated in the organism in the form of fat, especially in the abdominal area, strongly associated with a chronic inflammatory process involved in developing metabolic diseases. It is a subclinical and chronic inflammation, playing a key role in the pathophysiology of metabolic syndrome. Therefore, the probability of developing a metabolic disease may be related to a greater predisposition. It is believed that there may be a genetic predisposition in some people, but the leading causes are inadequate diet and physical inactivity. Despite not being a disease, metabolic syndrome is a set of indicators that coexist and characterize this predisposition and an increased risk for cardiometabolic pathologies and diabetes. These indicators or conditions are high blood pressure, hyperglycaemia, excess abdominal fat and adverse lipid levels. Even when the risk factors are only slightly elevated, the risk remains. The human being is not genetically adapted to live in an inflammatory state. In addition to this, a high caloric intake associated with a low energy expenditure typical of physical inactivity may be the explanatory basis of the metabolic syndrome.

7.6.1 How Is the Diagnosis Made?

According to the International Diabetes Federation [37], the diagnosis of metabolic syndrome is established when a person with abdominal adiposity fills two or more factors of the remaining four presented below.

- Abdominal Obesity: Abdominal perimeter that exceeds 102 cm in men and 88 cm in women.

- Triglycerides equal to or greater than 150 mg/dL.
- HDL cholesterol equal to or less than 40 mg/dL in men and 50 mg/dL in women.
- Blood pressure equal to or greater than 130–85 mmHg.
- Insulin resistance, fasting blood glucose equal to or greater than 100 mg/dL.

Diagnosis also can be made when at least three clinical findings out of the previous five are present [38].

It is estimated that more than 20% of the world's population meet the diagnostic criteria for metabolic syndrome. Individuals who test positive for metabolic syndrome are three times more likely to develop cardiovascular disease and five times more likely to develop type 2 diabetes mellitus [39].

Physical inactivity, so common in obese people, potentiates the metabolic syndrome's pathophysiological process. The amount and intensity of the exercises to deal with the metabolic syndrome can be expressed in the following recommendations:

150 min per week, with moderate to high-intensity exercises, such as the slow run or even most intense run modalities. For people with a lower level of physical condition, the exercise can be divided into 10–15 min with very interesting benefits.

These are, in fact, benefits when the exercise has adequate intensity and duration. However, the ideal type of exercise prescription is not yet fully established, although aerobic exercise seems to have a desirable effect on controlling the factors that integrate the metabolic syndrome.

Ostman et al. [39], in a meta-analysis of 16 randomized controlled trials to investigate whether aerobic exercise was reflected favourably in patients with metabolic syndrome, concluded that exercise training improves body composition, cardiovascular, and metabolic outcomes in patients with metabolic syndrome. Thus, in this context, aerobic exercise appears optimal.

Concerning the positive effects of running activities as a managing strategy for metabolic syndrome, it is important to point out that exercise should be combined with a proper diet. Running, even 5–10 min/day and at slow speeds <6 mph, is associated with markedly reduced risks of death from all causes and cardiovascular disease [11].

7.7 Benefits of Running in some Metabolic Diseases

1. Obesity
2. Dyslipidaemia
3. Type 2 Diabetes

7.7.1 Obesity

Based on the WHO definition, overweight and obesity are defined as abnormal or excessive fat accumulation that presents a health risk. A body mass index (BMI) over 25 is considered overweight, and over 30 is obese. According to the global burden of disease, the issue has grown to epidemic proportions, with over four million people dying each year due to being overweight or obese in 2017.

Obesity is the pandemic of the twenty-first century. It is a severe public health problem. The risk of chronic diseases increases with being overweight. It predisposes to metabolic syndrome, with all the risks that result, namely the predisposition to cardiovascular diseases and stroke, in addition to some forms of cancer. Obesity is a chronic disease with important metabolic implications. Clinically, the approach to this disease must also be made considering the comorbidities. Exercise and particularly running are a significant contribution to managing this disease when associated with diet, pharmacological therapy, and surgery in the most severe cases. The term obesity refers to excess fat and the exercise prescription, based on walking and running considering the intensity and duration of the exercise, providing a greater caloric expenditure.

Several methods are used to characterize obesity, such as skinfolds, electrical bioimpedance, underwater weighing, DEXA (dual-energy X-ray absorptiometry), with special interest for the latter method. A very practical approach for the quantitative characterization of obesity is the use of the body mass index (BMI), which limits the fact that it does not differentiate between fat mass and lean mass within the total body mass. It is an acceptable method to characterize overweight and obesity in practice, the quotient obtains its calculation between body mass in kilograms and the square of height in meters:

BMI expressed in kg/m^2 = weight (kg)/ $height^2$ (m)

Values below 18.5 characterize underweight; between 18.5 and 24.9 correspond to normality; between 25 and 29.9 overweight and obesity greater than 30 (class I = 30–34.9; class II = 35–39.9 and class III $\geq$ 40.0).

It is well established that obesity results from a positive energy balance associated with reducing physical activity. Most investigations focus on the mechanisms that regulate satiety and hunger centres in the central nervous system. Studies on the effects of leptin have influenced the understanding of obesity. The adipocyte produces leptin, and its role is to control body fat by acting on the hypothalamic centres that regulate satiety. In this context, leptin appears to be an appetite regulatory hormone synthesized by the adipocyte.

Although exercise does not cause significant weight loss hypothetically, the obese person benefits through increased cardiorespiratory fitness, glucose control, endothelial function, improvements in hyperlipidaemia, quality of life, and a reduction in future weight gain [40]. For weight loss in the obese, it is essential to have caloric restriction. At an early stage, calorie restriction is better than exercise for weight loss. However, exercise associated with caloric restriction improves body composition, increases fat loss, and decreases lean mass loss [41]. The great benefits of exercise in controlling obesity are not in the initial weight loss but in exercise maintenance to stabilize this weight loss [40]. The maintenance of successful weight loss requires a continuous reduction in energy consumption close to 320 kcal per day [42]. In conclusion, while food restriction is related to weight loss, exercise is the procedure for successfully maintaining lost weight [42]. According to ACSM, successful weight loss requires at least 250 min of moderate to vigorous exercise per week [43].

Manson et al. [44] demonstrated that the relationship between BMI and mortality had a J-shaped curve relationship for more than 20 years. Thus, the relationship between severe Class III obesity and the mortality rate is well established. For Class II obesity, moderate and slight mortality is attributed to Class I obesity. Interestingly, Class I obesity is not associated with a significant increase in the mortality rate for physically active individuals with satisfactory levels of physical condition.

Jogging exercises can be recommended and performed by individuals with Class I obesity. They tolerate this type of exercise well, where they burn more calories than if they just walked. Incidentally, running, even moderate leads to cardiorespiratory gains.

Running is a commonly adopted way to lose weight and fight obesity. Vincent [45] advocate that, at an early stage, intermittent walking and running programs are the most suitable for individuals who are unable to sustain running for long periods. For people with class III obesity, stationary cycling with intermittent intensities seems to be the most suitable for improving lipid oxidation before walking or running. Musculoskeletal pain should be avoided since fatigue can compromise the continuity of the exercise. Recovery time is critical to prevent overload injuries. In cases of obesity, those included in Class I are at lower risk for complying with a fitness program based on running.

7.7.2 Dyslipidaemia

Dyslipidaemia can be considered a metabolic disease. It is a disorder of lipoprotein metabolism. Dyslipidaemias are manifested by hyperlipemia in which cholesterol, triglycerides or both are elevated, associated with a low HDL level.

Cholesterol plays an important role as it is the essential component of cell membranes and brain cells. It also participates in the absorption of fat-soluble vitamins, such as Vit. D. The body produces cholesterol and obtains it through food.

Triglycerides are found in fat cells, adipocytes and can provide energy to ensure metabolic processes. Triglycerides are produced in the intestine and liver from fatty acids.

Lipoproteins are protein particles that carry lipids. There are different types of lipoproteins: chylomicrons, very-low-density lipoproteins (VLDL), low-density lipoproteins (LDL) and high-density lipoproteins (HDL).

The increase in the concentration of cholesterol and low-density lipoprotein (LDL-C) is associated with an increased risk of cardiovascular disease, a fact that has been well documented in studies by Lewington et al. [46]. Inversely, the high-density lipoprotein cholesterol (HDL-C), considered the "good cholesterol", is a favourable and significant predictor [47].

Regarding triglycerides, their increase was associated with an increased risk of ischemic heart disease, ischemic stroke and all-cause mortality [48].

Exercise benefits dyslipidaemia. Pederson and Saltin [49], analysing several meta-analyses, concluded that aerobic exercise could positively impact lipid profiles.

People with dyslipidaemia experience modest but beneficial effects of aerobic exercise training on blood lipids. Typically on the order of 5–10% decrease in LDL-C and an increase in HDL-C. The exercise benefits on blood lipids include: decrease concentrations of small dense LDL particles; increase of HDL-C concentration, and reduced postprandial lipemia. Also induced by exercise, the increase in lipoprotein enzyme activity is linked to the above-mentioned lipoproteins' changes.

There is evidence that running raises the value of HDL-C, which exhibited greater sensitivity to aerobic exercise than LDL-C and triglycerides. The majority of studies concerning the effect of running are consistent with the increase in HDL-C. It is believed that aerobic exercise effectively increases HDL if the caloric expenditure, intensity or distance covered are adequate [50]. In this context, reducing serum cholesterol can help reduce the risk of coronary heart disease. In addition, the aerobic impact associated with regular running improves the prognosis of cardiovascular disease.

The study results performed by Giuseppe Lippi et al. [50] show that middle-distance running elicits acute favourable changes of lipid profile with implications in public health promotion and improvement.

It is well established that dyslipidaemia characterizes the risk of cardiovascular disease. Exercise has a positive effect on individuals with dyslipidaemia, improving the lipid profile. In this sense, a reduction in serum cholesterol is beneficial for the risk of coronary heart disease. Regular exercise has a positive effect on individuals with dyslipidaemia and also positively influences the lipid profile [51].

Long-distance running improves the metabolism of carbohydrates, fats and proteins, thus preventing obesity, diabetes and cardiovascular diseases. Running at an intensity of 60–65% of VO₂max provides a maximum oxidation rate of fatty acids. Lipid oxidation is the predominant fuel source for the referred submaximal exercise intensities. Such a kind of running exercise has the principal effect on triglycerides by lowering them, and on HDL, by increasing it. Exercise does not have much impact on LDL unless combined with dietary changes and weight loss [52].

Hespanhol et al. [53] conducted randomized clinical trials with a minimum duration of 8 weeks of running training in a sample of healthy, untrained adults between 18 and 65 years of age. They obtained results demonstrating that endurance running produces substantial beneficial effects on body mass, body fat, VO₂max, triglycerides and HDL cholesterol. They also emphasize the importance of the duration of the running training, which is positively related to the health benefits achieved.

7.7.3 Type 2 Diabetes

Diabetes mellitus is a chronic metabolic disease. It is characterized by the inability to produce insulin, both quantitative and qualitative, or insulin resistance in tissues. As a result, hyperglycaemia arises.

Insulin is a hormone produced by the endocrine pancreas through β-cells. Insulin exerts its function in target organs, namely in muscle, adipose and hepatic tissue, indispensable for the physiological use of glucose by these tissues. In addition, type 2 diabetes is characterized by other metabolic impairments, particularly in the inflammatory homeostatic equilibrium and lipid metabolism. In this context, hyperlipidaemia (lipotoxicity) and hyperglycaemia (glucotoxicity) harm β-cell function and increase in insulin resistance.

Type 2 diabetes mellitus is a severe public health problem that has been increasing. It carries an increased risk of comorbidities. It has increased in various parts of the world in the past two decades and can be considered a worldwide public health problem. This disease is associated with numerous systemic complications that affect the retina, heart, brain, kidneys and nerves, with negative implications for proprioception, muscle strength and balance and coordination problems in patients with peripheral neuropathy. The evidence strongly supports the significant role of exercise in the treatment and control of diabetes mellitus and the associated complications.

Diabetes is a disease that presents itself in several ways. The exercise prescription and the option for a pharmacological treatment are influenced by the type of diabetes considered. Diabetes, once called adult or non-insulin-dependent diabetes, is no longer called this way. The designation type 2 diabetes is preferred. This is the most common form of the disease, with roughly 90–95% of diabetics with this type [54]. Diabetes affects a considerable number of the adult population, but it is no longer called adult diabetes because it has been observed with increasing frequency in adolescents, related to physical inactivity and overweight or obesity. The designation of non-insulin-dependent is also unsatisfactory because, in advanced cases of the disease, it may be necessary to carry out insulin therapy.

However, the pathophysiological explanation of type 2 diabetes is complex and multifactorial.

We can simplify it by saying that insulin resistance in peripheral tissues is a significant feature. When there is resistance to insulin action, the body may not use it properly, compromising carbohydrate metabolism in muscle and liver tissues. In most cases, insulin production is sufficient early in the course of the disease. However, it is a progressive disease. Over time, pancreatic beta cells cannot increase their secretory activity, which is ineffective in compensating for insulin resistance by developing hyperglycaemia. Treatment options based on diet and exercise are a good strategy, in addition to medication. It is a disease with a real genetic load. The likelihood is double for those who have diabetic parents. In addition to genetic influences, there are other risk factors: overweight, obesity, abdominal fat distribution and physical inactivity, lack of physical exercise, with an important role in insulin resistance.

Sedentary behaviours with low energy expenditure have been proven to influence cardiometabolic health and increase mortality and morbidity. Combating physical inactivity through exercise is a good and effective way to improve glycaemic control [54].

The recommendation of exercise in the treatment of diabetes is not new. It is believed that in classical antiquity, under the Hippocratic concept of medicine, exercise is recommended.

Exercise can be considered a major factor in the treatment of diabetes. Therefore, running programs has the potential to provide important benefits. For the diabetic individual, it is possible to be active and even to run. In fact, it may be helpful.

Three major metabolic dysfunctions characterize type 2 diabetes: the impairment in pancreatic beta-cell insulin secretion in response to a glucose stimulus; the reduced sensitivity to the action of insulin in major organs systems such as muscle, liver and adipose tissue; and the excessive hepatic glucose production in the basal state.

For the treatment of type 2 diabetes, exercise plays a key role in conjunction with a proper diet and medication. The benefits of exercise in diabetes include reduced hyperinsulinemia; improvement in insulin sensitivity, reduced body fat and normalization of dyslipoproteinemias.

In fact, exercise can play a role in the prevention or delay type 2 diabetes. Helmrich et al. [55] found a 6% decrease in age-adjusted risk for every increment of 500 kcal energy expenditure per week, studying an alumni group at the University of Pennsylvania.

Running and jogging is recommended for people with type 2 diabetes because it improves the sensitivity of muscle tissue to insulin action. This makes running especially helpful for people with type 2 diabetes to combat insulin resistance naturally. The American Diabetes Association encourages people with diabetes should perform aerobic exercise regularly. For type 2 diabetic adults, aerobic activity training should have a minimum duration of 10 min, progressing up to 30 min or more, most days of the week [53].

Regarding the use of short or long runs, we can mention that the short runs of 20–45 min can take place more times a week and with less glycaemic risk. Long runs range from 1 to 2 h, forcing you to go at a slower pace. They are only recommended for diabetic people who have adapted well to short runs, carrying a higher glycaemic risk. The use of short runs is also sufficient and recommended for type 2 diabetic individuals.

In nondiabetic individuals, metabolic-hormonal control allows maintenance of glucidic homeostasis. However, in diabetics, the response to exercise is not normal, and the balance between peripheral glucose use and hepatic metabolism is disturbed. The effect of diabetes on the response to exercise depends on several factors; among others, it is important to mention the medication, pre-effort blood glucose, volume and intensity of exercise.

Undoubtedly, exercise can be considered a major factor in the treatment of diabetes. The running programs have the potential to provide desirable benefits (Fig. 7.1).

In the domain of exercise, the prescription must be individualized according to the subject's physical and metabolic conditions. Running can be considered a multi-beneficial way to accom-

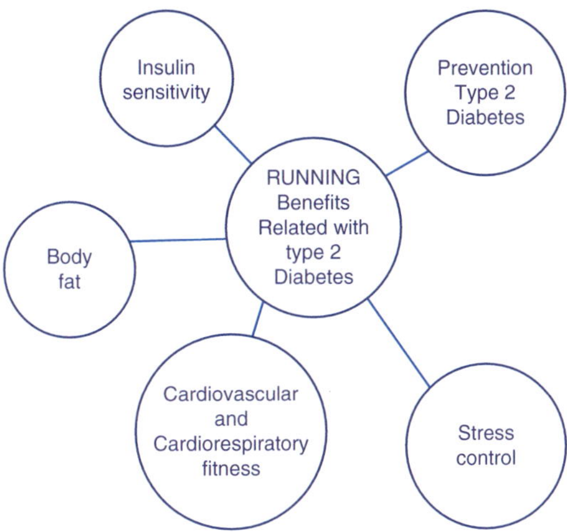

Fig. 7.1 Potential running programs benefits

plish exercise with influence in cardiorespiratory fitness, muscular strength and body composition. However, an exercise that benefits type 2 diabetes can take many forms. We refer to endurance and resistance training, which are the modalities that best affect type 2 diabetes. It is known that the effect of exercise on beneficial changes in glucose tolerance and insulin sensitivity is significantly in about 48–72 h after performing the exercise [56]. Logically, physical activity must be regular to achieve the desired goals. It is also known that patients with type 2 diabetes have lower levels of VO_2max, which will influence the running methodology adopted and the intensity of the training load.

With regard to resistance training, it is an excellent method to develop strength with evident effects on glucose tolerance and insulin sensitivity. However, it should be noted that people with type 2 diabetes must adapt the intensity and the way they run because they can develop autonomic neuropathy, with an influence on the heart rate, which is no longer a reliable indicator of intensity. For this reason, there are situations in which it is preferable to use the rate of perceived exertion to measure the intensity of the effort.

Exercise affects a patient's blood-sugar control positively. For those, exercise should be a primary means of blood-glucose control. On the priority list, it comes right after eating correctly. Epidemiologic studies revealed a high rate of type 2 diabetes, and heredity alone cannot account for such high prevalence rates [57]. Overweight, obesity and sedentarism play a crucial role. Exercise also prevents the development of diabetes complications.

Running can be a very recommended form of exercise prescription for people with type 2 diabetes. This is because running improves the body's sensitivity to insulin, which is extremely useful because it reduces insulin resistance. For the diabetic individual, it is possible to be active and even to run. It may be helpful.

In the domain of exercise, the prescription must be individualized according to the subject's physical and metabolic conditions. Running can be considered a multi-beneficial way to accomplish exercise with influence in cardiorespiratory fitness, muscular strength and body composition.

It is with some certainty that it can be said that running is a suitable form of exercise for people with diabetes, as long as their physical condition allows it. It improves insulin sensitivity, which is especially useful for people with type 2 diabetes because it helps to fight insulin resistance. Running is also an effective and controlled way to lose weight [58].

Sedentary behaviours with low energy expenditure have been proven to influence cardiometabolic health and increase mortality and morbidity. Combating physical inactivity through exercise is a good and effective way to improve glycaemic control [53].

Finally, it seems appropriate to say that physical exercise can be considered a "prescription". Considering the real benefits of exercise and its relative contraindications, we can safely say that if there were a pharmacological substance that would bring together all the benefits of exercise, those that are known and those that have not yet been discovered, if that were possible, it would be undoubtedly the best-selling medicine in the world.

References

1. Ledford H. The metabolic secrets of good runner—Chemical changes in runners linked to physical fitness. Nature. 2010.
2. Eckel RH, Alberti KG, Grundy SM, Zimmet PZ. The metabolic syndrome. Lancet (London, England). 2010;375(9710):181–3. https://doi.org/10.1016/S0140-6736(09)61794-3.
3. Igarashi Y, Nogami Y. Running to lower resting blood pressure: a systematic review and meta-analysis. Sports Med (Auckland, NZ). 2020;50(3):531–41. https://doi.org/10.1007/s40279-019-01209-3.
4. Macera CA, Hootman JM, Sniezek JE. Major public health benefits of physical activity. Arthritis Rheum. 2003;49(1):122–8. https://doi.org/10.1002/art.10907.
5. Warburton DE, Nicol CW, Bredin SS. Health benefits of physical activity: the evidence. CMAJ. 2006;174(6):801–9. https://doi.org/10.1503/cmaj.051351.
6. Hu FB, Willett WC, Li T, Stampfer MJ, Colditz GA, Manson JE. Adiposity as compared with physical activity in predicting mortality among women. N Engl J Med. 2004;351(26):2694–703. https://doi.org/10.1056/NEJMoa042135.
7. Hambrecht R, Niebauer J, Marburger C, Grunze M, Kälberer B, Hauer K, Schlierf G, Kübler W, Schuler G. Various intensities of leisure time physical activity in patients with coronary artery disease: effects on cardiorespiratory fitness and progression of coronary atherosclerotic lesions. J Am Coll Cardiol. 1993;22(2):468–77. https://doi.org/10.1016/0735-1097(93)90051-2.
8. Taylor RS, Brown A, Ebrahim S, Jolliffe J, Noorani H, Rees K, Skidmore B, Stone JA, Thompson DR, Oldridge N. Exercise-based rehabilitation for patients with coronary heart disease: systematic review and meta-analysis of randomized controlled trials. Am J Med. 2004;116(10):682–92. https://doi.org/10.1016/j.amjmed.2004.01.009.
9. Wolff I, van Croonenborg JJ, Kemper HC, Kostense PJ, Twisk JW. The effect of exercise training programs on bone mass: a meta-analysis of published controlled trials in pre- and postmenopausal women. Osteoporosis Int. 1999;9(1):1–12. https://doi.org/10.1007/s001980050109.
10. Feskanich D, Willett W, Colditz G. Walking and leisure-time activity and risk of hip fracture in postmenopausal women. JAMA. 2002;288(18):2300–6. https://doi.org/10.1001/jama.288.18.2300.
11. Lee DC, Pate RR, Lavie CJ, Sui X, Church TS, Blair SN. Leisure-time running reduces all-cause and cardiovascular mortality risk. J Am Coll Cardiol. 2014;64(5):472–81. https://doi.org/10.1016/j.jacc.2014.04.058.
12. Thune I, Furberg AS. Physical activity and cancer risk: dose-response and cancer, all sites and site-specific. Med Sci Sports Exerc. 2001;33(6 Suppl):S530–610. https://doi.org/10.1097/00005768-200106001-00025.
13. Lee IM. Physical activity and cancer prevention—data from epidemiologic studies. Med Sci Sports Exerc. 2003;35(11):1823–7. https://doi.org/10.1249/01.MSS.0000093620.27893.23.
14. Haskell WL, Lee IM, Pate RR, Powell KE, Blair SN, Franklin BA, Macera CA, Heath GW, Thompson PD, Bauman A. Physical activity and public health: updated recommendation for adults from the American College of Sports Medicine and the American Heart Association. Med Sci Sports Exerc. 2007;39(8):1423–34. https://doi.org/10.1249/mss.0b013e3180616b27.
15. Alzheimers Research and Prevention Foundation. Exercise and brain aerobics. The importance of physical exercise. www.alzheimersprevention.org.
16. Meng Q, Lin MS, Tzeng IS. Relationship between exercise and Alzheimer's disease: a narrative literature review. Front Neurosci. 2020;14:131. https://doi.org/10.3389/fnins.2020.00131.
17. Lee DC, Brellenthin AG, Thompson PD, Sui X, Lee IM, Lavie CJ. Running as a key lifestyle medicine for longevity. Prog Cardiovasc Dis. 2017;60(1):45–55. https://doi.org/10.1016/j.pcad.2017.03.005.
18. Schnohr P, Marott JL, Lange P, Jensen GB. Longevity in male and female joggers: the Copenhagen City heart study. Am J Epidemiol. 2013;177(7):683–9. https://doi.org/10.1093/aje/kws301.
19. Miller KR, McClave SA, Jampolis MB, et al. The health benefits of exercise and physical activity. Curr Nutr Rep. 2016;5:204–12. https://doi.org/10.1007/s13668-016-0175-5.
20. Garatachea N, Pareja-Galeano H, Sanchis-Gomar F, Santos-Lozano A, Fiuza-Luces C, Morán M, Emanuele E, Joyner MJ, Lucia A. Exercise attenuates the major hallmarks of aging. Rejuvenation Res. 2015;18(1):57–89. https://doi.org/10.1089/rej.2014.1623.
21. Srivastava AK. Challenges in the treatment of cardiometabolic syndrome. Indian J Pharmacol. 2012;44(2):155–6. https://doi.org/10.4103/0253-7613.93579.
22. Castro JP, El-Atat FA, McFarlane SI, Aneja A, Sowers JR. Cardiometabolic syndrome: pathophysiology and treatment. Curr Hypertens Rep. 2003;5(5):393–401. https://doi.org/10.1007/s11906-003-0085-y.
23. Mayo Clinic. Cardiometabolic program. Overview. http://www.mayoclinic.org/departments-centers/cardiovascular-diseases/overview/specialty-groups/cardiometabolic-program/overview.
24. Kemmler W, Scharf M, Lell M, Petrasek C, von Stengel S. High versus moderate intensity running exercise to impact cardiometabolic risk factors: the randomized controlled RUSH-study. Biomed Res Int. 2014;2014:843095. https://doi.org/10.1155/2014/843095.
25. Kemmler W, Lell M, Scharf M, Fraunberger L, von Stengel S. Hoch- versus moderat-intensive Laufbelastung—Einfluss auf kardio-metabolische Risikogrößen bei untrainierten Männern [High versus moderate intense running exercise—effects on cardio-

metabolic risk-factors in untrained males]. Deutsche medizinische Wochenschrift (1946). 2015;140(1):e7–e13. https://doi.org/10.1055/s-0040-100423.

26. Tuttor M, von Stengel S, Kohl M, Lell M, Scharf M, Uder M, Wittke A, Kemmler W. High intensity resistance exercise training vs high intensity (endurance) interval training to fight cardiometabolic risk factors in overweight men 30–50 years old. Front Sports Active Living. 2020;2:68. https://doi.org/10.3389/fspor.2020.00068.

27. Pedersen BK, Fischer CP. Beneficial health effects of exercise—the role of IL-6 as a myokine. Trends Pharmacol Sci. 2007;28(4):152–6. https://doi.org/10.1016/j.tips.2007.02.002.

28. Pedersen BK, Steensberg A, Fischer C, Keller C, Keller P, Plomgaard P, Febbraio M, Saltin B. Searching for the exercise factor: is IL-6 a candidate? J Muscle Res Cell Motil. 2003;24(2–3):113–9. https://doi.org/10.1023/a:1026070911202.

29. Lee JH, Jun HS. Role of myokines in regulating skeletal muscle mass and function. Front Physiol. 2019;10:42. https://doi.org/10.3389/fphys.2019.00042.

30. Ost M, Coleman V, Kasch J, Klaus S. Regulation of myokine expression: role of exercise and cellular stress. Free Radic Biol Med. 2016;98:78–89. https://doi.org/10.1016/j.freeradbiomed.2016.02.018.

31. Leal LG, Lopes MA, Batista ML. Physical exercise-induced Myokines and muscle-adipose tissue crosstalk: a review of current knowledge and the implications for health and metabolic disease. Front Physiol. 2018;9:1307. https://doi.org/10.3389/fphys.2018.01307.

32. Severinsen M, Pedersen BK. Muscle-organ crosstalk: the emerging roles of myokines. Endocr Rev. 2020;41(4):594–609. https://doi.org/10.1210/endrev/bnaa016.

33. Pedersen BK, Akerström TC, Nielsen AR, Fischer CP. Role of myokines in exercise and metabolism. J Appl Physiol. 2007;103(3):1093–8. https://doi.org/10.1152/japplphysiol.00080.2007.

34. Nielsen S, Pedersen BK. Skeletal muscle as an immunogenic organ. Curr Opin Pharmacol. 2008;8(3):346–51. https://doi.org/10.1016/j.coph.2008.02.005.

35. Huh JY. The role of exercise-induced myokines in regulating metabolism. Arch Pharm Res. 2018;41(1):14–29. https://doi.org/10.1007/s12272-017-0994-y.

36. Marcucci-Barbosa LS, Martins-Junior F, Lobo LF, et al. 10 km running race induces an elevation in the plasma myokine level of nonprofessional runners. Sport Sci Health. 2020;16:313–21. https://doi.org/10.1007/s11332-019-00608-3.

37. International Diabetes Federation: the IDF consensus worldwide definition of the metabolic syndrome. http://www.idf.org/webdata/docs/Metabolic_syndrome_definition.pdf.

38. Huang PL. A comprehensive definition for metabolic syndrome. Dis Model Mech. 2009;2(5–6):231–7. https://doi.org/10.1242/dmm.001180.

39. Ostman C, Smart NA, Morcos D, Duller A, Ridley W, Jewiss D. The effect of exercise training on clinical outcomes in patients with the metabolic syndrome: a systematic review and meta-analysis. Cardiovasc Diabetol. 2017;16(1):110. https://doi.org/10.1186/s12933-017-0590-y.

40. Swift DL, Johannsen NM, Lavie CJ, Earnest CP, Church TS. The role of exercise and physical activity in weight loss and maintenance. Prog Cardiovasc Dis. 2014;56(4):441–7. https://doi.org/10.1016/j.pcad.2013.09.012.

41. Miller CT, Fraser SF, Levinger I, Straznicky NE, Dixon JB, Reynolds J, Selig SE. The effects of exercise training in addition to energy restriction on functional capacities and body composition in obese adults during weight loss: a systematic review. PLoS One. 2013;8(11):e81692. https://doi.org/10.1371/journal.pone.0081692.

42. Hill JO, Thompson H, Wyatt H. Weight maintenance: what's missing? J Am Diet Assoc. 2005;105(5 Suppl 1):S63–6. https://doi.org/10.1016/j.jada.2005.02.016.

43. Donnelly JE, Blair SN, Jakicic JM, Manore MM, Rankin JW, Smith BK, American College of Sports Medicine. American College of Sports Medicine position stand. Appropriate physical activity intervention strategies for weight loss and prevention of weight regain for adults. Med Sci Sports Exerc. 2009;41(2):459–71. https://doi.org/10.1249/MSS.0b013e3181949333.

44. Manson JE, Willett WC, Stampfer MJ, Colditz GA, Hunter DJ, Hankinson SE, Hennekens CH, Speizer FE. Body weight and mortality among women. N Engl J Med. 1995;333(11):67. https://doi.org/10.1056/NEJM199509143331.

45. Vincent HK. Advising the obese patient on starting a running program. Curr Sports Med Rep. 2015;14(4):278. https://doi.org/10.1249/JSR.0000000000000171.

46. Lewington S, Whitlock G, Clarke R, Sherliker P, Emberson J, Halsey J, Qizilbash N, Peto R, Collins R. Blood cholesterol and vascular mortality by age, sex, and blood pressure: a meta-analysis of individual data from 61 prospective studies with 55,000 vascular deaths. Lancet (London, England). 2007;370(9602):1829–39. https://doi.org/10.1016/S0140-6736(07)61778-4.

47. Toth PP, Barter PJ, Rosenson RS, Boden WE, Chapman MJ, Cuchel M, D'Agostino RB Sr, Davidson MH, Davidson WS, Heinecke JW, Karas RH, Kontush A, Krauss RM, Miller M, Rader DJ. High-density lipoproteins: a consensus statement from the National Lipid Association. J Clin Lipidol. 2013;7(5):484–525. https://doi.org/10.1016/j.jacl.2013.08.001.

48. Nordestgaard BG, Benn M, Schnohr P, Tybjaerg-Hansen A. Nonfasting triglycerides and risk of myocardial infarction, ischemic heart disease, and death in men and women. JAMA. 2007;298(3):299–308. https://doi.org/10.1001/jama.298.3.299.

49. Pedersen BK, Saltin B. Evidence for prescribing exercise as therapy in chronic disease. Scand J Med Sci Sports. 2006;16(Suppl 1):3–63. https://doi.org/10.1111/j.1600-0838.2006.00520.x.

50. Lippi G, Tarperi C, Salvagno GL, Benati M, Danese E, Montagnana M, Moghetti P, Schena F. Comparison of plasma lipids changes after middle-distance running in euglycemic and diabetic subjects. J Public Health Emerg. 2019;3:10. https://doi.org/10.21037/jphe.2019.06.01.
51. Wang Y, Xu D. Effects of aerobic exercise on lipids and lipoproteins. Lipids Health Dis. 2017;16:132. https://doi.org/10.1186/s12944-017-0515-5.
52. Górecka M, Krzemiński K, Buraczewska M, et al. Effect of mountain ultra-marathon running on plasma angiopoietin-like protein 4 and lipid profile in healthy trained men. Eur J Appl Physiol. 2020;120:117–25. https://doi.org/10.1007/s00421-019-04256-w.
53. Hespanhol Junior LC, Pillay JD, van Mechelen W, et al. Meta-analyses of the effects of habitual running on indices of health in physically inactive adults. Sports Med. 2015;45:1455–68. https://doi.org/10.1007/s40279-015-0359-y.
54. Colberg SR, Sigal RJ, Yardley JE, Riddell MC, Dunstan DW, Dempsey PC, Horton ES, Castorino K, Tate DF. Physical activity/exercise and diabetes: a position statement of the American Diabetes Association. Diabetes Care. 2016;39(11):2065–79. https://doi.org/10.2337/dc16-1728.
55. Helmrich SP, Ragland DR, Leung RW, Paffenbarger RS Jr. Physical activity and reduced occurrence of non-insulin-dependent diabetes mellitus. N Engl J Med. 1991;325(3):147–52. https://doi.org/10.1056/NEJM199107183250302.
56. Way KL, Hackett DA, Baker MK, Johnson NA. The effect of regular exercise on insulin sensitivity in type 2 diabetes mellitus: a systematic review and meta-analysis. Diabetes Metab J. 2016;40(4):253–71. https://doi.org/10.4093/dmj.2016.40.4.253.
57. Mambiya M, Shang M, Wang Y, Li Q, Liu S, Yang L, Zhang Q, Zhang K, Liu M, Nie F, Zeng F, Liu W. The play of genes and non-genetic factors on type 2 diabetes. Front Public Health. 2019;7:349. https://doi.org/10.3389/fpubh.2019.00349.
58. Williams PT. Greater weight loss from running than walking during a 6.2-yr prospective follow-up. Med Sci Sports Exerc. 2013;45(4):706–13. https://doi.org/10.1249/MSS.0b013e31827b0d0a.

Running Activities in Times of COVID-19 Pandemic

8

Helena Herrero and Del Coso Juan

8.1 Introduction

Coronavirus disease 2019 (COVID-19), produced by the SARS-CoV-2 virus, is an illness that can cause a severe acute respiratory syndrome and other life-threatening conditions in a relatively high proportion of infected individuals [1]. COVID-19 become a pandemic in March 2020 and hit most countries that were not prepared to contain a virus that readily spreads from person to person predominantly to and from the respiratory route and through droplets [2]. To reduce the spread of SARS-CoV-2 during the first wave of COVID-19, governments of most countries declared home confinement measures in addition to other isolation and hygiene measures and social distancing [3]. During home confinement, a strict quarantine was set to prohibit athletes from practising any form of exercise outside of their own residence. This entailed the suspension of most professional activities for weeks, including professional and amateur sports competitions [4–6]. The COVID-19 pandemic produced the suspension of major international events in 2020 such as the Olympic Games and the UEFA Euro. After the first wave of COVID-19, other waves impacted most countries at the end of 2020 and through 2021, although the timing and duration of these waves were associated with intrinsic factors of each country and the season, as the spread of COVID-19 is somewhat associated with ambient temperature [7]. However, in these posterior waves of COVID-19, the health authorities of most countries reduced the pressure of lockdowns and most sports competitions have been maintained. This has entailed the coexistence of sports training and competition with a pandemic. The objective of this document is to provide basic information to prepare sports activities that include running actions in times of COVID-19 pandemic. It is important to note that the guidelines provided in this manuscript are formulated for healthy athletes; the recommendation for older athletes, athletes with a known pathology or other populations may be different.

8.2 Home Training Due to Home Confinement

The virus that causes COVID-19 spreads mainly from person to person through respiratory droplets, aerosols, and fomites, so governments of several countries declared confinement measures at different points of the COVID-19 pandemic to reduce the propagation of the virus and to avoid

H. Herrero (✉)
Medical Department,
Royal Spanish Federation of Football, Madrid, Spain
e-mail: hherrero@rfef.es

D. C. Juan
Centre for Sport Studies,
Rey Juan Carlos University, Fuenlabrada, Spain
e-mail: juan.delcoso@urjc.es

infection. Although several countries have implemented mass vaccination protocols in 2021, it is still probable that home confinement measures are set in other countries with lower access to vaccines or in certain regions with high rates of COVID-19 pandemic. Or it is still probable that some exercise routines, such as team training, are limited to reduce the propagation of the virus among athletes of team sports. If this is the case, athletes have multiple forms of home training to mitigate the detraining effect of confinement on physical conditioning. However, the success of these routines may differ depending on the sports discipline and the characteristics of each residence. If there is low availability of space and instruments, home training programs should include strength-based activities with body loads, proprioception activities, and intermittent routines performed with low range displacements. If available, exercise on a treadmill can aid to perform long-duration endurance exercise activities or even to perform high-intensity intermittent activities. In the case of athletes involved in sports that entail any form of running, the treadmill is the machine of choice, although cardiovascular fitness can be maintained by using a bike trainer or a rowing machine in the case of lack of availability of a treadmill. If possible, it is recommended that athletes perform at home the critical movements and actions associated with the sports discipline. This is an easy task for endurance running athletes, but it may be more difficult to implement for team sport athletes. For the latter, including accelerations, decelerations, sprints, changes of direction, and kicking and throwing the ball to pass or shot to goal may be key to reduce the detraining effects produced by home confinement and to facilitate the recovery of fitness once home confinement is suspended [8]. The schedule of training, the number of sessions per week and the duration of the training practices should be maintained, when feasible, to mimic pre-confinement routines, as this may help to reduce the detraining effect of confinement [9] and to instil the idea in the athlete that they have maintained physical fitness despite the conditions. In any case, it is important to note that despite home training during confinement, some reduction of physical capabilities is expected to happen, particularly in elite and professional athletes [10].

8.3 Return to Training and Competition After Home Confinement

One of the most important aspects of exercise in the times of COVID-19 is the return to training and competition after home confinement. This is because, despite the efforts of athletes to maintain physical condition during home confinement, many may show some detraining signs resulting in an increased risk of injury on return to training [11]. Although the return to training and competition may differ depending on the duration and conditions of home training, it should entail a retraining protocol that may last from days to weeks. Athletes should be aware that confinement and its detraining outcomes may have reduced their ability to perform sport-specific running actions and high-intensity exertions over time. Moreover, they should pay greater attention than usual to workload, perception of exertion, and to signs and symptoms of injury [12]. Overall, the return to training and competition should entail two phases: initially, a short re-training phase that includes a broad-spectrum assessment of the athlete's physical condition (power, endurance, joint mobility and body composition; Fig. 8.1). The tests employed in this initial phase should be specific for the running characteristics of each sport discipline and they should be aimed to ascertain if the player is ready for sport-specific training, to injury prevention and to determine the best manner of physical re-adaptation of players. After the re-training phase has been completed, athletes should start a re-adaptation phase with the objective of recovering full performance of specific running activities in terms of intensity, duration and number of repetitions [13]. It is highly important that athletes are progressively exposed to increased volume and intensity during this phase and to avoid participation in competition without proper re-adaptation to exercise. Regarding return to

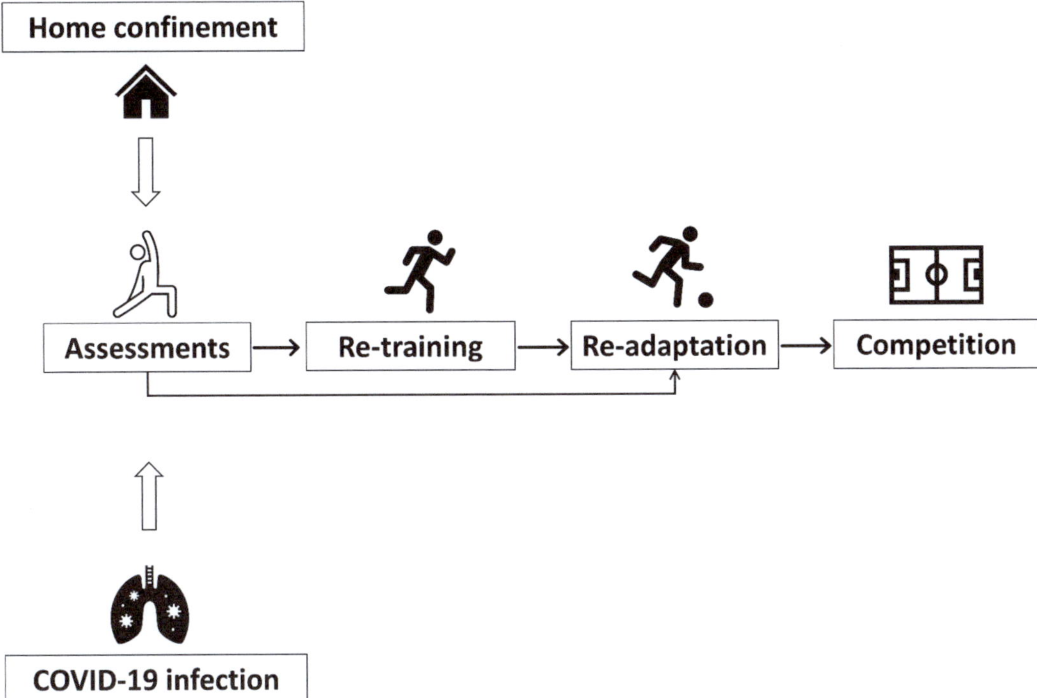

Fig. 8.1 Phases to return to training and competition in running activities during COVID-19 pandemic

competition, again, this may depend on the duration and conditions of home confinement, but the athlete should be enrolled in competitive activities once there is the certainty that he/she is fully re-adapted to overcome the stress imposed by competition. In this regard, friendly competitions or simulated competitions may help to re-adapt athletes to the competitive demands. If the time to restart competition is long enough to allow re-adaptation, sports performance can be unaffected by the lockdown [14]. Last, the avoidance of a congested calendar after the return to the sports competition may be key to avoid injury or to produce an excessive workload in athletes, particularly if home confinement has lasted more than 14 days.

8.4 Training and Competition After COVID-19 Infection

The conditions of return to training and competition in athletes that had suffered COVID-19 infection may present differences as the symptoms of this condition present a high interindividual variability [15]. In fact, even today, there is still much to learn about the various ways the SARS-CoV-2 virus affects athletes' health. For those athletes that have suffered symptoms of low magnitude or for those who have been asymptomatic, the recommendations are the same as in the previous section, as it is expected that they had been confined at home and have been able to exercise regularly. For those athletes with moderate symptoms of COVID-19, the return to exercise should be gradual, giving priority to recovery of the illness rather than a fast return to normal training and competition. As in the case of home confinement, the first measure is to test the athlete's physical condition, but in this case, only testing with low physical demands should be performed (joint mobility, body composition, etc.; Fig. 8.1). Depending on the results of this testing, re-training may considerably differ, but efforts should be directed to recover basic physical condition. Once this phase is finished, a re-adaptation phase should be implemented, as explained in the prior section. For those who have had severe symptoms of COVID-19, the return to training and competition can be a challenging task, particularly for those who have experienced several days of fever and or diarrhoea, muscle pain or pneumonia. In this case, the return to the competition may last several weeks and should include an initial rehabilitation phase with the main purpose of preparing the athlete to exercise. In this case, testing athlete's physical fitness is secondary as it is expected that the athlete has experienced a severe reduction of his/her physical fitness. Only when the players can perform light-intensity exercise without muscle pain and with a normal pattern of breathing, the two steps re-training/re-adaptation program should be implemented.

8.5 Training and Competition with Mask

After the resumption of sports competitions due to COVID-19 pandemic, professional and high-performance athletes have undergone constant PCR and antigen testing to detect athletes infected with COVID-19 in an early phase, even before there are any symptoms. However, due to the impossibility of affording the cost of this continuous testing, amateur athletes have been recommended or even mandated to train and compete with a mask to reduce the likelihood of COVID-19 transmission. This recommendation is present in several countries despite the World Health Organization does not recommend wearing a mask for vigorous physical activity because it can make breathing more difficult [16] and because sweat can make the mask wet, which impacts breathing and may promote the growth of microorganisms. The use of a face mask while exercising is mostly recommended for those exercise and sport situations where it is unfeasible to keep a distance of at least 1 m from a teammate or opponent. Hence, this has been recommended mainly to team sports athletes, although there are several situations in individual sports where social distance cannot be maintained. The close proximity of participants and increased respiration rate due to the demands of exercise may facilitate the infection, although there is evidence to demonstrate that the likelihood of transmission in team sports competition is low [17]. Additionally, health authorities emphasise the importance of wearing masks mainly in indoor settings, and in crowded outdoor settings while, in most policies raised by health authorities, exercising outdoors does not require the use of a mask (Fig. 8.2). In any case, if the running athlete is mandated to use a mask during training or competition, or whether it is convenient to use a mask to reduce the likelihood

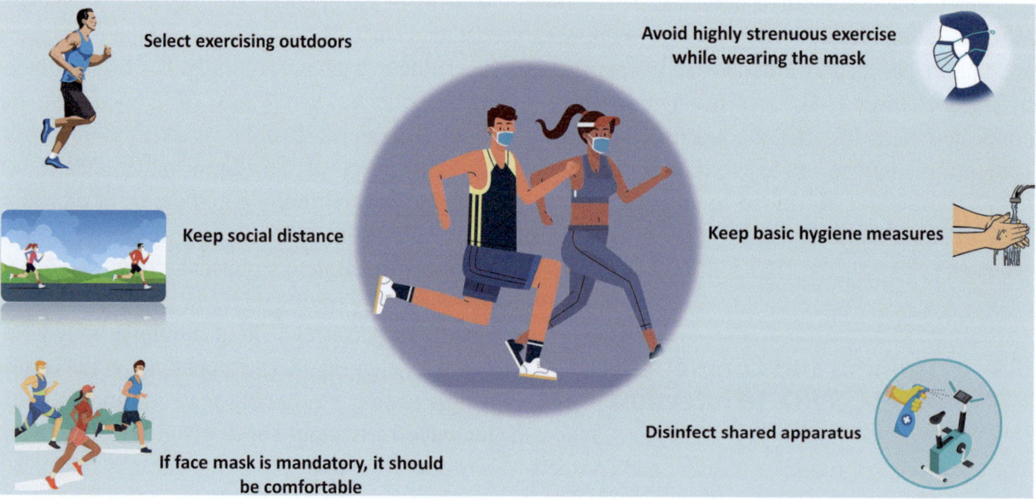

Fig. 8.2 Basic hygiene measures to reduce the likelihood of COVID-19 infection during running activities in sport

of transmission, it is important to note that the activity should be of moderate intensity. In the case of needing executing high-intensity exercise while wearing a mask, the use of a disposable surgical face mask may be recommended because it does not affect performance [18], while N95-type masks can produce some ventilatory alterations [19]. In fact, it has been suggested that filtering facepiece respirators (FFRs) such as N95, FFP1 and FFP2 should not be used for exercise purposes [20]. From a practical perspective, the training of tactics in team sport athletes can be performed while wearing the mask while physical conditioning can be done individually while maintaining physical distance. In any case, if the athlete cannot wear a mask during training due to difficulty of breathing, it is recommended to change the activity to an outdoors location and while maintaining physical distance. Additionally, other side effects while wearing the face mask such as dizziness and light-headedness are clear symptoms indicating that the exercise activity should be suspended.

8.6 Other Important Measures to Be Taken into Account While Exercising During COVID-19 Pandemic

Apart from the guidelines exposed above, other measures must be contemplated while exercising in times of COVID-19 pandemic. These measures are also applicable to avoid the transmission of other viruses and, in general, can be maintained as general rules for exercise in any situation. For example, it is recommended that the athlete washes his/her hands before and after the exercise session while the use of clean sports clothing in each session can be an additive hygiene measure. Carrying a hand sanitiser and a towel during the exercise session can be recommended for those activities that require constant touching of an apparatus/device. Avoiding touching the face while exercising is also a useful recommendation to reduce the likelihood of infection.Sharing beverage cans or sports equipment such as shirts, towels, etc., with other athletes can be a source of contagion and therefore they should be prohibited. In the case of using materials such as weights, exercise mats, etc., this should be disinfected before and after each use.

References

1. Randelli PS, Compagnoni R. Management of orthopaedic and traumatology patients during the coronavirus disease (COVID-19) pandemic in northern Italy. Knee Surg Sports Traumatol Arthrosc. 2020;28(6):1683–9. https://doi.org/10.1007/s00167-020-06023-3.
2. Collignon P. COVID-19 and future pandemics: is isolation and social distancing the new norm? Intern Med J. 2021;51(5):647–53. https://doi.org/10.1111/imj.15287.
3. Sarto F, Impellizzeri FM, Spörri J, Porcelli S, Olmo J, Requena B, et al. Impact of potential physiological changes due to COVID-19 home confinement on athlete health protection in elite sports: a call for awareness in sports programming. Sports Med. 2020;50(8):1417–9. https://doi.org/10.1007/s40279-020-01297-6.
4. Eirale C, Bisciotti G, Corsini A, Baudot C, Saillant G, Chalabi H. Medical recommendations for home-confined footballers' training during the COVID-19 pandemic: from evidence to practical application. Biol Sport. 2020;37(2):203–7. https://doi.org/10.5114/biolsport.2020.94348.
5. Bisciotti GN, Eirale C, Corsini A, Baudot C, Saillant G, Chalabi H. Return to football training and competition after lockdown caused by the COVID-19 pandemic: medical recommendations. Biol Sport. 2020;37(1):313–9. https://doi.org/10.5114/biolsport.2020.96700.
6. de Souza DB, González-García J, Campo RI.-D, Resta R, Buldú JM, Wilk M, Coso J. Players' physical performance in LaLiga when the competition resumes after COVID-19: insights from previous seasons. Biol Sport. 2020;37(1):2–7. https://doi.org/10.5114/biolsport.2020.96856.
7. Christophi CA, Sotos-Prieto M, Lan FY, Delgado-Velandia M, Efthymiou V, Gaviola GC, et al. Ambient temperature and subsequent COVID-19 mortality in the OECD countries and individual United States. Sci Rep. 2021;11(1):8710. https://doi.org/10.1038/s41598-021-87803-w.
8. de Albuquerque Freire L, Tannure M, Sampaio M, Slimani M, Znazen H, Bragazzi NL, et al. COVID-19-related restrictions and quarantine COVID-19: effects on cardiovascular and Yo-Yo test performance in professional soccer players. Front Psychol. 2020;11:589543. https://doi.org/10.3389/fpsyg.2020.589543.
9. Mosqueira-Ourens M, Sánchez-Sáez JM, Pérez-Morcillo A, Ramos-Petersen L, López-Del-Amo A, Tuimil JL, Varela-Sanz A. Effects of a 48-day home quarantine during the covid-19 pandemic on the first outdoor running session among recreational runners in

Spain. Int J Environ Res Public Health. 2021;18(5):1–11. https://doi.org/10.3390/ijerph18052730.

10. Fikenzer S, Fikenzer K, Laufs U, Falz R, Pietrek H, Hepp P. Impact of COVID-19 lockdown on endurance capacity of elite handball players. J Sports Med Phys Fitness. 2021;61(7):977–82. https://doi.org/10.23736/S0022-4707.20.11501-9.

11. Silva JR, Brito J, Akenhead R, Nassis GP. The transition period in soccer: a window of opportunity. Sports Med. 2016;46(3):305–13. https://doi.org/10.1007/s40279-015-0419-3.

12. Pol R, Hristovski R, Medina D, Balague N. From microscopic to macroscopic sports injuries. Applying the complex dynamic systems approach to sports medicine: a narrative review. Br J Sports Med. 2019;53(19):1214–20. https://doi.org/10.1136/bjsports-2016-097395.

13. Herrero-Gonzalez H, Martín-Acero R, Del Coso J, Lalín-Novoa C, Pol R, Martín-Escudero P, et al. Position statement of the Royal Spanish Football Federation for the resumption of football activities after the COVID-19 pandemic (June 2020). Br J Sports Med. 2020;54(19):1133–4. https://doi.org/10.1136/bjsports-2020-102640.

14. Brito de Souza D, López-Del Campo R, Resta R, Moreno-Perez V, Del Coso J. Running patterns in LaLiga before and after suspension of the competition due to COVID-19. Front Physiol. 2021;12:666593. https://doi.org/10.3389/fphys.2021.666593.

15. Chen L, Zheng S. Understand variability of COVID-19 through population and tissue variations in expression of SARS-CoV-2 host genes. Inform Med Unlocked. 2020;21:100443. https://doi.org/10.1016/j.imu.2020.100443.

16. The World Health Organization. Coronavirus disease (COVID-19): masks; 2021. https://www.who.int/news-room/q-a-detail/coronavirus-disease-covid-19-masks. Accessed 14 June 2021

17. Jones B, Phillips G, Kemp S, Payne B, Hart B, Cross M, Stokes KA. SARS-CoV-2 transmission during rugby league matches: do players become infected after participating with SARS-CoV-2 positive players? Br J Sports Med. 2021. https://doi.org/10.1136/bjsports-2020-103714.

18. Shaw K, Butcher S, Ko J, Zello GA, Chilibeck PD. Wearing of cloth or disposable surgical face masks has no effect on vigorous exercise performance in healthy individuals. Int J Environ Res Public Health. 2020;17(21):1–9. https://doi.org/10.3390/ijerph17218110.

19. Epstein D, Korytny A, Isenberg Y, Marcusohn E, Zukermann R, Bishop B, et al. Return to training in the COVID-19 era: the physiological effects of face masks during exercise. Scand J Med Sci Sports. 2020;31(1):70–5. https://doi.org/10.1111/sms.13832.

20. Hamuy Blanco J, Janse van Rensburg DC. Should people wear a face mask during exercise: what should clinicians advise? I BJSM blog—social media's leading SEM voice. Br J Sports Med (Blog); 2020. https://blogs.bmj.com/bjsm/2020/06/12/should-people-wear-a-face-mask-during-exercise-what-should-clinicians-advise/. Accessed 14 June 2021

Shoes for Running 9

Bermon Antoine, Turner Christopher,
and Bermon Stéphane

9.1 Running Shoes: Main Characteristics and Their Evolution

9.1.1 History

Humans are thought to have been running for approximately two million years [1] at least in part for the practice of persistent hunting. More "recently", the first Olympic Games took place in 776 BC at Olympia where athletes ran the "stade" or "stadion", a 180 m sprint, naked, barefoot and on an earthen track. The longest Olympic race was the dolichus (4800 m) introduced in 720 BC.

According to legend, the first marathon was run by Philippides a Greek messenger, who ran naked and therefore probably barefoot, from the town of Marathon, to the city of Athens in 490 BC. Despite the myth having no connection to the ancient Olympics, a marathon race of approximately 40 km was included in the first modern Olympic Games that were staged in Athens in 1896.

Even though the oldest shoes are thought to date from the eighth millennium BC, the first running shoes are surprisingly recent as they date only from the nineteenth century. Leisure running was unknown at the time. The working class had no leisure time and the wealthy saw running as undignified. The rebirth of competitive running began when gentlemen aristocrats bet large sums of money to decide who employed the fastest footmen. Initially, like gentlemen's dress shoes which were worn by their liveried footmen as part of their uniform, the first running shoes were made of leather that had the disadvantage to stretch when wet.

The first rubber soles were invented in 1832 by Wait Webster and replaced wooden outsoles. Twenty years later, Joseph Willian Foster, founder of the company known today as Reebok, added spikes to improve grip.

At the end of the nineteenth century, sneakers were born when an industrial rubber sole was connected to an upper part made of canvas. They became popular among adults and allowed silent walking hence their name derived from the verb "to sneak".

At the beginning of the twentieth century, Adi Dassler started handcrafting running spikes and transformed them from casual looking shoes with nails coming out of the outsole front to track shoes as we know them today (except they were

B. Antoine
Centre Hospitalier Universitaire de Besançon,
Service de Médecine Physique et Réadaptation,
Besançon, France

T. Christopher
Heritage Department, World Athletics,
Monaco City, Monaco
e-mail: chris.turner@worldathletics.org

B. Stéphane (✉)
Health and Science Department, World Athletics,
Monaco City, Monaco
e-mail: stephane.bermon@worldathletics.org

© The Author(s), under exclusive license to Springer-Verlag GmbH, DE, part of Springer Nature 2022
G. L. Canata et al. (eds.), *The Running Athlete*, https://doi.org/10.1007/978-3-662-65064-6_9

Fig. 9.1 The Karhu footwear (credit Karhu)

made of kangaroo leather). He also designed shoes for the high jumpers with spikes in both the back and the front of the outsole. The Dassler brothers (later founders of Adidas and Puma) made short distance running spikes shoes (up to 800 m) that were internationally recognized by athletes during the 1920s.

The Finnish running shoe manufacturer Karhu, whose emblem was three-stripes on the sides of their footwear, was one of the world's leading athletics shoemakers (see Fig. 9.1). At the 1952 Helsinki Olympics, Karhu's shoes were worn by 15 Olympic champions. Adi Dassler bought from Karhu the rights to use the three-stripes which had originally been invented to add support to the foot and would later become the symbol of Dassler's brand.

In the 1960s, New Balance started to produce the Trackster, a light multiple width shoe with the famous ripple sole supposed to improve traction, reduce fatigue and prevent shin splints. In parallel, Onitsuka Tiger (renamed ASICS in 1977) developed a shoe with a multiple layered sole to give extra cushion. The "Cortez" was released ahead of the 1968 Mexico City Olympic Games. The shoe was developed in association with Bill Bowerman, joint founder with Phil Knight of Blue Ribbon Sports (renamed Nike in 1971), who were Onitsuka Tiger's distributor in the USA. The two companies ended their relationship when Blue Ribbon launched an identical version of the same shoe. Bowerman's air-cushioned heel was the precursor to the air-cushioned soles still used today. In this same period, a plastic plate (also called a traction plate) was added under the toe box of track spiked shoes, allowing spikes to become removable and changeable.

In 1968, Puma launched "Brush" spikes, track shoes that came with 68 needle-like spikes on the front of the sole that provided better grip on synthetics tracks, than the traditional 4 or 6 nail spikes. Following a brace of world record-breaking sprint performances in the USA, the design was declared illegal prior to the Mexico City Games.

In 1974, the Nike Waffle Trainer patented by Bowerman with a lighter traction sole for track shoes built out of a waffle pan hit the market. Brooks started to sell the first pronation controlling shoes since pronation was thought at that time to cause injuries.

During the 1980s, Adidas implanted the first piece of technology in its Micropacer by fitting a pedometer into the tongue of the shoe. Asics launched their first model with cushioning compound made of silicone known as GEL, displacing more impact than air-cushioned heel and still used today.

The end of the twentieth century is marked mostly by an improvement of cushioning technology with which Nike will break in 2004 when releasing its Nike Free, a training shoe supposed to strengthen the foot by reducing weight and cushioning, therefore recreating barefoot running feelings. Nike was quickly followed by Vibram and its more extreme FiveFingers minimalist shoe.

In 2009, the Hoka One One broke with the minimalist trend by proposing a lightweight strongly cushioned running shoe.

In 2013, New Balance presents the first 3D-printed sole adapted to the runner's foot characteristics as well as 3D-printed spike plates. Adidas Boost Technology improved their cushioning system by using a chemical compound (developed by the chemical company BASF) that compresses under pressure and bounce back to provide energy return with every stride, therefore surpassing the ethylene vinyl acetate commonly used since the 1970s.

The latest breakthrough in that domain belongs to Nike with its advanced footwear technology relying on a combination of a thick and light midsole made of polyamide elastomer and a long stiff plate that is proven to be an important

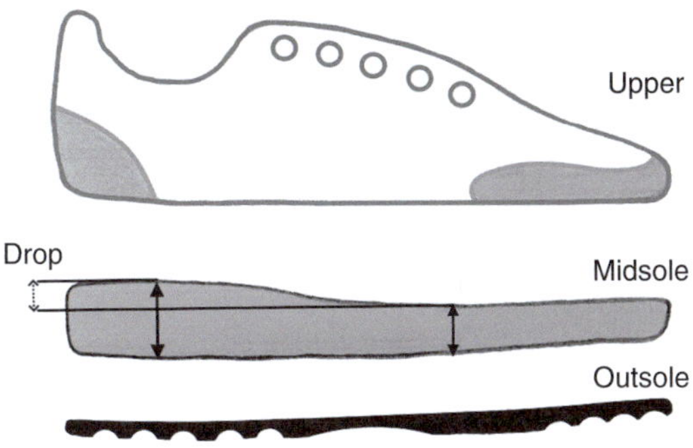

Fig. 9.2 Different parts of the shoe

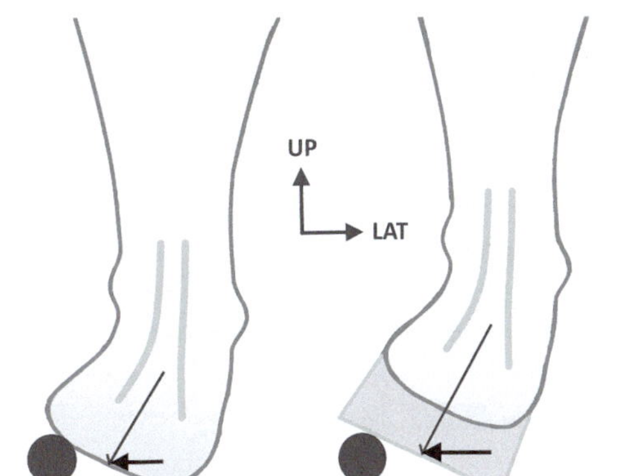

Fig. 9.3 Frontal plane lever and shoe height

factor contributing to the latest improvement in road race performances [2].

9.1.2 Main Technical Characteristics (Fig. 9.2)

– **Weight:**
Light shoes (under 440 g a pair) have been proven to improve running economy and athletic performance [3]. Therefore, shoes manufacturers have spent a lot of time and money across the years to develop light foam and materials. Typical weight for a standard road running shoe is around 210 g, whereas typical weight of a sprint spiked shoe is around 150 g.

– **Flexibility:**
Depending on their different features, some shoes will be more flexible than others, thus allowing deformation in the frontal plane (pronation/supination of the forefoot) and in the sagittal plane (curving of the shoe in a plantar/dorsiflexion manner).

– **Protection from the ground and cushioning:**
To protect our feet from ground irregularities (and coldness) is probably the reason why we made shoes in the first place. As they evolved into running-specific shoes, protection from ground force reaction became important and cushioning midsole were introduced. Modern shoes have a wide range of midsole height, from no midsole at all in minimalist shoes to

more than 3 cm (at rearfoot) as found in the advanced footwear technology. This protection comes with two major setbacks. The first one being a decreased proprioception since sensory feedback from the ground is diminished and the second one is an increased lever in the frontal plane in case of non-horizontal landing (see below).

– **Drop:**
Drop is defined by the extra height under the heel compared to the forefoot (see Fig. 9.2). As with midsoles, its height varies from zero in minimalist shoes to more than 1 cm. It is interesting to note that midsole height and drop are not necessarily linked. While most highly cushioned shoes also provide an important drop (more than 6 mm) some have no drop. Drop and cushioning are important parameters regarding running style. On the one hand, a low cushioning, low drop shoe will promote a forefoot strike like observed in barefoot or minimalist runners. On the other hand, a thick midsole under the heel will promote a heel strike by absorbing impact force and by increasing the amount of plantar flexion needed to land on the forefoot.

A thick midsole under the heel will increase the lever force (black arrow) applied on the ankle in the frontal plane if the foot lands on an irregular or inclined surface (see Fig. 9.3). This is an important factor to consider since, combined to the reduced proprioception, it can lead to higher stress on the passive and active stabilizer of the ankle and cause (or at least increase the rate of) injuries [4].

- **Adherence:**
Adherence is to the outsole what cushioning is to the midsole. At one end of the spectrum, indoor running outsole will favour a low weight and a high responsiveness by being thin and mostly flat. On the other end of that spectrum, trail running shoes will favour protection, traction and durability, with a thicker outsole, deep grooves and rubber crampons. Road running shoes are in between those two extremes with relatively light and moderately deep grooved outsoles to ensure adherence even on a wet ground.

Track shoes feature different types of spikes to maximize adherence on the track, depending on their intended use (see below).

- **Stability technology:**
As found in some foot orthosis, some shoes directly include stability features like heel cups, heel counter, arch supports and pronation/supination control elements.

Regarding the upper part of the shoe, laces ensure a proper fit and middle foot straps can add an extra level of stability and secure the feet when high decelerations and/or ground reaction forces are expected (jumping and throwing events mostly).

- **Passive assistance:**
Recent road running shoes feature advanced footwear technology designed to passively improve running performance by restoring energy from the strike to the push of the foot on the ground. This technology includes a thick and light midsole made of polyamide block elastomer in which a long stiff plate (often made of carbon) is inserted.

Some track shoes are also equipped with a shoe-long plate designed to store energy when stretched and restore it at the push. This stiff element also limits excessive deformation of the shoe under the huge strains developed by sprinters or jumpers.

9.2 Non-spiked Running Shoes

Non-spiked running shoes are the most polyvalent and therefore the most popular outside of the track. They are composed of a rubber outsole, a cushioning midsole of various thicknesses and a soft upper part (see Fig. 9.2).

9.2.1 Road Running

When it comes to road running shoes, they are two schools. Shod runners who wear regular running footwear believe that the shoe must assist the foot and help to improve performance. Minimalist runners on the other hand believe that footwear should only provide protection from the ground and ensure adherence while interfering as little as possible with the biomechanics, as observed in barefoot runners.

9.2.1.1 Cushioned or Regular Shoes

This type of shoe features a cushioning midsole most often associated with a moderate to important drop therefore allowing the runner to land on the heel. It also sometimes features stability technology designed mostly to control pronation and support the medial arch. To improve performance by decreasing the running energy cost, passive assistance technology (stiff plate) can be embedded in the midsole.

Even though technological advancements have allowed lighter and more efficient (air, gel, new foams, etc.) cushioning materials, they still come at the price of an increased weight compared to minimalist shoes. Other drawbacks of these technologies are less proprioception and a repeated rear foot strikes resulting in an increased load on the knee and hip [5, 6].

It is interesting to note that these technologies, coupled with the well-worked aesthetic of running shoes, has progressively found its way out of the running industry and into the daily footwear of a lot of people.

9.2.1.2 Minimalist Shoes

Since the beginning of the twenty-first century, there has been a counter-current to this technological escalade and a growing demand for simpler and less assistive shoes. Minimalist shoes are designed to be as light and flexible as possible. To interfere as little as possible with foot biomechanics, they either provide a large toe box or different compartments for each toe. To be placed on the highest minimalist ranking, the shoe should also provide little to no cushioning, drop and stability control technology.

On hard surfaces, these shoes favour a more natural forefoot strike, therefore putting more stress on calves, Achilles' tendon, intrinsic foot muscle and plantar fascia.

9.2.1.3 Adherence on the Road

Besides those differences, road running shoes share the same adherence features with a mostly flat outsole with grooves of different depths to ensure adherence on wet road just like a standard road tire would do.

9.2.2 Mountain and Trail Running

Shoes designed to run in the wild provide less cushioning and less drop to limit the frontal plane lever discussed earlier and to avoid an excessive plantar flexion while going downhill. They are harder to provide more protection from the irregular ground and sometimes feature a plastic rock plate under the forefoot or in the whole length of the shoe. The outsole features lugs designed to provide maximal adherence on uneven terrain like an off-road tire.

It is interesting to note that pronation control is way less common in trail running shoes, probably to provide more manoeuvrability. Other interesting features are a waterproof and reinforced upper for protection.

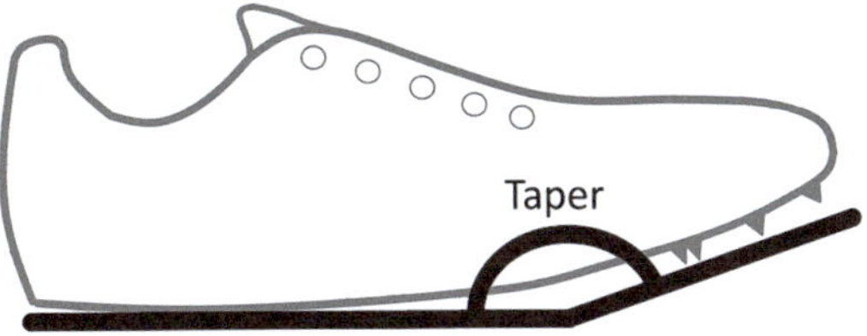

Fig. 9.4 Taper on spiked shoes

9.3 Spiked Running Shoes

Spikes are used to maximize adherence on the track and on cross country terrain. They are attached to the outsole of the shoe, under the forefoot, via a plastic traction plate.

A specificity of spiked shoes is the angle between the forefoot and the rest of the shoe called "taper" (see Fig. 9.4).

9.3.1 Track Running Shoes

Sprints spiked shoes are super light and rigid shoes with a high number of spikes (6–11). They are designed to maximize the speed by enhancing forefoot traction on the synthetic track. They provide no cushioning since the rear foot does not even touch the ground. The shorter the distance run, the more rigid the shoe and the higher the taper. Low shoe weight is important as weight is negatively correlated with performance in sprints.

Above 100 m races, plates will become less rigid, allowing the runner to turn on the curved track. With increasing distance, shoes will provide athletes with more cushioning, lower number of spikes and reduced taper.

Steeplechase shoes basically look like long distance running spike with technology offering better water proofing and water draining.

9.3.2 Cross Country Shoes

While track shoes look very different from their road running counterparts, cross country shoes look very much like road running footwear with 4–6 spikes under the forefoot. These shoes provide a moderate amount of cushioning, a more flexible spike plate and sometimes include stability control elements.

In cross country shoes, the removable/replaceable aspect of the spikes is very important since spikes will be chosen depending on the characteristics of the ground and weather conditions (dry, wet, muddy, snowy).

9.3.3 Different Types of Spikes

Spikes number, shape and length varies greatly depending on distance and terrain.

Sprint shoes can feature up to 11 spikes where most long-distance running shoes provide only 4. That part of each spike which projects from the sole or the heel shall not exceed 9 mm except in the high jump and javelin throw, where it shall not exceed 12 mm. The spike must be so constructed that it will, at least for the half of its length closest to the tip, fit through a square sided 4 mm gauge. Cross country spikes can be 15 mm long.

Spikes come in four main different forms (Fig. 9.5):

- Pyramids spikes (a) are the most polyvalent, they are used at a 6 mm length on the track and up to 15 mm on muddy/wet cross-country races.

- Needles/Pins spikes (b) are intended to penetrate deeper into the ground thus providing more traction for both track and cross country.
- Tartans spikes (c) are somewhere between the two. They are mostly used on synthetic tracks.
- Christmas trees spikes (d) were initially intended to not penetrate the track surface, they are very commonly used because they offer both great traction and energy return.

Different shapes are thought to provide different amounts of bounce or to be less aggressive for the track surface. While there is very little evidence, one study [7] showed that Christmas trees generated the greatest amount of energy return followed by pyramids. This study also showed that all spikes penetrated the ground.

Recently, Asics was released and was awarded for a track shoe of a new kind. Instead of featuring spikes, it is equipped with a carbon fibre reinforced thermoplastic outsole that looks a lot like a honeycomb. This new technology is built to avoid penetrating the track and is supposed to offer better propulsion at every step and therefore higher speed on sprints.

9.3.4 Spiked Shoes to Run… and Jump

High jump shoes are very close to running flats with spikes under the forefoot and under the heel to maximize performance. They sometimes have an asymmetrical design depending on the foot

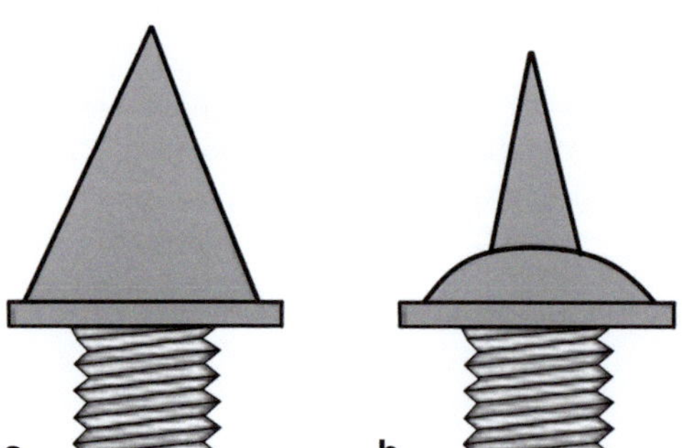
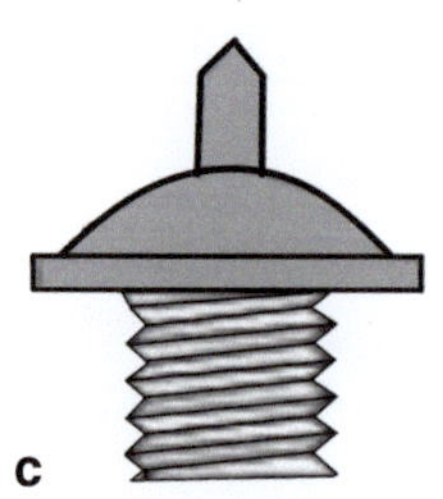

Fig. 9.5 Different types of spikeskes

from which the athlete jumps. Long jump and pole vault footwear look like sprint spiked shoes with a little cushioning under the mid foot and the heel and little taper. Triple jump shoes provide a little more cushioning and almost no taper, therefore looking like middle-distance running shoes.

Most jumping shoes have a midfoot strap over the laces to improve stability, as jumping is associated with very high and brief energy transfer from the horizontal plan to the vertical plan.

9.4 Running Shoes and Injuries

A review of factors influencing running-related injuries among US military recruits [5] suggest that transitioning to forefoot striking may reduce risk for hip, knee and tibial injuries, but increases the risk for calf, Achilles' tendon, foot and ankle injuries. Therefore, it seems that no ideal foot strike pattern exists when it comes to preventing running overuse injuries. The same study found minimal evidence associating running surfaces and injury risk and no evidence supporting a "running shoe prescription per foot type", but reiterated that low physical fitness, history of prior injury and high weekly running distance are factors that increase injury risks, but not shoes specifications.

Currently, minimalist shoe wear does not appear to be associated with greater or lesser injury risk when compared with traditional running shoe wear. However, the very definition of a minimalist shoe remains unclear as some studies used cushioned shoes with 4 mm heel drop that might be likely to promote a rearfoot striking pattern without really offering enough protection for such a pattern.

Another interesting finding regarding footwear and injuries among US military is that, examining historical records, Knapik et al. [8] found no difference in injury risk before 1982 when physical training was performed in boots and after it started to be performed in running shoes.

Similarly, in a non-systematic narrative review of current literature, Malisoux and al [9]. concluded that there is very little evidence supporting the prescription of an "appropriate footwear" in order to prevent injuries and that it is possible that the role of running shoe technology in injury prevention has been largely overrated.

Giving a closer look at minimalist shoes and striking pattern, Boyer et al. [10] found that forefoot striking in regular shoes helps decrease vertical load rates that are known to be associated with overuse running-related injuries, but increased anteroposterior and lateral load rates (more plantar flexion and inversion mainly) which might induce injuries too.

Rice et al. [11] conducting a similar study, including an additional minimal footwear group reported decreased vertical, lateral and medial load rates and equal anteroposterior load rates in the minimalist group compared to both forefoot strike and rearfoot strike cushioned groups. In that minimalist group, a sub analysis between no midsole and minimal midsole found that running with a forefoot strike in full minimal shoes (without cushioning) results in the lowest vertical load rates.

A systematic review of the role of footwear constructions in running-related injuries and performance was realized by Sun et al. [12]. Evidence about the genesis of injuries was scarce. However, two interesting points were noted. Regarding midsole hardness and its relationship with injuries, only a few longitudinal studies were conducted. Among these, Theisen et al. [13] randomly assigned soft and hard midsole shoes to 247 runners to wear for 5 months and found no difference in injury rate. Malisoux et al. [14] conducted a randomized controlled trial including 848 runners assigned either a "hard" or a "soft" pair of shoes (both 34 mm at heel with 10 mm drop) and found a lowered injury rate only among light runners assigned a soft pair of shoes over a 6-month follow-up.

Very few studies have explored the link between heel drop and injury rate. However, Malisoux et al. [15] conducted a randomized controlled trial, using first-time injury as the primary outcome and found no significant difference on injury rate among different heel drops.

It seems that the injury rate is lower among runners who use multiple pairs of shoes and practice different sports [16]. Indeed, a prospective study on 264 amateur runners, despite longer and more frequent training sessions and twice more competition time concluded that using different pairs of shoes and practising other sports is a protecting factor. This finding, which has some significant practical implications, suggests that submitting the musculoskeletal system to different types of constraints might decrease the probability of crossing the running-related injury threshold by either diminishing the repetitive aspect of the applied strain and/or by increasing more widely the body's capacity to sustain and adapt to that strain (the stress response).

9.5 Running Shoes and Performance

9.5.1 World Records

2020 has been a prolific year in terms of world records and personal best broken in distance running events. Almost all these records on the track or on the road (5 km, 10 k, half-marathon and marathon) have been broken by athletes using the advanced footwear technology, which includes a stiff plate embedded in a light and thick midsole.

9.5.2 What Science Says

Three publications [2, 17, 18] recently confirmed that this spectacular improvement in race times in road races is likely technological and not physiological.

Although the mechanisms underlying such performance enhancement are still to be fully elucidated, some of the purported performance enhancement mechanisms associated with the advanced footwear technology are summarized here-after.

The sole rebound is one of them and refers here to the "bounciness" experienced largely from the midsole when it returns to its original shape after being compressed. One may argue that sole rebound does not provide much benefit to the runner as it does not provide horizontal force assisting in anterior propulsion but rather vertical force. However, if the compression energy return permits a greater vertical height, then a greater anterior displacement is possible during the flight time. Increases in the stride length of approximately 5 cm, while keeping the stride frequency constant, have been reported in athletes running with an AFT shoe model [17]. Therefore, fewer steps will be required to complete a race.

Furthermore, the concave sole construction in AFT shoes may orient the ground reaction force (GRF) vector in a more anterior direction if the sole rebound is provided by the anterior part of the shoe sole.

A stiff yet pliable sole/midsole could contribute to forward propulsion by recoiling just before the shoe leaves the ground. A relatively stiff sole/midsole may bend near the metatarsophalangeal joints' axes during maximal forefoot loading and store potential energy in the form of elastic strain energy. This energy can then be released as the sole recoils late in the contact phase, thus performing work that propels the runner. Stefanyshyn and Fusco [19] reported that shoes with higher flexural stiffness increased the maximum running speed for all runners in their tests, regardless of their length, weight, shoe size and fitness level. These results are confirmed by Hoogkamer et al. who confirmed that elite athletes running 5-min trial at 14 km/h, 16 km/h and 18 km/h lowered their energy cost of running by 4% when using AFT shoes [20]. Although the mechanism for this lowered energy cost is not fully understood, it is likely achieved through a slightly reduced work of muscles that must contract to help tendons to operate as springs.

A sole with an external curvature and an embedded curved stiff plate may provide leverage that is generated when the centre of pressure rolls over to forefoot. As such, the stiff sole tends to push the heel upwards and forwards. Nigg et al. mention this and refer to it as a "teeter-totter" effect [21]. Nigg explains two mechanisms for this effect. First, a stiff curved plate moves the resultant GRF anteriorly during the stance phase of running. Secondly, a bending point (around which the teeter-totter effect takes place) allows the shoe sole to act as a fulcrum and a curved shoe sole. The mechanism could be beneficial for the propulsion, although a very stiff sole is required to generate a significant force. Nigg et al. report improvements of 4–6% but do not present any models or results that support this statement quantitatively [21]. They also state that a significant contribution (1–3%) may be due to longer ground contacts facilitated by the curved sole.

9.6 Running Shoe Rules

World Athletics, the international governing body for Track and Field events, recently updated its shoe rules [22].

The shoe total thickness measured at 12% (heel) and 75% (forefoot) of the shoe internal length must not exceed the value presented in Table 9.1. Moreover, non-spiked shoe should not include more than one stiff plate. Spiked shoes can include one stiff plate in addition to the traction plate used to attach the spikes.

Table 9.1 Shoe sole thickness table according to Rule 5 of the World Athletics technical rules

Event	Maximal thickness at 12% and 75% of shoe internal length	Further rules requirements
Field events (except triple jump)	20 mm	For all field events, the sole at the centre of the athlete's forefoot must not be higher than the sole at the centre of the athlete's heel
Triple jump	25 mm	
Track events up but not including 800 m	20 mm	
Track events from 800 m and above	25 mm	
Cross country	25 mm	
Road events (running and race walking)	40 mm	

9.7 Running and Foot Orthosis

9.7.1 Running Economy and Performance

A 2019 systematic review and meta-analysis on foot orthosis and running economy and performance [23] conclude that foot orthosis may alter running economy by adding weight and diminishing the spring effect of longitudinal foot arch. In this same study, shock absorbing insoles showed negative but statistically non-significant effect and carbon insole showed positive non-significant effect on running economy. Regarding performance, data was insufficient to make a conclusion.

Foot orthoses have also been shown to impact the longitudinal arch motion, specifically limiting its deformation and recoil during running. In a recent study McDonald et al. [24] showed that blocking arch deformation with semi-rigid foot orthoses negatively impacts running economy by 6%.

9.7.2 Patellofemoral Pain

Foot orthosis is frequently prescribed as part of the treatment for patellofemoral pain in both runners and non-runners. No systematic review of literature or meta-analysis was found on the subject.

A cross-controlled study evaluated frontal and transverse plane kinematics of the lower limb during a run with and without a rearfoot medially wedged prefabricated foot orthosis [25]. No difference was found on the hip and knee parameters in both the subjects with a highly "everted" calcaneus and those with an "inverted" one.

Braga et al. [26] found statistically significant improvement in "lower limb mechanical pattern associated with injuries" (mainly hip adduction and ankle eversion) when an orthosis with 7° medial wedges at forefoot and rearfoot was used on runners with increased foot pronation. These changes however were only evaluated in the very short term.

Sinclair et al. [27] showed that semi-custom insoles improve patellofemoral pain specially in runners qualified as "weak and tight" (mainly with weak quadriceps and hip abductors and tight gastrocnemius and rectus femoris).

To conclude, the 2018 consensus on patellofemoral pain [28] recommends the use of prefabricated foot orthosis for short-term relief and suggests a prescription based on a foot posture and mobility analysis, in combination with other therapies.

Based on these results, it seems logical to prescribe a medially wedged orthosis (both at front and rear foot) in patients with patellofemoral pain who show an excess of foot eversion in association with a rehabilitation programme to strengthen the quadriceps and hip abductors and to stretch calves and rectus femoris.

9.7.3 Other Pathologies

Arch supporting foot orthosis is also commonly prescribed in medial tibial stress syndrome. A review of the current literature [29] shows no particular improvement in pain.

Plantar fasciitis is also a frequent running pathology that leads to the prescription of orthosis. A 2018 systematic review and meta-analysis [30] found foot orthosis to be efficient a reducing pain in the medium term with no effect on short- and long-term pain and no difference between customized and prefabricated orthoses.

For this same pathology, a 2019 systematic review of mechanical treatment [31] suggests the use of contoured full-length insoles and find them to be more efficient when associated with night splints and rocker shoes.

9.8 Trends and Perspectives

Humans have been running for a very long time and since the nineteenth century mostly for either leisure and/or in search for performance. Since that time, injury prevention and technological advancement regarding shoes have never stopped. It seems however that things are taking two different paths.

9.8.1 The Technological Path

One is the advancement of footwear technology that as we saw plays an important role in performance enhancing but still fails to show consistent improvement with regard to injury prevention.

As one does not stop progress, designers and engineers are actively working on tomorrow's running shoe that might feature the following technologies:

- Self-lacing that adapts to the users will and/or to the detected use of the shoe.
- Heating systems.
- Adaptive/dynamic cushioning depending on the user's will and/or the detected pattern of running.
- Connected shoes that might feature a whole range of different technologies, from pedometer to running style analysis and that might very well be self-recharging when used.
- Shoes that could be recycled to meet growing sustainability objectives.

9.8.2 The (Almost) Self-sufficient Path

On the other hand, more and more runners are being seduced by a de-escalation of the assistance provided by shoes and by an almost self-sufficient foot naturally equipped for both performance and sustaining repetitive stress injury free. The evolution of that path could be the development of lighter and more flexible shoes and an improvement in scientific knowledge concerning the proper way to transition towards this kind of footwear. This de-escalation could also lead to a popularization of barefoot running. Historically barefoot runners have won major international titles or set records. These include in the 1960s and 1970s, Abebe Bikila, Ron Hill, and Bruce Tulloh. Today this minimalist path shows interesting results regarding performance and running economy but very little actual record breaking. Talking about barefoot running, a recent review of the literature [28] associated it with more foot pathologies but less deformity, an equal rate of injuries (but different ones than shod runners) and not enough data on performance. It is anecdotal but interesting to note that recently, after being criticized for wearing forbidden Nike Vaporfly during a 5 km track event, the South African runner Mbuleli Mathanga decided to go barefoot on his next 10 km event and finished second, breaking the provincial record.

When it comes to injury prevention, minimalist shoes show interesting perspectives but mixed results, probably due to a lack of proper transitioning from regular shoes and unclear recommendations on how to transition.

References

1. Bramble DM, Lieberman DE. Endurance running and the evolution of Homo. Nature. 2004;432:345–52.
2. Bermon S, Garrandes F, Szabo A, Berkovics I, Adami PE. Effect of advanced shoe technology on the evolution of road race times in male and female elite runners. Front Sports Act Living. 2021;3:653173.
3. Moore IS. Is there an economical running technique? A review of modifiable biomechanical factors affecting running economy. Sports Med. 2016;46:793–807.
4. Altman AR, Davis IS. Barefoot running: biomechanics and implications for running injuries. Curr Sports Med Rep. 2012;11:244–50.
5. Molloy JM. Factors influencing running-related musculoskeletal injury risk among U.S. military recruits. Mil Med. 2016;181:512–23.
6. Kulmala J-P, Avela J, Pasanen K, Parkkari J. Forefoot strikers exhibit lower running-induced knee loading than rearfoot strikers. Med Sci Sports Exerc. 2013;45:2306–13.
7. Bahamonde R, Streepey J, Goyke L, Myers A, Mikesky A. Energy return of different shapes of track spikes. Conference: International Society of Sport Biomechanics. 2014.
8. Knapik JJ, Pope R, Orr R, Grier T. Injuries and footwear (part 1): athletic shoe history and injuries in relation to foot arch height and training in boots. J Spec Oper Med. 2015;15:102–8.
9. Malisoux L, Theisen D. Can the "appropriate" footwear prevent injury in leisure-time running? Evidence versus beliefs. J Athl Train. 2020;55:1215–23.
10. Boyer ER, Rooney BD, Derrick TR. Rearfoot and midfoot or forefoot impacts in habitually shod runners. Med Sci Sports Exerc. 2014;46:1384–91.
11. Rice HM, Jamison ST, Davis IS. Footwear matters: influence of footwear and foot strike on load rates during running. Med Sci Sports Exerc. 2016;48:2462–8.
12. Sun X, Lam GWK, Zhang X, Wang J, Fu W. Systematic review of the role of footwear constructions in running biomechanics: implications for running-related injury and performance. J Sports Sci Med. 2020;19:20–37.
13. Theisen D, Malisoux L, Genin J, Delattre N, Seil R, Urhausen A. Influence of midsole hardness of standard cushioned shoes on running-related injury risk. Br J Sports Med. 2014;48:371–6.
14. Malisoux L, Delattre N, Urhausen A, Theisen D. Shoe cushioning influences the running injury risk according to body mass: a randomized controlled trial involving 848 recreational runners. Am J Sports Med. 2020;48:473–80.
15. Malisoux L, Chambon N, Urhausen A, Theisen D. Influence of the heel-to-toe drop of standard cushioned running shoes on injury risk in leisure-time runners: a randomized controlled trial with 6-month follow-up. Am J Sports Med. 2016;44:2933–40.
16. Malisoux L, Ramesh J, Mann R, Seil R, Urhausen A, Theisen D. Can parallel use of different running shoes decrease running-related injury risk? Scand J Med Sci Sports. 2015;25:110–5.
17. Muniz-Pardos B, Sutehall S, Angeloudis K, Guppy FM, Bosch A, Pitsiladis Y. Recent improvements in Marathon run times are likely technological, not physiological. Sports Med. 2021;51:371–8.
18. Senefeld JW, Haischer MH, Jones AM, Wiggins CC, Beilfuss R, Joyner MJ, Hunter SK. Technological advances in elite marathon performance. J Appl Physiol (1985). 2021;130(6):2002–8. https://doi.org/10.1152/japplphysiol.00002.2021.
19. Stefanyshyn D, Fusco C. Increased shoe bending stiffness increases sprint performance. Sports Biomech. 2004;3:55–66.
20. Hoogkamer W, Kipp S, Frank JH, Farina EM, Luo G, Kram R. A comparison of the energetic cost of running in marathon racing shoes. Sports Med. 2018;48:1009–19. Erratum in: Sports Med. 2017 Dec 16.
21. Nigg BM, Cigoja S, Nigg SR. Teeter-totter effect: a new mechanism to understand shoe-related improvements in long-distance running. Br J Sports Med. 2020;7:bjsports-2020-102550.
22. World Athletics. Technical rules. Amendments to Rule 5. https://www.worldathletics.org/about-iaaf/documents/technical-information.
23. Crago D, Bishop C, Arnold JB. The effect of foot orthoses and insoles on running economy and performance in distance runners: a systematic review and meta-analysis. J Sports Sci. 2019;37:2613–24.
24. McDonald KA, Stearne SM, Alderson JA, North I, Pires NJ, Rubenson J. The role of arch compression and metatarsophalangeal joint dynamics in modu-

lating plantar fascia strain in running. PLoS One. 2016;11:e0152602.

25. Almonroeder TG, Benson LC, O'Connor KM. The influence of a prefabricated foot orthosis on lower extremity mechanics during running in individuals with varying dynamic foot motion. J Orthop Sports Phys Ther. 2016;46:749–55.

26. Braga UM, Mendonça LD, Mascarenhas RO, Alves COA, Filho RGT, Resende RA. Effects of medially wedged insoles on the biomechanics of the lower limbs of runners with excessive foot pronation and foot varus alignment. Gait Posture. 2019;74:242–9.

27. Sinclair J, Janssen J, Richards JD, Butters B, Taylor PJ, Hobbs SJ. Effects of a 4-week intervention using semi-custom insoles on perceived pain and patello-femoral loading in targeted subgroups of recreational runners with patellofemoral pain. Phys Ther Sport. 2018;34:21–7.

28. Collins NJ, Barton CJ, van Middelkoop M, Callaghan MJ, Rathleff MS, Vicenzino BT, et al. 2018 consensus statement on exercise therapy and physical interventions (orthoses, taping and manual therapy) to treat patellofemoral pain: recommendations from the 5th international patellofemoral pain research retreat, Gold Coast, Australia, 2017. Br J Sports Med. 2018;52:1170–8.

29. Menéndez C, Batalla L, Prieto A, Rodríguez MÁ, Crespo I, Olmedillas H. Medial tibial stress syndrome in novice and recreational runners: a systematic review. Int J Environ Res Public Health. 2020;17:7457.

30. Whittaker GA, Munteanu SE, Menz HB, Tan JM, Rabusin CL, Landorf KB. Foot orthoses for plantar heel pain: a systematic review and meta-analysis. Br J Sports Med. 2018;52:322–8.

31. Schuitema D, Greve C, Postema K, Dekker R, Hijmans JM. Effectiveness of mechanical treatment for plantar fasciitis: a systematic review. J Sport Rehabil. 2020;29:657–74.

Wearable Tech for Long-Distance Runners

10

Andrea Aliverti, Michele Evangelisti, and Alessandra Angelucci

10.1 Introduction

"Wearable" means whatever a subject can wear, as sweaters, hats, pants, glasses, bras, socks, watches, patches, or devices fixed on the belt, without encumbering daily activities or restricting mobility [1]. Some authors include smartphones in the definition [2], however this extension of the concept of wearability will be considered in a specific subsection of this chapter. Recent advances in sensors, miniaturized processors, body area networks (BANs), and wireless data transmission technologies allow the assessment of environmental, physical and physiological parameters in different environments, without restriction of activity. Specifically, BANs are systems composed of a network of wearable devices that can be implanted, attached in fixed positions, or carried by the person [3]. It is needless to say that such technological solutions provide a great opportunity to track different types of data outdoors, which is of interest for long-distance runners. In fact, more than 90% of regular runners use a Global Positioning System (GPS)-equipped watch or some other run tracking device [4]. Despite the widespread use, fewer than 10% of commercially available wearables are validated against an accepted gold standard.

In order to determine the outline of this chapter, an ultra-trail running professional coach, Michele Evangelisti [5], was interviewed and provided information on the most relevant aspects of monitoring of training, competition, and recovery with wearables. The expert coach identified unobtrusiveness and lightness as a common characteristic that wearables should have to be used during long-distance running, with lightness being particularly relevant for increasing distances.

The rest of the chapter illustrates the state of the art of commercial and research monitors intended for use during long-distance running, divided by the type of parameter monitored. The use of smartphones to monitor running parameters is then discussed, followed by an overview of future trends in the field of monitoring with wearables and a conclusion by the authors.

10.2 Heart Rate Monitors

Techniques to measure heart rate can be based on different technologies, such as electrocardiography, photoplethysmography, or accelerometry.

A. Aliverti (✉) · A. Angelucci
Department of Electronics, Information and Bioengineering (DEIB), Politecnico di Milano, Milan, Italy
e-mail: andrea.aliverti@polimi.it; alessandra.angelucci@polimi.it

M. Evangelisti
Como, Italy
e-mail: michele.evangelisti@oyl.run

Measuring the electrical activity of the heart by means of an electrocardiogram (ECG) can provide important information about the heartbeat [2]. Photoplethysmography (PPG) is an optical technique used to detect changes in the blood flow volume in a peripheral vascular bed by illuminating a suitable portion of human skin at two wavelengths, typically 660 nm (red) and 940 nm (infrared) and collecting the light reflected or transmitted through the tissue [6].

The pulsatile component of the PPG signal is called 'AC' component and its fundamental frequency, typically 1 Hz, depends on the heart rate; the advantage of PPG is that it can be performed in several body locations, most notably at the fingertip or on the wrist like in the case of smartwatches and fitness trackers.

Seismocardiography (SCG) refers to the measurement of body and in particular thoracic vibration induced by the heart's contraction and ejection of blood from the ventricles [7]. Nowadays, it is possible to record SCG by placing a MEMS accelerometer on the chest of a person, however accelerometers are sensitive to movement and for this reason SCG is not indicated for monitoring during highly dynamic activities such as running.

Wearable technology devices that return heart rate include chest straps, smart bras, earbuds, and sensors placed around the forearm or wrist like smartwatches or fitness trackers. A list of commercial-level and research-level heart rate monitors is reported in Table 10.1, along with the other measured parameters. It must be noted that over 10,000 wearables are now available on the market, therefore this selection only presents the most relevant ones from a technological point of view. Furthermore, many companies offer both smartwatches and fitness trackers, the latter being a simplified version of the watches.

Several devices have been validated and the results have been published; some of the most relevant ones are selected in this chapter. The accuracy of Apple Watch 1 and Fitbit Charge HR has been compared with the Oxycon Mobile Spirometer [9]. One of the most recent works on the Apple Watch is a poster published in May 2021 assessing the Series 6 pulse oximetry and ECG algorithm in children [10]. Also the accuracy of the heart rate estimation of the Garmin Forerunner 235 has been evaluated [17]. Other types of devices, such as the Ōura Ring, have been studied in comparison with a commercial ECG monitor [13].

Despite the number of studies that compare heart rate monitors with gold standards, validation of heart rate responses in wearable technology devices is generally composed of laboratory-based protocols that are steady state or periodic (e.g., running on a treadmill with a fixed speed) in nature and, as a result, high accuracy measures are returned. However, there is a need to understand device validity in applied settings that include varied intensities of exercise. Navalta et al. [44] performed a validation study on PPG-based devices (namely, smartwatches and activity trackers) that were compared to ECG-based devices (in the case of the study, chest straps) during trail running. All PPG-based devices displayed poor heart rate agreement during variable intensity trail running. Until technological advances occur in PPG-based devices allowing for acceptable agreement, heart rate in outdoor environments should be obtained using an ECG-based chest strap that can be connected to a wristwatch or other comparable receiver.

10.3   Running Parameters

Traditional external workload metrics, such as running distance (e.g., kilometers), duration (e.g., hours and minutes), and speed (e.g., minutes per kilometer), can be tracked by smartwatches equipped with GPS technologies [4].

Several wearable devices, including inertial sensors (inertial measurement units, or IMUs, comprising accelerometers, gyroscopes, and magnetometers), and insole pressure measurement systems, provide running technique infor-

Table 10.1 Selection of the most relevant commercial and research-level heart rate monitors

Device name	Company	Type	Main features	Data transmission	Reference
Apple Watch (Series 6 and previous versions; Series 7 coming this autumn)	Apple, USA	Smartwatch	Single-lead ECG (used for HR extraction), PPG sensor (used for SpO_2 and HR extraction), accelerometer, gyroscope, altimeter, GPS/GNSS	Cellular (possible to call and send messages without smartphone around), Bluetooth, Wi-Fi	[8–11]
Ōura Ring	Ōura Health Oy, Finland	Smart ring	PPG, accelerometer	Bluetooth Low Energy	[12, 13]
Fitbit Smartwatches (Sense and previous versions)	Fitbit, US	Smartwatch	Multipurpose electrical sensors (compatible with ECG and EDA app), PPG (used for continuous HR, HRV, RR, SpO_2 monitoring), temperature sensor, microphone, accelerometer, gyroscope, GPS + GLONASS, altimeter	Bluetooth Low Energy, Wi-Fi	[14, 15]
Fitbit Trackers (Charge 5 and previous versions)	Fitbit, US	Wristband	Multipurpose electrical sensors (compatible with ECG and EDA app), PPG (used for continuous HR and SpO_2 monitoring), temperature sensor, accelerometer, GPS + GLONASS	Bluetooth Low Energy	[14, 15]
Running smartwatches (Forerunner® 945, previous versions)	Garmin Ltd., US	Smartwatch	PPG (used for HR, RR, and SpO_2 monitoring), temperature sensor, accelerometer, gyroscope, GPS + GLONASS, Galileo satellite navigation, compass, barometric altimeter	Cellular (LTE, no calls/ messages, only tracking), Bluetooth, ANT+, Wi-Fi	[16, 17]
HRM-Pro™, versions dedicated to specific sports (HRM-Run™)	Garmin Ltd., US	Chest strap	PPG (used to extract HR), accelerometer	Bluetooth Low Energy, ANT+	[16, 18]
O2Ring and its version for children	Viatom Technology, China	Smart ring	PPG (used to extract HR and SpO_2)	Bluetooth	[19]
HeartView (wearable), DualEK (wearable and portable)	Viatom Technology, China	Chest strap or patch ECG (with electrodes)	Single-lead ECG	Bluetooth	[19]
Hexoskin	Carre Technologies inc (Hexoskin), Canada	Garment	Single-lead ECG (used for HR, HRV, stress, QRS event, HRZ, resting HR, HR recovery monitoring), respiratory inductance plethysmography (used for RR, Minute ventilation, VO_2max monitoring), accelerometer	Bluetooth Low Energy	[20–22]

(continued)

Table 10.1 (continued)

Device name	Company	Type	Main features	Data transmission	Reference
Astroskin	Carre Technologies inc (Hexoskin), Canada	Garment	3-lead ECG (used for HR, HRV, stress, QRS event, HRZ, resting HR, HR recovery monitoring), respiratory inductance plethysmography (used for RR, Minute ventilation, VO$_2$max monitoring), accelerometer, blood pressure, skin temperature, PPG (used for SpO$_2$ estimation)	Bluetooth Low Energy	[20, 21, 23]
Performer	L.I.F.E. Italia Srl, Italy	Garment	2-lead ECG on the back, 1-lead ECG on the chest (used to extract HR), 3 sensors for respiratory mechanics (used to extract RR), finger PPG sensor (used to extract SpO$_2$), 9 inertial measurement units, temperature sensor, near-infrared spectroscopy (NIRS)	Cellular (4G, 5G), Wi-Fi	[21, 24, 25]
Huawei Watch 3, previous versions	Huawei, China	Smartwatch	PPG (used to extract HR and SpO$_2$), accelerometer, gyroscope, geomagnetic sensor, temperature sensor, barometric pressure sensor, GPS + GLONASS, Galileo satellite navigation, BeiDou satellite navigation	UMTS (3G), LTE (4G), Bluetooth/BLE	[26, 27]
Galaxy Watch4, previous versions	Samsung, South Korea	Smartwatch	PPG (used to measure HR and SpO$_2$), electrical sensors (for ECG and EDA measurement), atmospheric pressure sensor, GPS	Bluetooth, LTE, Wi-Fi	[28, 29]
GTR 2 Sport, previous versions	Amazfit, China	Smartwatch	PPG (used to measure HR and SpO$_2$), accelerometer, gyroscope, geomagnetic sensor, GPS + GLONASS, barometric pressure sensor	Bluetooth, WLAN 2.4 GHz	[30, 31]
Polar Vantage M2, other and previous versions	Polar Electro Oy, Finland	Smartwatch	PPG-based proprietary technology (Precision Prime™ used to measure HR), activity recognition, filtering of movement signal, GPS + GLONASS	Bluetooth	[32]
Polar H10	Polar Electro Oy, Finland	Chest strap	Electrical sensors (for continuous ECG monitoring)	Bluetooth, ANT+	[32, 33]
Polar Verity Sense	Polar Electro Oy, Finland	Chest strap	PPG (used to detect HR)	Bluetooth, ANT+	[32]
Mi Smart Band 6	Xiaomi, China	Wristband	PPG (used to detect HR and SpO$_2$), accelerometer, gyroscope	Bluetooth, NFC	[34]
Summit 2+, previous versions	Montblanc, Germany	Smartwatch	PPG (used to detect HR), accelerometer, gyroscope, barometric pressure sensor, GPS/GNSS, microphone	Bluetooth Low Energy, NFC, 3G, LTE, Wi-Fi	[35]

Table 10.1 (continued)

Device name	Company	Type	Main features	Data transmission	Reference
Elemnt Rival Multisport GPS Watch	Wahoo Fitness, US	Smartwatch	PPG (used to detect HR), accelerometer, GPS + GLONASS, GPS-based compass	Bluetooth, ANT+	[36]
TICKRx, other versions	Wahoo Fitness, US	Chest strap	PPG (used to detect HR)	Bluetooth, ANT+	[36]
E4	Empatica, Italy/US	Wristband	PPG (used to detect HR, HRV, SpO_2), EDA sensor, accelerometer, infrared thermophile	Bluetooth	[37, 38]
Zephyr™ Strap/ BioModule	Medtronic, US–Zephyr™ Performance Systems	Chest strap	HR, HRV, RR, accelerometer, GPS	Bluetooth	[39, 40]
Zio XT and Zio AT	iRhythm Technologies Inc., US	Patch ECG	Single-lead ECG	Bluetooth	[3, 41]
Oxitone 1000M	Oxitone Medical, Israel	Wristband	PPG (used to detect HR and SpO_2), temperature sensor	Bluetooth	[42, 43]

mation (e.g., cadence, stride length, contact time). Specifically, cadence (i.e., step or stride frequency) and step or stride length determine running velocity, while the contact time is known to affect the energetics and mechanics of running [45]. Another parameter of interest in the vertical ground reaction force (GRF), which can not only be obtained by means of sensorized insoles, as evidence shows high correlations between the peak accelerations obtained from wrist-worn accelerometers and peak vertical ground reaction force (GRF), resultant GRF, and peak loading rates [46]. One of the most relevant studies in the field is the work by Ruder et al. [47] during the 2016 Boston Marathon, in which the relationship between foot strike patterns and landing impacts during a marathon was studied.

The combination of traditional metrics and specific running mechanics may be beneficial for both performance and injury risk. Theoretically, wearable devices can objectively track training loads in real-time, and identify training load errors prior to the occurrence of running-related injuries.

Still, recently published studies suggest that accurately estimating running kinematics with wearable devices is not trivial, largely due to sensor noise and signal drift [4]. Such estimations require complex computational methods that are not always commercially available; furthermore, the algorithms on each commercial device are different and not published by the owning company, thus making it virtually impossible to know how a given output is obtained.

10.4 Other Physiological Parameters

10.4.1 Respiratory Rate

Several types of sensors can be used for respiratory monitoring, with different measured parameters and levels of invasiveness. A typical approach in wearables is to measure respiratory rate from the PPG signal obtained at the wrist or in other body locations [48], but this estimation has the same limitations of PPG-based heart rate

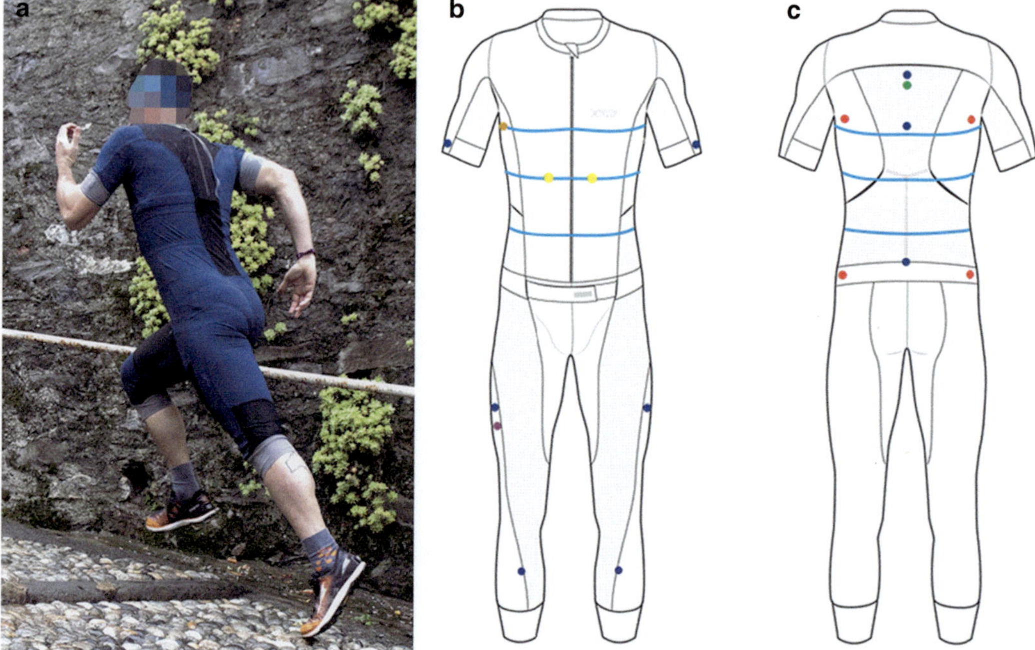

Fig. 10.1 L.I.F.E. Italia Srl's compression garment (Performer); (**a**) Expert coach Michele Evangelisti [5] wearing the garment; (**b**) front; and (**c**) back of the garment. The blue dots are IMUs, the yellow dots represent a single-lead ECG, the orange dot is a temperature sensor at the armpit, the purple dot is a near-infrared spectroscopy (NIRS) embedded device, the red dots represent a 2-lead ECG and the green dot is the logger with the GPS sensor. The light blue stripes are resistive respiratory sensors

estimations, i.e., it is not highly accurate during dynamic activities when there is movement artifact. Furthermore, methods to estimate respiratory rate from PPG are computationally expensive to be performed on an embedded system such as a wearable device.

Another approach for measuring breathing consists in deriving breathing parameters from body surface (namely, chest wall) motion detection, i.e., by inferring changes in thoracic volume from geometrical changes at discrete locations on the torso.

Respiratory inductance plethysmography (RIP) involves the use of two transducer bands placed around the subject to monitor excursions of the chest and abdomen over time. As the subject breathes, the volumes of the two compartments change, and these changes are reflected in alterations in the self-inductance of the coils [49]. RIP has been validated against spirometry during incremental running exercises and respiratory rate showed a low mean absolute percent error of 2.74 [50]. Hexoskin® and Astroskin® (Carré Technologies, Montréal, Canada [20]) are currently available on the market and base their respiratory rate and minute ventilation measurements on the use of RIP. These devices were validated in multiple studies that involved tests in different postures and with incremental levels of exercise; volume measurements have low variability and good agreement and consistency [22].

Several wearable sensors are based on the principle that the resistance measured varies with the movement of the torso. An example is L.I.F.E. Italia Srl's compression garment [24], already illustrated in Table 10.1 and shown in detail in Fig. 10.1. The respiratory sensors are based on the variation of the resistance and are positioned on the anterior external surface of the garment as follows: two are thoracic, one is central and two are abdominal [25, 51].

Another wearable respiratory monitor is a thoracic band called AirGo™, which measures the thoracic circumference changes with a stretchable knitted matrix of nylon and spandex with a built-in silver-coated yarn; the system also embeds an Inertial Measurement Unit for motion detection and human activity recognition. The device was validated during incremental dynamic activities against a gold standard metabolic cart [52] and tested on 20 subjects during daily life, associating respiratory parameters with the performed activity [53]. Measurements of respiratory rate and tidal volume using a disposable Band-Aid®-like sensor have been evaluated. This sensor allows to derive the two aforementioned parameters by simply measuring the local strain of the ribcage and abdomen during breathing, thus providing an interesting perspective given the unobtrusiveness of the solution [54].

Other fields of research and development regard sensors based on variations of capacitance [55] or on the signal coming from Inertial Measurement Units [56], but the results have been mostly referred to static postures and these solutions are not yet used for monitoring during sport activities.

10.4.2 Electrodermal Activity and Sweat

As it is reported in Table 10.1, many devices have electrical sensors that can be used for ECG and/or for electrodermal activity (EDA) monitoring. EDA, which can also be called galvanic skin response, is a measure of the changes in electrical conductance of the skin. EDA, as a reflection of autonomic innervation of sweat glands, provides a quantitative functional measure of sudomotor activity. When subjects undergo physical activity, EDA is increased as sweating rate increases, as a product of an initial recruitment of sweat glands and then an increased sweat secretion per gland [57]. Sweat comes through varying numbers of ducts in the sweat glands at different levels. The sweat ducts can be thought of as a set of variable resistors wired in parallel: the more the sweat level rises and the more ducts that are filled up,

the lower the resistance in that variable set of parallel resistors. In this manner, changes in the level of sweat in the ducts produce observable variations in EDA [58]. For this reason, this parameter can be used during long-distance running to estimate how much an athlete is sweating.

10.4.3 Glucose

Continuous blood glucose monitors (CGM) can provide real-time insights into how our bodies react when creating the energy needed to perform an activity such as running. They can be used to assess if the nutrition plan is correct or needs modifications, since each body reacts differently to food inputs. One example is the Supersapiens system by TT1 Products, Inc. [59], based on the Abbott's Libre Sense Glucose Sport Biosensor. This sensor is applied on the upper arm and gives real-time feedback on blood glucose concentration, enabling to detect hyper- or hypoglycemia. The drawback of this technology is that part of it must be implanted in the skin; however, the benefit of real-time glucose monitoring is high for professional athletes or amateur runners who want to maximize their performance starting from their nutrition. Expert coach Michele Evangelisti selected real-time glucose concentration as the most interesting parameter not currently measured by standard smartwatches, due to its ability to provide feedback on the possibility to further increase the level of effort of an athlete.

10.5   Environmental Parameters

In addition to physiological and running parameters, wearables also have the capability to sense environmental parameters, such as position, temperature, humidity, atmospheric pressure, and altitude, among others. Sensor fusion is often performed in these cases, where sensor fusion is defined as the combining of sensory data or data derived from sensory data such that the resulting information is better than what would be possible when these sources were used individually.

In Table 10.1, different positioning technologies are mentioned. These technologies are based on Global Navigation Satellite Systems (GNSS), which are constellations of satellites orbiting around the Earth and allowing for a very precise positioning [60]. GNSS-based technologies include GPS (Global Positioning System), GLONASS, Galileo, and BeiDou. Specifically, GLONASS is based on a Russian satellite constellation and BeiDou on a Chinese one, while Galileo is Europe's constellation. The actual constellations are expected to expand in the next few years and the International GNSS Service (IGS) is committed to the creation of a multi-GNSS service.

Smartwatches and wearables for sports often include temperature sensors. Such sensors might be used for environmental or body temperature, but it must be noted that in catalogs usually simply the term "temperature" is present, and it is not always specified which one is monitored. It might be of interest to focus on body temperature detection as well in running wearables to provide more information on the state of the runner, especially in highly demanding conditions such as long-distance running. Environmental temperature, on the other hand, can be obtained from the Internet using GNSS information if an Internet connection is available or if the wearable is synchronized with an Internet-connected device, as it is often the case with smartwatches synchronized with smartphones. Also, socks for foot temperature monitoring have been developed [61], mostly for diabetic patients, but this technology might be easily extended to other fields as well.

Humidity and atmospheric pressure are usually monitored together with temperature. Atmospheric pressure allows to estimate altitude, especially if this information can be associated with GNSS geo-localization data. These sensors are typically referred to as altimeters in commercial devices, as it is reported in Table 10.1. Changes in altitude are used to compute the difference in height during a training session or a competition; as expert coach Michele Evangelisti reports, it is a necessary information to evaluate the performance of an athlete during trail running.

Recent developments in sensor technology allow to obtain very small integrated devices that can measure multiple parameters, such as temperature, humidity, and pressure. This is for instance the case of the Bosch Sensortec's BME680 integrated sensor [62], which has a volume of $3.0 \times 3.0 \times 0.93$ mm^3 includes volatile organic compounds (VOCs) sensing capabilities. The most recent sensor of that series, BME688 [63], can also detect volatile sulfur compounds (VSCs) and other gases such as carbon monoxide and hydrogen with the same volume as BME680. Such advancements will allow to easily integrate gas scanning in wearables without an increase in volume.

Figure 10.2 represents real sample data from a male amateur athlete during an ultra-trail running competition. The parameters represented are altitude, speed, heart rate, cadence, and temperature.

10.6 Smartphone Apps

As it was mentioned in the Introduction, some authors consider smartphones as wearables.

There are indeed several smartphone apps specifically designed for sports, even though the smartphone per se is a non-specific sporting good [64]. Some apps can be paired with other wearables, while other apps can function independently and provide running-related features. An example of such apps is Adidas Running by runtastic GmbH [65]. While regular runners tend to use dedicated devices, self-standing apps can be used as a first approach to the sport or to integrate the information obtained from wearables.

Smartphones have indeed several embedded sensors, such as GNSS systems and inertial measurement units. Several algorithms have been developed in the literature to extract activity-related parameters and perform human activity recognition [66]. Also, access to the Internet allows apps to download real-time data concerning the environment given the position in addition to the information coming from the embedded sensors. Furthermore, microprocessors inside of smartphones have general more

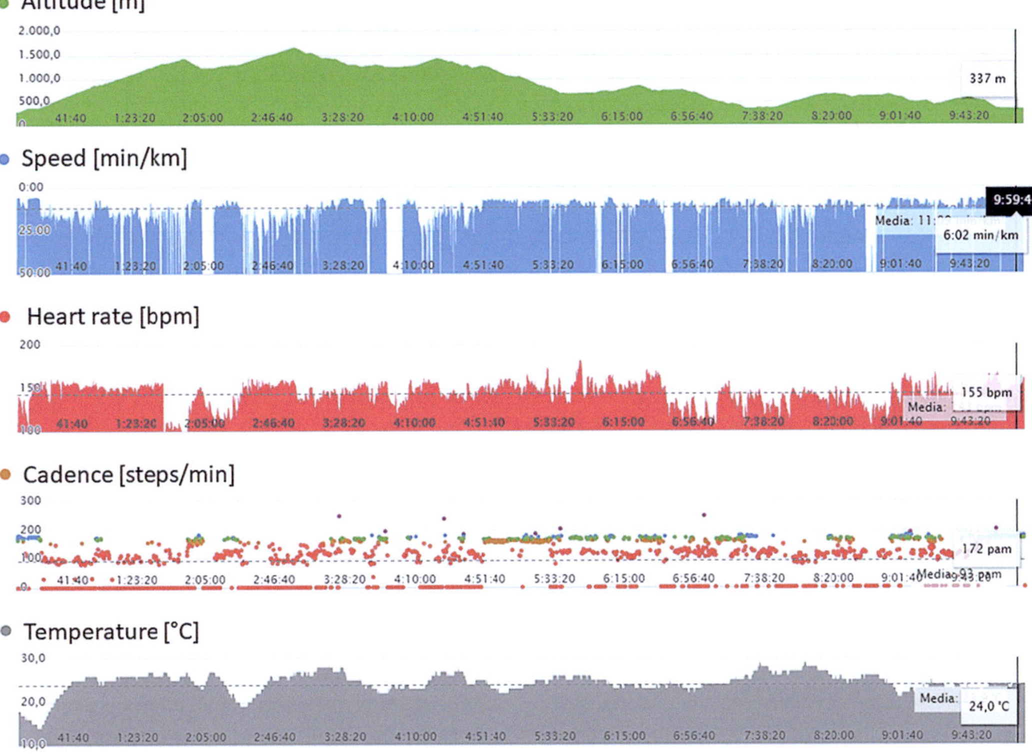

Fig. 10.2 Variations in time of various parameters (altitude, speed, heart rate, cadence, temperature) detected by a Garmin smartwatch of a male athlete (55 years old, height 176 cm, weight 80 kg) during an ultra-trail running competition

computational powers than those used in smartwatches or fitness trackers.

In smartphones, there is the additional possibility to manually insert data from the users. For instance, users are usually asked to manually insert their weight and height in order to estimate calories expenditure.

One function of interest that can be found in some apps is the menstrual tracking feature, which is useful both for women practicing sports and coaches of women athletes who want to fine tune the training to the specifics of the woman's body.

Finally, smartphone apps might be designed with social network features, thus enabling to connect with friends or other athletes in the community, such as the Adidas Running app.

10.7 Future Trends

10.7.1 Transcutaneous Partial Pressure of CO_2

Another parameter that could be of interest for runners, especially trail runners that perform their sports activity at high altitudes with decreasing levels of partial pressure of O_2 and CO_2 in the air, is the arterial partial pressure of CO_2 ($PaCO_2$), which can be estimated by means of the transcutaneous partial pressure of CO_2 ($PtCO_2$). Transcutaneous measurements make use of the fact that the CO_2 gas diffuses through body tissue and skin and can be detected by a sensor at the skin surface. By warming the sensor, a local hyperemia is induced, which increases the supply of arterial blood to the dermal capillary bed below the sensor. In general, this value correlates well with the corresponding $PaCO_2$ value also during exercise [67, 68]. Wearable devices for the measurement of $PtCO_2$ by means of optical sensors are currently a field of research [69]. Another use case of this type of monitoring might be for long-distance running in big cities, where CO_2 levels in the atmosphere are often high and above the recommended thresholds.

10.7.2 Environmental Pollution and Air Quality

Several parameters are indicators of pollution, such as ambient carbon dioxide (CO_2), carbon monoxide (CO), nitrogen dioxide (NO_2), and particulate matter of different sizes (PM_{10}, $PM_{2.5}$, PM_1, and $PM_{0.1}$, where the x in PM_x indicates the diameter of the pollutant in micrometers). Most environmental monitors collect data in fixed stations; however, air quality may change drastically even from one street to the other. One example is shown in Fig. 10.3. On the left side,

the path followed by a person during the day in the city of Milan is shown: the area is nearby Politecnico di Milano, the red arrow indicates the location of an ARPA (Regional Environmental Protection Agency) station measuring various parameters and the green arrow indicates a parking lot which is typically crowded and full of cars. The ARPA station is located nearby a green area. On the left side, data coming from a wearable $PM_{2.5}$ sensor are reported in time. The values measured in correspondence of the red arrow (ARPA station) are consistent with the official ones; however, it can be seen that they are not representative of the situation: air pollution is much worser (and above the suggested threshold of 25 ppb [70]) in correspondence of the parking lot.

This is a rationale behind using environmental monitoring during long-distance running: there is an increasing interest toward these solutions to provide the users with real-time information that can be used to make choices on where exactly to run in cities. This is even more relevant if the exposure to a given pollutant is taken into consideration: exposure is computed as minute ventila-

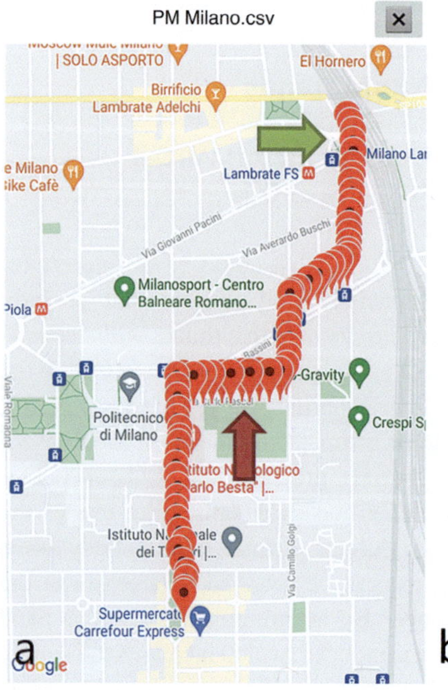

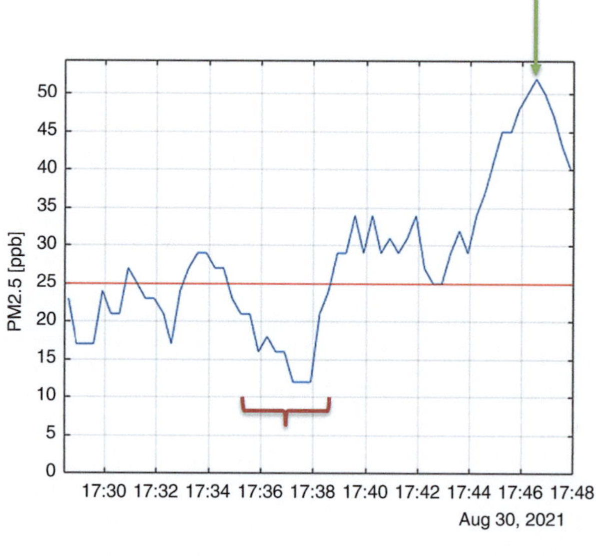

Fig. 10.3 $PM_{2.5}$ values acquired in different points of the city of Milan (sampling rate = 1 min) (**a**) and plotted over time (**b**). The red arrow indicates the location of the ARPA station nearby a green area, while the green arrow indicates a parking lot

tion per parameters concentration, so athletes doing efforts for a long time are more exposed to pollutants [71].

10.8 Conclusions

This chapter illustrated the main parameters that can be detected by means of wearable devices specifically designed for sport activities, with a focus on long- distance running. Most recent advancements in heart rate monitoring, physiological monitoring, running parameters assessment, and environmental monitoring are discussed and presented. The presented solutions are already on the market or still at research level. The use of smartphone apps is also briefly illustrated. Some future perspectives are also discussed, especially related to the impact of the external environment on the physiological parameters of the athlete, which is part of a broader debate on the effect of environmental pollution on human health in general.

Despite the widespread use of wearable devices, not all of them have been validated against a gold standard and some validation studies did not show high accuracy. Furthermore, the technology behind most commercially available solutions is rarely published, thus making it almost impossible for the final user to know how reliable their devices are. However, it is generally true that, the more a measurement is "close" to the signal we want to measure (e.g., electrical measurement of heart activity to estimate heart rate, analysis of thoracic movement to estimate respiratory rate), the higher the accuracy of the estimated parameter is. Also, it is worth mentioning that some high-profile companies publish their data and make them available for the scientific community.

References

1. Aliverti A. Wearable technology: role in respiratory health and disease. Breathe. 2017;13(2):e27–36.
2. Nasiri S, Khosravani MR. Progress and challenges in fabrication of wearable sensors for health monitoring. Sensors Actuat A Phys. 2020;312:112105.
3. Angelucci A, Aliverti A. Telemonitoring systems for respiratory patients: technological aspects. Pulmonology. 2020;26(4):221–32.
4. Moore IS, Willy RW. Use of wearables: tracking and retraining in endurance runners. Curr Sports Med Rep. 2019;18(12):437–44.
5. Michele Evangelisti [Internet]. [cited 2021 Nov 13]. https://www.micheleevangelisti.com/.
6. Jubran A. Pulse oximetry. Crit Care. 2015;19(1):1–7.
7. Inan OT, Migeotte P-F, Park K-S, Etemadi M, Tavakolian K, Casanella R, et al. Ballistocardiography and seismocardiography: a review of recent advances. IEEE J Biomed Health Inform. 2014;19(4):1414–27.
8. Apple Official Website [Internet]. [cited 2021 Oct 1]. https://www.apple.com/it/.
9. Bai Y, Hibbing P, Mantis C, Welk GJ. Comparative evaluation of heart rate-based monitors: Apple Watch vs Fitbit Charge HR. J Sports Sci. 2018;36(15):1734–41.
10. Littell LM, Roelle L, Dalal A, Van Hare G, Orr W, Miller N, et al. Assessment of applewatch series 6 pulse oximetry and ecg algorithm in children. J Am Coll Cardiol. 2021;77(18_Supplement_1):482.
11. Perez MV, Mahaffey KW, Hedlin H, Rumsfeld JS, Garcia A, Ferris T, et al. Large-scale assessment of a smartwatch to identify atrial fibrillation. N Engl J Med. 2019;381(20):1909–17.
12. Ōura Ring Official Website [Internet]. [cited 2021 Oct 1]. https://ouraring.com/.
13. Kinnunen H, Rantanen A, Kentt T, Koskim ki H. Feasible assessment of recovery and cardiovascular health: accuracy of nocturnal HR and HRV assessed via ring PPG in comparison to medical grade ECG. Physiol Meas. 2020;41(4):04NT01.
14. Fitbit Official Global Website [Internet]. [cited 2021 Oct 1]. https://www.fitbit.com/global/it/home.
15. Mishra T, Wang M, Metwally AA, Bogu GK, Brooks AW, Bahmani A, et al. Pre-symptomatic detection of COVID-19 from smartwatch data. Nat Biomed Eng. 2020;4(12):1208–20.
16. Garmin Official US Website [Internet]. [cited 2021 Oct 1]. https://www.garmin.com/en-US/.
17. Støve MP, Haucke E, Nymann ML, Sigurdsson T, Larsen BT. Accuracy of the wearable activity tracker Garmin Forerunner 235 for the assessment of heart rate during rest and activity. J Sports Sci. 2019;37(8):895–901.
18. Cassirame J, Vanhaesebrouck R, Chevrolat S, Mourot L. Accuracy of the Garmin 920 XT HRM to perform HRV analysis. Australas Phys Eng Sci Med. 2017;40(4):831–9.
19. Viatom Official Website [Internet]. [cited 2021 Oct 1]. https://www.viatomtech.com/.
20. Hexoskin Official Website [Internet]. [cited 2021 Oct 1]. https://www.hexoskin.com/.
21. Angelucci A, Cavicchioli M, Cintorrino IA, Lauricella G, Rossi C, Strati S, et al. Smart textiles and sensorized garments for physiological monitoring: a review of available solutions and techniques. Sensors (Switzerland). 2021;21(3):1–23.

22. Villar R, Beltrame T, Hughson RL. Validation of the hexoskin wearable vest during lying, sitting, standing, and walking activities. Appl Physiol Nutr Metab. 2015;40(10):1019–24.

23. Villa-Colín J, Shaw T, Toscano W, Cowings P. Evaluation of astroskin bio-monitor during high intensity physical activities. In: Memorias del Congreso Nacional de Ingeniería Biomédica; 2018. p. 262–265.

24. L.I.F.E. Official Website [Internet]. [cited 2021 Oct 1]. https://www.x10x.com/.

25. Sarmento A, Vignati C, Paolillo S, Lombardi C, Scoccia A, Nicoli F, et al. Qualitative and quantitative evaluation of a new wearable device for ECG and respiratory Holter monitoring. Int J Cardiol. 2018;272:231–7.

26. Huawei Official Global Website [Internet]. [cited 2021 Oct 1]. https://consumer.huawei.com/en/.

27. Guo Y, Wang H, Zhang H, Liu T, Liang Z, Xia Y, et al. Mobile photoplethysmographic technology to detect atrial fibrillation. J Am Coll Cardiol. 2019;74(19):2365–75.

28. Galaxy Watch4 User Manual [Internet]. [cited 2021 Oct 1]. https://www.samsung.com/it/support/model/SM-R895FZKAITV/.

29. Hwang J, Kim J, Choi KJ, Cho MS, Nam GB, Kim YH. Assessing accuracy of wrist-worn wearable devices in measurement of paroxysmal supraventricular tachycardia heart rate. Korean Circ J. 2019;49(5):437–45.

30. Amazfit Official Shop [Internet]. [cited 2021 Oct 1]. https://amazfit.shop/.

31. Zhang S, Xian H, Chen Y, Liao Y, Zhang N, Guo X, et al. The auxiliary diagnostic value of a novel wearable electrocardiogram-recording system for arrhythmia detection: diagnostic trial. Front Med. 2021;8(June):1–7.

32. Polar Official Global Website [Internet]. [cited 2021 Oct 1]. https://www.polar.com/en.

33. Gilgen-Ammann R, Schweizer T, Wyss T. RR interval signal quality of a heart rate monitor and an ECG Holter at rest and during exercise. Eur J Appl Physiol. 2019;119(7):1525–32.

34. Xiaomi Official Global Website [Internet]. [cited 2021 Oct 1]. https://www.mi.com/global.

35. Montblanc Official Website [Internet]. [cited 2021 Oct 1]. https://www.montblanc.com/en-ua.

36. Wahoo Fitness Official European Website [Internet]. [cited 2021 Oct 1]. https://eu.wahoofitness.com/.

37. Empatica Official Website [Internet]. [cited 2021 Oct 1]. https://www.empatica.com/.

38. McCarthy C, Pradhan N, Redpath C, Adler A. Validation of the Empatica E4 wristband. In: 2016 IEEE EMBS international student conference (ISC). IEEE; 2016. p. 1–4.

39. Zephyr™ Performance Systems Official Website [Internet]. [cited 2021 Oct 1]. https://www.zephyranywhere.com/.

40. Nazari G, Bobos P, MacDermid JC, Sinden KE, Richardson J, Tang A. Psychometric properties of the Zephyr bioharness device: a systematic review. BMC Sports Sci Med Rehabil. 2018;10(1):6.

41. Steinhubl SR, Mehta RR, Ebner GS, Ballesteros MM, Waalen J, Steinberg G, et al. Rationale and design of a home-based trial using wearable sensors to detect asymptomatic atrial fibrillation in a targeted population: the mHealth Screening to Prevent Strokes (mSToPS) trial. Am Heart J. 2016;175:77–85.

42. Oxitone Official Website [Internet]. [cited 2021 Oct 1]. https://www.oxitone.com/.

43. Guber A, Epstein Shochet G, Kohn S, Shitrit D. Wrist-sensor pulse oximeter enables prolonged patient monitoring in chronic lung diseases. J Med Syst. 2019;43(7):230.

44. Navalta JW, Montes J, Bodell NG, Salatto RW, Manning JW, DeBeliso M. Concurrent heart rate validity of wearable technology devices during trail running. PLoS One. 2020;15(8):e0238569.

45. Morin J-B, Samozino P, Zameziati K, Belli A. Effects of altered stride frequency and contact time on leg-spring behavior in human running. J Biomech. 2007;40(15):3341–8.

46. Rowlands AV, Stiles VH. Accelerometer counts and raw acceleration output in relation to mechanical loading. J Biomech. 2012;45(3):448–54.

47. Ruder M, Jamison ST, Tenforde A, Mulloy F, Davis IS. Relationship of foot strike pattern and landing impacts during a Marathon. Med Sci Sports Exerc. 2019;51(10):2073–9.

48. Karlen W, Raman S, Ansermino JM, Dumont GA. Multiparameter respiratory rate estimation from the photoplethysmogram. IEEE Trans Biomed Eng. 2013;60(7):1946–53.

49. Sackner MA, Watson H, Belsito AS, Feinerman D, Suarez M, Gonzalez G, et al. Calibration of respiratory inductive plethysmograph during natural breathing. J Appl Physiol. 1989;66(1):410–20.

50. Harbour E, Lasshofer M, Genitrini M, Schwameder H. Enhanced breathing pattern detection during running using wearable sensors. Sensors. 2021;21(16):5606.

51. Contini M, Sarmento A, Gugliandolo P, Leonardi A, Longinotti-Buitoni G, Minella C, et al. Validation of a new wearable device for type 3 sleep test without flowmeter. PLoS One. 2021;16(4):e0249470.

52. Antonelli A, Guilizzoni D, Angelucci A, Melloni G, Mazza F, Stanzi A, et al. Comparison between the Airgo™ device and a metabolic cart during rest and exercise. Sensors. 2020;20:3943.

53. Angelucci A, Kuller D, Aliverti A. A home telemedicine system for continuous respiratory monitoring. IEEE J Biomed Health Inform. 2021;25:1247–56.

54. Chu M, Nguyen T, Pandey V, Zhou Y, Pham HN, Bar-Yoseph R, et al. Respiration rate and volume measurements using wearable strain sensors. NPJ Digit Med. 2019;2(1):1–9.

55. Naranjo-Hernández D, Talaminos-Barroso A, Reina-Tosina J, Roa LM, Barbarov-Rostan G, Cejudo-Ramos P, et al. Smart vest for respiratory rate monitoring of

copd patients based on non-contact capacitive sensing. Sensors (Switzerland). 2018;18(7):1–24.

56. Cesareo A, Nido SA, Biffi E, Gandossini S, D'Angelo MG, Aliverti A. A wearable device for breathing frequency monitoring: a pilot study on patients with muscular dystrophy. Sensors. 2020;20(18):5346.

57. Posada-Quintero HF, Reljin N, Mills C, Mills I, Florian JP, VanHeest JL, et al. Time-varying analysis of electrodermal activity during exercise. PLoS One. 2018;13(6):e0198328.

58. Lara ÓD, Pérez AJ, Labrador MA, Posada JD. Centinela: a human activity recognition system based on acceleration and vital sign data. Pervasive Mob Comput. 2012;8(5):717–29.

59. Supersapiens [Internet]. [cited 2021 Nov 13]. https://www.supersapiens.com/en-EN/.

60. Li X, Zhang X, Ren X, Fritsche M, Wickert J, Schuh H. Precise positioning with current multi-constellation global navigation satellite systems: GPS, GLONASS, Galileo and BeiDou. Sci Rep. 2015;5(1):1–14.

61. Reyzelman AM, Koelewyn K, Murphy M, Shen X, Yu E, Pillai R, et al. Continuous temperature-monitoring socks for home use in patients with diabetes: observational study. J Med Internet Res. 2018;20(12):e12460.

62. Gas Sensor BME680 | Bosch Sensortec [Internet]. [cited 2021 Nov 13]. https://www.bosch-sensortec.com/products/environmental-sensors/gas-sensors/bme680/.

63. Gas Sensor BME688 | Bosch Sensortec [Internet]. [cited 2021 Nov 13]. https://www.bosch-sensortec.com/products/environmental-sensors/gas-sensors/bme688/.

64. Janssen M, Scheerder J, Thibaut E, Brombacher A, Vos S. Who uses running apps and sports watches? Determinants and consumer profiles of event runners' usage of running-related smartphone applications and sports watches. PLoS One. 2017;12(7):e0181167.

65. Adidas Running [Internet]. [cited 2021 Nov 13]. https://www.runtastic.com/en/.

66. Qi W, Su H, Aliverti A. A smartphone-based adaptive recognition and real-time monitoring system for human activities. IEEE Trans Human Mach Syst. 2020;50(5):414–23.

67. Eberhard P. The design, use, and results of transcutaneous carbon dioxide analysis: current and future directions. Anesth Analg. 2007;105(Suppl. 6):48–52.

68. Contini M, Angelucci A, Aliverti A, Gugliandolo P, Pezzuto B, Berna G, et al. Comparison between PtCO2 and PaCO2 and derived parameters in heart failure patients during exercise: a preliminary study. Sensors. 2021;21(19):6666.

69. Grangeat P, Gharbi S, Accensi M, Grateau H. First evaluation of a transcutaneous carbon dioxide monitoring wristband device during a cardiopulmonary exercise test. Proc Annu Int Conf IEEE Eng Med Biol Soc EMBS. 2019:3352–5.

70. World Health Organization. WHO. Air quality guidelines for particulate matter, ozone, nitrogen dioxide and sulphur dioxide. Global update 2005. World Health Organization. 2006;38:E90038. https://www.euro.who.int/__data/assets/pdf_file/0005/786.

71. Slezakova K, Pereira MC, Morais S. Ultrafine particles: levels in ambient air during outdoor sport activities. Environ Pollut. 2020;258:113648.

Return to Running After Anterior Cruciate Ligament Reconstruction

11

Alessandro Compagnin, Marco Gastaldo, and Francesco Della Villa

11.1 Introduction

Anterior cruciate ligament (ACL) injury is one of the most common and devastating knee injuries in pivoting and contact sports [1]. With an annual incidence of 68.6 injuries per 100,000 person-years, more than 200,000 ACL tears occur in the United States annually (incidence rates of ACL injuries vary between populations being studied) [1, 2]. ACL surgery is often performed as a first-line treatment after the injury. Anterior cruciate ligament reconstruction (ACLR) is the predominant method of surgery in current practice and hundreds of thousands of these surgeries are performed every year [3]. Despite being one of the most studied and discussed topic in the Sport Medicine literature, no consensus has been reached about many aspects of this recovery process and it is well established that outcomes after ACLR are far from being perfect [4].

Because of the nature of the majority part of the sports that are practised nowadays (e.g., football, basketball, etc.), running is considered a foundational task for every athlete involved in those activities (most sports require the player to be able to run) [5]. Return to running after ACLR is considered by the athlete and the rehabilitation team a milestone in the recovery process but, although the importance of running is well established in the rehabilitation community, there is not a consensus on the pathway that has to be followed to allow a patient to return to run successfully. A correct and thorough running implementation can represent a useful training stimulus for the athlete but on the other hand, if introduced too early or with the wrong progression, it can pose a threat for the athlete's recovery process. To bridge the gap between research and practice, and to optimize the outcomes after ACLR, there is the need to provide clinicians who work with ACL injured patients a clear criteria-based progression for running implementation. To guide the patient in the return to running process, a deep understanding of the running demands (biomechanics, forces to be absorbed, different running types, etc.) and the patient's injury and current condition (injured body area, biological healing time, athlete's physical profile, etc.) is paramount.

Therefore, the aim of this chapter is to provide a clear understanding of the running biomechanics, analysing the features of the ACL injury and

A. Compagnin
Isokinetic Medical Group, FIFA Medical Centre of Excellence, London, UK
e-mail: a.compagnin@isokinetic.com

M. Gastaldo
Isokinetic Medical Group, FIFA Medical Centre of Excellence, Turin, Italy
e-mail: m.gastaldo@isokinetic.com

F. Della Villa (✉)
Education and Research Department, Isokinetic Medical Group, FIFA Medical Centre of Excellence, Bologna, Italy
e-mail: f.dellavilla@isokinetic.com

consequential reconstruction, to present an easily implementable checklist to utilize before allowing a patient to resume running and to provide a clear rehabilitation progression to help the clinician achieving those goals with ACL reconstructed patients.

11.2 The Demands of Running

The decision of introducing a new task during the rehabilitation process of an athlete should be based on the analysis of that task demands and the athlete's envelope of function as described by Dye in 1996 [6]. The envelope of function can be defined as the range of load that can be applied across an individual joint in a given period without reaching a supraphysiologic overload or structural failure [6]. In other words, this envelope indicates the load tolerance of a specific joint determined by multifactorial elements (anatomical, kinematic, physiological, etc.). Indicators that an imposed task demand is superior to the athlete's joint envelope of function can include pain, discomfort, functional instability, effusion, warmth and tenderness [6]. Failure to match the task demands and the athlete's load tolerance may result in exposing the athlete to tasks which they are not prepared for. In order to prevent this to happen, it is important to fully understand and quantify the demands of the task in terms of level of loading that may be placed on the body, and estimate the load tolerance of the athlete. Starting from the analysis of the task, the level of loading of a specific activity can be considered as:

- Peak loading (e.g., peak ground reaction forces) [5];
- Volume load (e.g., load times repetition) [5];
- Rate of loading (e.g., time over which it is delivered/experienced) [5].

Biomechanical studies reported that for every step taken during running, the weight acceptance of the athlete is estimated to be around 2–3 times the athlete's body mass [7]. It is important to stress that the ground reaction forces are variables depending on many modifiable and non-

modifiable factors such as running biomechanics (modifiable) [8, 9], footwear and orthotics (modifiable) [8, 9], environment and running surface (modifiable) [8], athlete strength (modifiable), running speed (modifiable), athlete biological characteristics like age, height, etc. (non-modifiable). A variation on the previously mentioned parameters could potentially change both the force-time and centre of pressure patterns, influencing in this way the task total load.

11.3 Criteria to Return to Running After ACLR

Once the task demands are clear, it is time to better understand the load tolerance of the athlete to implement the right activity at the right moment. An accurate assessment of many aspects of the athlete is paramount to create the "athlete profile" and understand if the patient is ready or not to resume running. There is not a consensus in the current literature about which criteria should be achieved before allowing a patient to return to running. This lack of support by the current evidence on the topic leaves the clinicians without safety guidelines to follow in order to implement a safe and effective return to run process.

In a recent scoping review, Rambaud et al. (2018) reported that in deciding when a patient is ready to return to running, time-based criteria after ACLR was the most cited criterion [10]. The median time from which return to run was permitted was 12 postoperative weeks [10]. There are many consequences of different nature that can arise from this choice:

1. *Risk of harming the athlete*: the ability to perform specific tasks like running is not only related to the time that has passed since the surgery, but more specifically to the patient's function [4], strength, mobility, movement quality, etc. If a patient's load tolerance is not trained enough to accept the force generated by the running, the athlete will be exposed to overload and risk of injury;
2. *Athlete's psychological failure*: when a specific time frame has been communicated to

the athlete, that date becomes a fixed point for the patient. Depending on multifactorial elements (e.g., surgery characteristics, rehabilitation process, level of the athlete, etc.), every patient has a different rate of progression and will achieve different training targets at different times. If the criteria to allow a patient to return to run post ACLR is only based on time, and the patient is not able to run at that specific moment in time, the athlete will likely think that they are "failing" since running is still not possible [5].

Time-based criteria are selected arbitrary and do not reflect the athlete's function and readiness to run. Fewer than 1/5 studies reported clinical, strength or performance-based criteria for return to running, even though the best evidence recommends performance-based criteria combined with time-based criteria to commence running activities following ACLR [7]. A shift from time-based to criteria-based rehabilitation approach is necessary during the ACL rehabilitation process. Different papers presented multiple objective criteria that should be achieved before allowing a patient to return to run. The criteria presented in those articles can be divided into five different categories (see Table 11.1 for more information about the Return to Running criteria after ACLR) that are suggested to be achieved before allowing a patient to return to run:

11.3.1 Joint Homeostasis

Goals: Absent or minimal pain on a numerical rating scale (NRS, ≤3) during walking [11, 12], zero or trace effusion [4].

The "quiet knee" is a knee that does not show (or shows in minimal part) typical signs of inflammation (pain, redness, heat, swelling and loss of function). Pain and swelling can result in arthrogenic muscle inhibition (AMI), a complex and multifactorial neurophysiological phenomenon that hinders optimal recruitment of the knee spanning muscles (especially knee extensors), limiting the ability of the athlete to produce force. Before starting to run it is suggested to minimize pain and

reach a minimal activity related effusion status (≤1 cm change at the supra patella circumferential measurement in response to activity) [18].

11.3.2 Knee Mobility

Goals: Full knee extension (straight knee, equal to the other side) and ≥120°/130° of knee flexion [4, 5, 11, 12].

Restoring joint range of motion (flexion and extension) is paramount during the rehabilitation process. Even small knee extension deficits as little as 3° appear to adversely affect post-surgical subjective and objective outcomes after ACL reconstruction [19, 20]. Depending on the running speed and technique, knee flexion angles can reach values as high as 120° of knee flexion. Restoring knee joint mobility is critical for the recovery of optimal gait and running biomechanics.

11.3.3 Gait Biomechanics

Goals: Walk on a treadmill for at least 10 min without pain or swelling [12] and with an optimal biomechanics.

Abnormal gait patterns have been associated with muscle weakness, decreased functional performance, low patient satisfaction outcomes after surgery and with post-operative complications including osteoarthritis [5]. Abnormal gait patterns often become further exacerbated when the patient returns to running. Re-establishing normal gait early and safely after surgery is a key priority [5].

11.3.4 Strength

Open kinetic chain strength goals: Limb Symmetry Index (LSI) for Quadriceps and Hamstrings Strength ≥70% assessed by isometric or isokinetic knee extension and flexion [4, 5, 10].

Closed kinetic chain strength goals: Single leg closed kinetic chain peak strength of at least 1.25 times body mass on leg press [4, 5].

Table 11.1 Recommended criteria for running implementation after anterior cruciate ligament reconstruction. Each outcome measure, specific test and published reference are included, as well as the goal to achieve in order to allow the patient to start to run

Category	Outcome measure	Test	Goal
Joint homeostasis	Pain	Numeric Rating Scale (NRS)	Absent or minimal pain on a numerical rating scale (NRS) (NRS <3) during walking [11, 12]
	Effusion	Stroke test	Zero or trace effusion [4]
		Knee circumference measurements [13]	Minimal activity related effusion (<1 cm change patella)
Knee mobility	Extension	Prone hang test [14]	Straight knee (0°). The heel height difference is measured (approximately 1 cm = 1°) Equal to the other side
		Supine with a long-arm goniometer	Straight knee (0°). Bony landmarks: greater trochanter, lateral femoral condyle, and lateral malleolus.
	Flexion	Supine/prone with long arm goniometer	Knee flexion ≥120°/130° [4, 5, 11, 12]
Gait biomechanics	Walk assessment	Visual or 2D assessment of walking gait	Walk on a treadmill for at least 10 min without pain or swelling [12] and with an optimal biomechanics
	Thigh muscles, open kinetic chain (OKC) strength	Isokinetic strength assessment of quadriceps and hamstrings muscles [4]	Limb Symmetry Index (LSI) for Hamstrings and Quadriceps strength ≥70% [1, 4, 5]
		Isometric strength assessment of quadriceps and hamstrings muscles [4]	Limb Symmetry Index (LSI) for Hamstrings and Quadriceps strength ≥70% [1, 4, 5]
	Closed kinetic chain (CKC) muscle strength	Leg Press Test [4]: 90° knee flexion and seat at 45°, maximal weight achieved for 8 RM test	Single limb closed kinetic chain peak strength of at least 1.25 times body mass on single limb leg press [4, 5] or 1.5 × BM predicted 1 RM [4]
Strength	Calf capacity	Single leg heel raises [15]: The athlete stands on one foot on the edge of a step and performs a heel raise through full ROM. Heel raises are performed at 1 repetition every 2 s [4]. The test is concluded when the subject is unable to move through the full range or slows below the cadence [4]	Greater than 20 reps and within 5 repetitions versus the other side [4]
	Gluteal muscle capacity	Single leg bridge test (variation [16]): The athlete is supine with 90° knee flexion angle, one foot on the floor and arms crossed on the chest. The subject lifts the hips from the floor to neutral hip position and then returns down to the ground [4]. The test is concluded when the subject cannot reach the height or gives up [4]	Greater than 20 reps and within 5 repetitions versus the other side [4]
Functional outcomes	Movement quality assessment	Single Leg Squat [4, 17]: Squat to at least 60° of knee flexion, minimal trunk motion, minimal pelvic motion and no hip adduction nor internal rotation [4]	Good movement quality (no zeros and score greater than 6) [4]

Accessory muscles strength: Optimal work capacity of glutes and calf muscles [11, 12] (single leg calf raises [15] and glutes bridges [16] capacity test greater than 20 reps and within 5 repetitions versus the other side [4]).

Lower limb weakness (in particular of the knee extensor compartment) alters biomechanics, reduce functional performance, and may be linked to poorer return to sport outcomes [4]. Extensive research indicates that most patients are unable to sufficiently restore quadriceps strength after ACLR at the moment of return to running and sport.

Restoring isolated or analytic strength of knee flexors and extensors muscles is only one of the many components of the strength recovery process after ACLR. Running and most of functional activities are performed in a closed kinetic chain fashion and for this reason, assessing and recovering closed kinetic chain strength is necessary to allow a gradual progression to more demanding activities. The ability to perform functional tasks involves the neuromuscular system in the production, transmission and dissipation of the ground reaction forces via the neuromuscular system. Inability to absorb those forces through the neuromuscular system (e.g. insufficient functional eccentric muscle strength of the lower limb) would result in movement compensations and/or overreliance/acceptance of the passive restraints such as ligament, joint complexes, and fascial system, potentially resulting in overload and/or acute injuries. Developing and testing the athlete's lower limbs ability to produce and accept force can provide the necessary foundation and understanding on when the athlete may be ready to return to run ACLR.

Optimization of the gluteal musculature is needed to be able to control movement patterns, especially during single-leg activities [21]. The tri-planar function of the gluteus maximus and gluteus medius (abduction, extension, external rotation) serves to control femoral adduction and internal rotation and to produce hip extension, thereby protecting the knee from high-risk positions that increase ACL strain [21]. Deficits in hip muscle strength after ACLR have been reported in the literature [22]. Weakness of the gluteal muscles can contribute to altered movement patterns which increase knee and ACL loading and are thought to be important risk factors for ACL injury [4]. A lack of appropriate neuromuscular control and strength of the gluteal muscles has been related to high-risk landing strategies during single-leg activities [21]. A strong focus on addressing dysfunction of the gluteal muscles during mid-stage rehabilitation, as well as considering the trunk, pelvic and hip musculature in general is recommended [4].

Calf muscle strength is important for load acceptance and propulsion. Biomechanical studies found that the soleus muscle contributes the most muscle force production during running at speeds up to 7 m s^{-1} [23] and that the ankle eccentrically accepts up to 50% of the impact forces from landing [24]. During single-leg drop landing task, the muscles that generated the greatest posterior shear force have been reported to be the soleus, medial hamstrings, and biceps femoris [25]. Since ACL injury occurs promptly after initial contact, the soleus may be particularly important for reducing the likelihood of ACL injury, as it makes a more substantial contribution to the posterior shear joint reaction force during the first 25% of the landing phase [25]. Reduced ankle plantarflexion strength is likely to affect both load absorption and propulsion significantly during running gait [18] and potentially lead to compensatory strategies in functional activities such as running.

11.3.5 Functional Outcomes

Goal: "Good" movement quality on Single Leg Squat testing [4, 17].

Squatting movement can be defined as the foundational exercise for the majority part of the most practised sport nowadays. The Single Leg Squat (SLS) variation involves a triple extension-flexion movement performed on one leg. To optimally execute a single leg squat, the athlete must have sufficient lower limb strength (the movement involves supporting and moving the whole bodyweight up and down), balance (in order to stay on one leg), and neuromuscular control to execute the movement with a proper technique (motor strategy and lower limb/pelvis/trunk alignment). Poor single leg squat performance is associated with poor biomechanics in more complex tasks as this forms the motor pattern foundation for many tasks involving single leg stance and triple flexion and extension [4].

11.4 Return to Running Progression

Return to running after anterior cruciate ligament reconstruction (ACLR) is a long and intricate process that has to be carefully pre-planned and tailored on the patient in order to optimize the outcomes of ACLR rehabilitation. As previously stated in the preceding section, the patient must achieve many different goals to be allowed to resume running in a safe and effective way. To achieve those goals with the patient, the medical team has to identify the rehabilitation priorities (in accordance with the current rehabilitation stage) and address them keeping in mind the upcoming new priorities. In this section, the authors will present a series of rehabilitation strategies (Table 11.2) that can be implemented to achieve the previously mentioned goals and bring the patient back to running.

11.4.1 Pain (Knee Homeostasis): Absent or Minimal Pain on a Numerical Rating Scale (NRS)

The sources of pain after ACLR can be many and dealing with pain is one of the priorities of the early-stage rehabilitation [27, 38]. When a specific load imposed to a joint is greater than the joint's loading ability to handle that load, pain is one of the body signals that indicate this overload. Pain can be used to determine task and exercise progression, as these factors will relate to the loading stress experienced by the knee [39]. Dealing with pain is necessary during the ACLR

rehabilitation process and, in order to do so, clinicians can utilize different strategies (Fig. 11.1): (a) temporary decrease the load imposed to the joint; (b) increase the load tolerance of the joint; (c) temporary decrease the pain by the use of cryotherapy [40] or transcutaneous electrical nerve stimulation (TENS) [29, 41]. Progression to a more demanding task or exercise is allowed only when increase of pain is not reported by the patient (numeric rating scale) as response to previous tasks [5].

11.4.2 Effusion (Knee Homeostasis): Zero or Trace Effusion [4] with Minimal Activity Related Effusion (<1 cm Change Patella)

Swelling and subsequent changes in knee circumferences after a task are signs of joint overload. Managing swelling is necessary during the ACLR rehabilitation process and these are some suggested strategies by the authors that can be implemented by the clinicians to do so (Fig. 11.2): (a) use of cryotherapy (there is a debate in the literature regarding the effects of ice on joint swelling and more evidence is needed); (b) joint elevation; (c) stimulate the "muscle pump action" with movement and exercise therapy; (d) knee load management and monitoring [5]; (e) lymphatic massage; (f) hydrotherapy [26].

11.4.3 Knee Extension and Flexion (Mobility): Straight Knee (0°) or Equal to the Other Side and Knee Flexion ≥120°/130°

Regaining full knee extension is a priority in the early-stage of rehabilitation as it can reduce pain, stimulate joint homeostasis, help to prevent the formation of scar tissue and capsular retractions that may limit joint mobility [40] and allow the patient to regain a normal gait pattern. Even small knee extension deficits as little as 3°

Table 11.2 Suggested rehabilitation strategies in order to achieve the desired goal and allow the patient to return to run. Each outcome measure, specific goal, and published reference for supporting the suggested rehabilitation strategies are listed in the following table

Outcome measure	Goals	Rehabilitation strategies
Pain	Absent or minimal pain on a numerical rating scale (NRS) (NRS <3) during walking [11, 12]	Temporary decrease the load imposed to the joint
		Increase the load tolerance of the joint
		Temporary decrease the pain using cryotherapy
Effusion	Zero or trace effusion [4]	Use of cryotherapy (there is a debate in the literature regarding the effects of ice on joint swelling and more evidence is needed)
		Joint elevation
	Minimal activity related effusion (<1 cm change patella)	Stimulate the "muscle pump action" with movement and exercise therapy
		Knee load management and monitoring
		Lymphatic massage
		Hydrotherapy [26]
Knee extension and flexion	Full knee extension (straight knee, equal to the other side) [5]	Low-load—long-duration mobility exercises performed multiple times per day [27]
		Manual therapy (patella mobility and terminal knee extension or flexion)
	Knee flexion >120°/130°	Active exercise at the end of the range of motion
Walking biomechanics	Walk on a treadmill for at least 10 min without pain or swelling [12] and with optimal biomechanics	Full knee extension [5, 11, 28]
		Good quadriceps muscle activation (no quadriceps lag on active straight leg raise [5])
		Minimal and non-reactive swelling to activities [5, 11, 28]
		Good neuromuscular control of the gait movement [5, 11, 28]
		Gradual increase of the knee walking volume
Quadriceps muscle strength	Limb Symmetry Index (LSI) >70% on Isokinetic/Isometric Test [4, 5]	Quadriceps analytic exercises (OKC and safe implementation) [28, 29]
		Use of modalities (neuromuscular electrical stimulation (NMES [4] and blood flow restriction training (BFR) [4, 29–31]
		Implementation of a periodized approach to strength training [29]
		Cross-education phenomenon [32–35]
		Gradual implementation of CKC exercises targeting the thigh musculature
Hamstrings muscle strength	Limb Symmetry Index (LSI) >70% on Isokinetic/Isometric Test [4, 5]	Low-intensity hamstring exercises initiated in the early-stage of rehabilitation followed by progressive overload [36]
		Use of modalities (NMES and BFR) [36]
		Medial to lateral hamstrings balance [36] training all hamstrings functions (hip extension, knee flexion and knee rotations)
		Full ROM strengthening exercises (especially deep knee flexion angles where the strength deficit is more accentuated)
		Progressive eccentric strength training [36]
Calf strength and capacity	Greater than 20 reps and within 5 repetitions versus the other side [4]	Calf muscle strengthening (Gastrocnemius and Soleus)
		Anti-pronation muscles strengthening

(continued)

Table 11.2 (continued)

Outcome measure	Goals	Rehabilitation strategies
Glutes strength and capacity	Greater than 20 reps and within 5 repetitions versus the other side [4]	Restore optimal lumbo-pelvic stability and balance [37]
		Strengthen the gluteus muscles [55] training all the most important gluteal functions (hip extension, abduction, and external rotation);
		Re-integrate the gluteal muscle into the motor pattern [37]
Closed kinetic chain muscle strength	Single leg closed kinetic chain peak strength of at least 1.25 times body mass on single limb leg press [4, 5]	Closed kinetic chain strength training
		Strengthening of all the lower limb muscles contributing to CKC force production
Movement quality	Good Single Leg Squat Movement Assessment [4, 17]	See Fig. 11.10 for an example of task intensity progression suggested to gradually reach the demands of the single leg squat

Fig. 11.1 Examples of suggested rehabilitation strategies to implement to decrease pain. *TENS* Transcutaneous electrical nerve stimulation

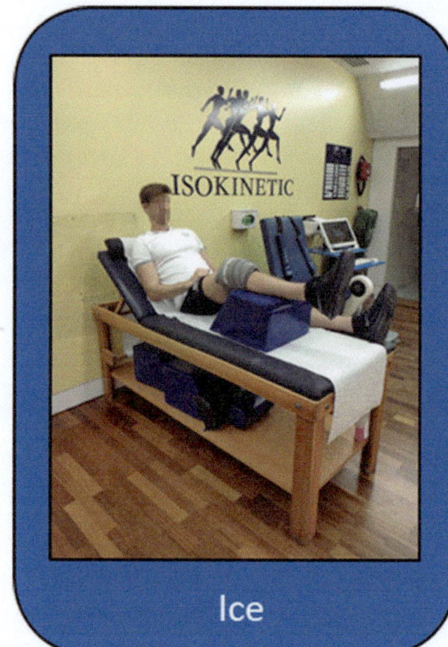

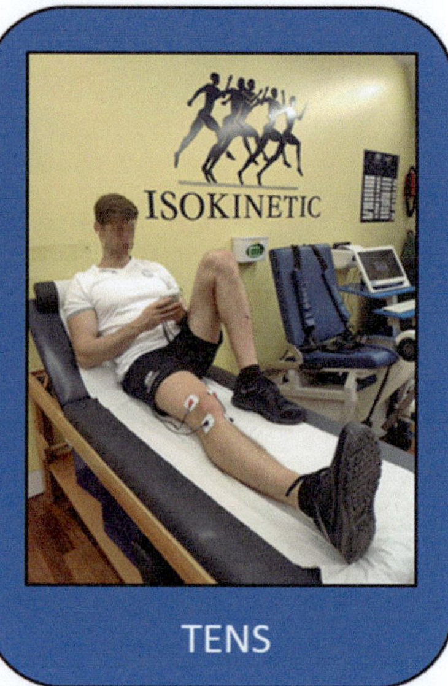

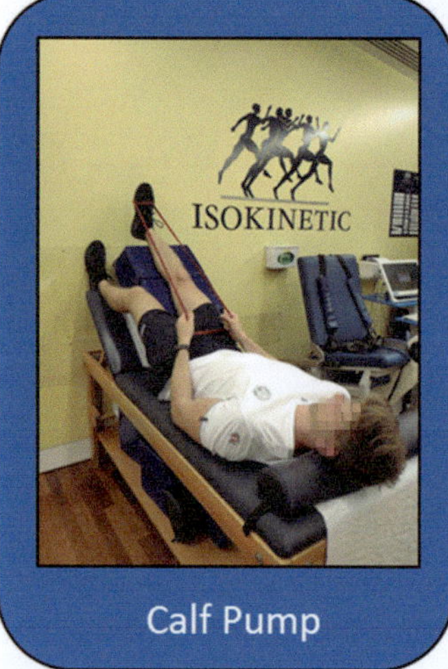

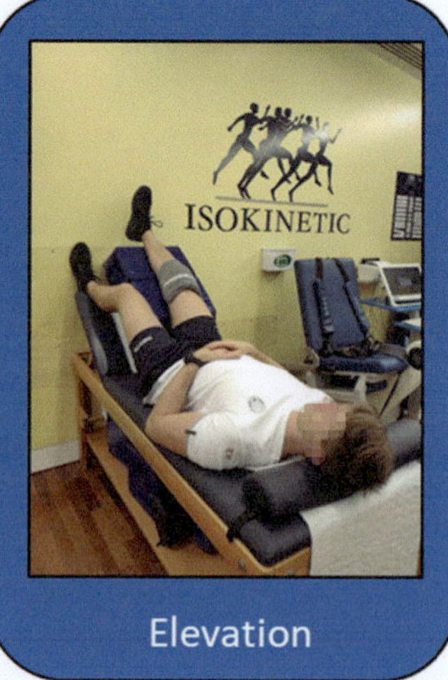

Fig. 11.2 Examples of suggested rehabilitation strategies to implement to manage the swelling

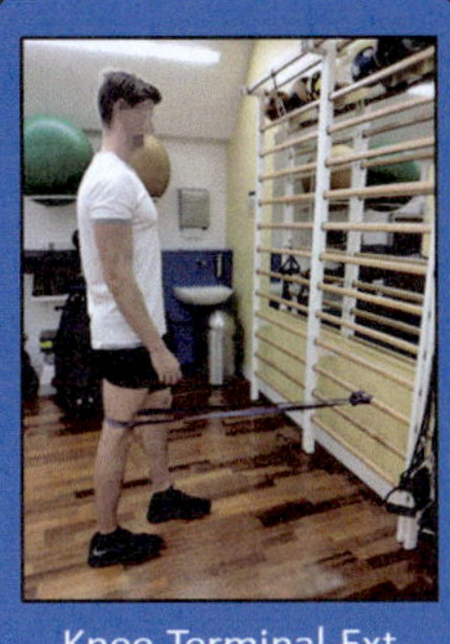

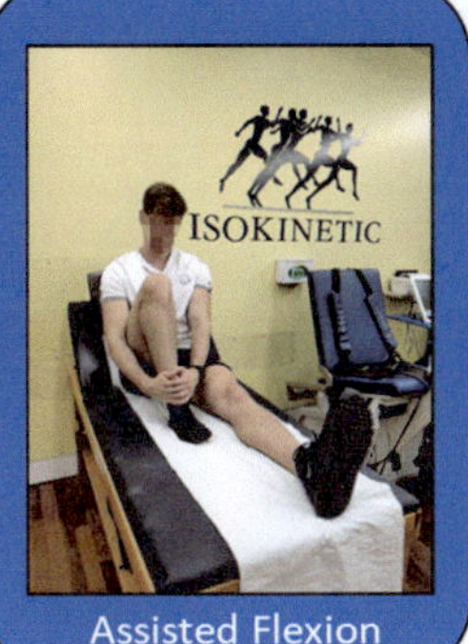

Fig. 11.3 Examples of suggested rehabilitation strategies to implement to recover the knee full mobility. The first couple of exercises can be used to recover the passive and active knee extension. The second couple of exercises can be used to improve the passive knee flexion. *Ext* Extension

appear to adversely affect post-surgical subjective and objective outcomes after ACL reconstruction [19, 20]. Although the initial primary focus of the early-stage rehabilitation is on obtaining full knee extension, the recovery of knee flexion is still key in order to maximize the outcomes after ACLR, but it is achieved in a much more gradual manner [27]. To recover full knee mobility after ACLR the following strategies can be implemented (Fig. 11.3): (a) low-load—long-duration mobility exercises performed multiple times per day [27]; (b) manual therapy (patella mobility and terminal knee extension or flexion); (c) active exercise at the end of the range of motion.

11.4.4 Walking Biomechanics: Walking on a Treadmill for At Least 10 min Without Pain or Swelling [12] and With Optimal Biomechanics

Untreated abnormal gait patterns can hinder the return to run process and the global outcomes of ACLR rehabilitation. Walking is a functional activity that can be optimally restored only if full knee extension is achieved [5, 11, 28], good quadriceps muscle activation is reached (no quadriceps lag on active straight leg raise [5]), swelling is minimal and non-reactive to activities [5, 11, 28] and good neuromuscular control of the gait movement is showed [5, 11, 28]. To restore walking, the previously stated criteria should be met. Allowing a patient to start to walk with suboptimal biomechanics can lead to increase in running compensatory strategies at the time of return to running. Once these goals have been achieved, walking tolerance can be developed gradually increasing the walking volume (Fig. 11.4).

11.4.5 Quadriceps Muscle Strength: Limb Symmetry Index (LSI) >70% on Isokinetic/Isometric Test for Quadriceps Strength [4, 5]

After ACLR, strength recovery is one of the main obstacles that patients encounter during the rehabilitation process. The traumatic effects of injury and subsequent surgery result in large deficits in the thigh muscle (especially knee extensors muscles) volume, neural activation and strength. Failure to achieve less than a 20% difference versus the contralateral limb is common at 6-month post-ACLR [42]. Only 29% of patients achieved a limb symmetry index (LSI) greater than 90%, when the reconstructed limb was compared to pre-surgery strength values at 6-month post-ACLR (note, pre-surgery, not pre-injury), whilst 57% were able to restore the injured limb's strength to within 10% of the uninjured limb (conventional LSI) [43].

Fig. 11.4 Examples of suggested rehabilitation strategies to implement to improve walking mechanics and tolerance

To improve the outcomes of ACLR rehabilitation process, full strength recovery has to be achieved. Limiting the strength loss between injury and surgery with the implementation of pre-operative treatment is showing promising results in the literature [28, 29, 44]. Following a structured rehabilitation plan after ACLR is necessary to completely restore the strength-velocity curve after the surgery. Strategies that can be implemented after ACLR in order to recover the strength are the following (Fig. 11.5): (a) quadriceps analytic exercises (OKC and safe implementation) [28, 29]; (b) use of modalities (neuromuscular electrical stimulation (NMES) [4, 28, 29, 45] and blood flow restriction training (BFR) [4, 29–31]; (c) implementation of a periodized approach to strength training [29]; (d) cross-education phenomenon [32–35]; (e) gradual implementation of CKC exercises targeting the thigh musculature.

defined as ACL agonists (their contraction decreases ACL strain). The hamstrings muscles have many functions but of particular interest after ACLR (especially with hamstrings graft) is the function of the medial hamstrings in preventing medial condyle lift-off and dynamic knee valgus, a known ACL injury risk factor [36, 46]. Strategy to restore hamstrings strength before allowing a patient to return to run include (Fig. 11.6): (a) low-intensity hamstring exercises initiated in the early-stage of rehabilitation followed by progressive overload [36]; (b) use of modalities (NMES and BFR) [36]; (c) medial to lateral hamstrings balance [36] without neglecting any of the hamstrings functions (hip extension, knee flexion and knee rotations); (d) full ROM strengthening exercises (especially deep knee flexion angles where the strength deficit is more accentuated); (e) progressive eccentric strength training [36].

11.4.6 Hamstrings Muscle Strength: Limb Symmetry Index (LSI) >70% on Isokinetic/Isometric Test for Hamstrings Strength [4, 5]

The hamstrings muscle complex is vitally important for the knee since this muscle group can be

11.4.7 Calf Strength and Capacity: Greater than 20 Reps and Within 5 Repetitions Versus the Other Side [4]

Plantarflexion strength and work capacity importance has been extensively discussed in the previous section and enhancing the strength of the calf

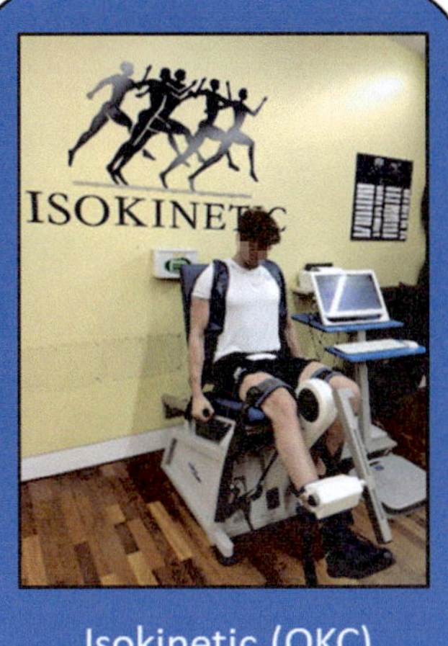

Fig. 11.5 Examples of suggested rehabilitation strategies to implement to improve quadriceps strength. *Iso* isometric, *Ext* extension, *NMES* neuromuscular electrical stimulation, *OKC* open kinetic chain, *CKC* closed kinetic chain

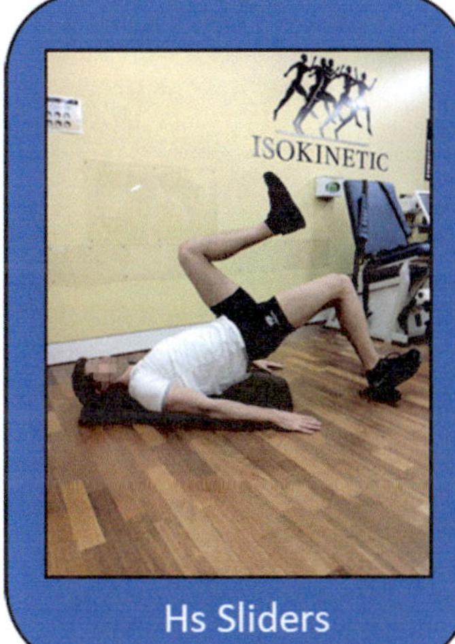

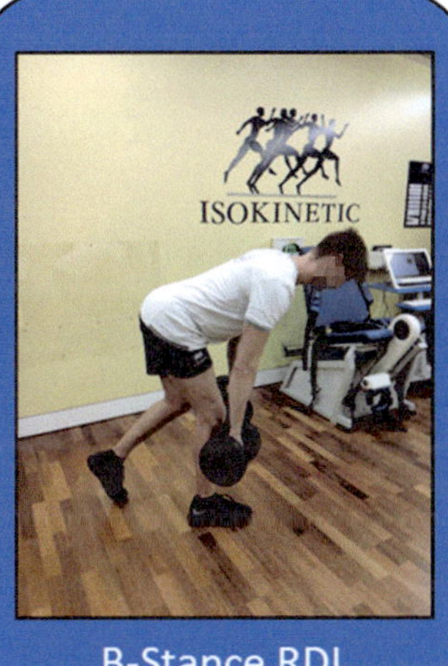

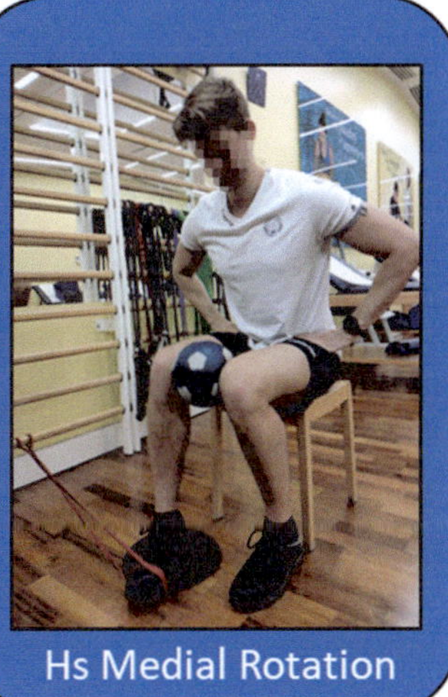

Fig. 11.6 Examples of suggested rehabilitation strategies to implement to improve hamstrings strength. *Hs* hamstrings, *RDL* Romanian Deadlift

muscles has positive effects on the return to running process. The authors suggest not to limit the assessment and treatment intervention to calf muscle (gastrocnemius and soleus) only, but to extend it to the whole ankle complex, especially to the muscles that control the pronation of the foot (intrinsic foot muscles, tibialis posterior and flexor hallucis longus). Here there are some examples of suggested intervention to maximize the function of the ankle complex before allowing a patient to return to run (Fig. 11.7).

11.4.8 Glutes Strength and Capacity: Greater than 20 Reps and Within 5 Repetitions Versus the Other Side [4]

The glutes complex is of vital importance for the lower limb health, and especially for the knee. In a healthy individual, the gluteus maximus is considered the biggest (volume) and strongest muscle of the body. The gluteus maximus is prone to inhibition after injury and weakness of the glu-

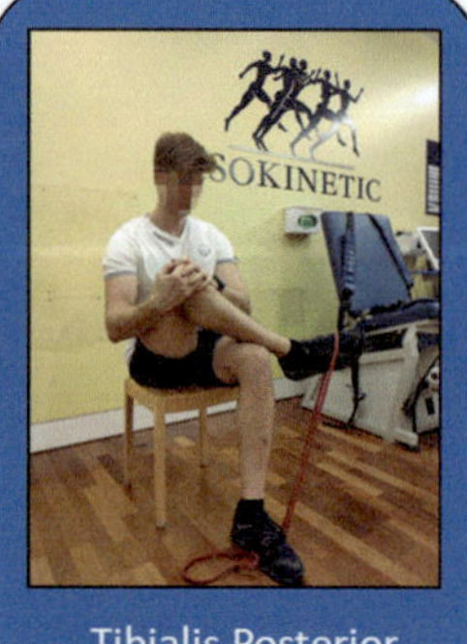
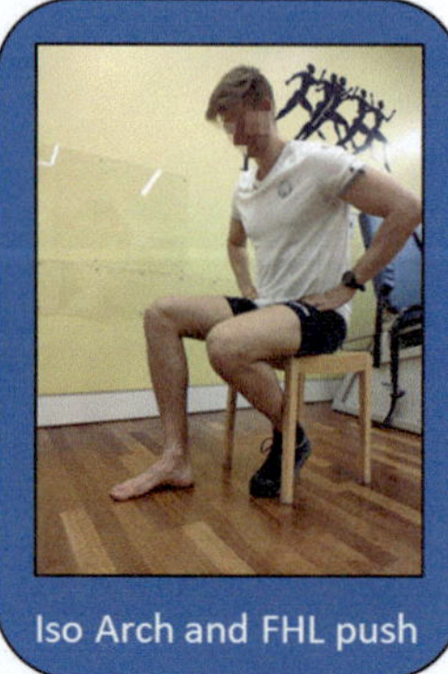

Fig. 11.7 Examples of suggested rehabilitation strategies to implement to improve the ankle complex strength and stability. The first couple of exercises can be used to improve calf strength while the second couple are suggested with the goal of increasing the foot stability and prevent excessive pronation during functional exercises. *Gastro* Gastrocnemius, *Iso* Isometric, *FHL* Flexor Hallucis Longus

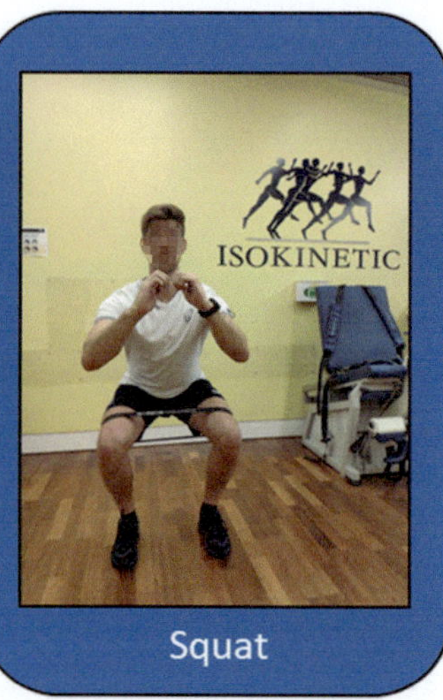

Fig. 11.8 Examples of suggested rehabilitation strategies to implement to improve gluteal strength. The suggested exercises follow a specific progression to maximize the gluteal strength: (1) restore optimal lumbo-pelvic stability and balance; (2) analytic gluteus strengthening; (3) re-integrate the gluteal muscles into the motor pattern

teal muscles can contribute to altered movement patterns which increase knee and ACL loading and are thought to be important risk factors for ACL injury [4]. Restoring the strength of the gluteal complex is of fundamental importance to optimize the return to run process after ACLR. In order to do so, the authors suggest the following indications (Fig. 11.8): (a) restore optimal lumbo-pelvic stability and balance [37]; (b) strengthen the gluteus muscles [37] training all the most important gluteal functions (hip extension, abduction and external rotation); (c) re-integrate the gluteal muscle into the motor pattern [37].

11.4.9 Movement Quality: Good Single Leg Squat Movement Assessment

Functional exercises and performance can be considered as the expression of one's ability of utilizing their neuromuscular system to complete

Fig. 11.9 Suggested exercise progression to safely achieve an optimal movement quality on Single Leg Squat. This exercise progression gradually increases the task demand exercise after exercise, allowing the patient to gradually adapt to the increased load of the new task. Do not proceed with this exercise sequence if the patient reports an adverse reaction (pain, increase of swelling, instability, etc.) to the previous task implementation

a specific task. Reasons for sub-optimal movement quality that differs from the "good quality" execution have to be found assessing the neuromuscular system and all its sub-components (intramuscular coordination, intermuscular coordination, balance, mobility, etc.). Once the reason for the dysfunctional movement has been found, it can be addressed through neuromuscular training strategies by the clinician. From the task demand intensity stand of view, a possible exercise progression for a correct Single Leg Squat implementation is illustrated in Fig. 11.9.

11.4.10 Lumbo-Pelvic, Trunk and Core Stability

Even if core stability is not included in the criteria checklist for the return to run process, it does not mean that it is not important. Core stability and lumbo-pelvic control are important in the rehabilitation process after ACLR especially during functional movements. An ipsilateral trunk lean may increase ACL loading as a result of a lateral shift in centre mass, achieving a resultant vector line lateral to the knee joint and causing a knee abduction moment [47]. During the rehabilitation of an ACL reconstructed patient, the approach to core training will evolve from "simple" to "complex". The focus will shift from "local stabilizers" (analytic core work) to "global stabilizers" (functional core work), to conclude with "load transfer" tasks (higher power functional exercises with trunk control) (Fig. 11.10). All the core functions should be trained during the recovery process (e.g. anti-extension, anti-lateral flexion, rotational exercises, etc.).

Once the patient has achieved all the previously stated criteria for return to running, a gradual running implementation is suggested. Depending on the patient readiness, many parameters can be modified to increase or decrease the task demands for the athlete (running surface, speed, volume, frequency, shoes, etc.).

Inefficient running biomechanics play an important role in the development and incidence of running injuries [48]. After anterior cruciate ligament (ACL) reconstruction (ACLR), athletes demonstrate significant alterations in surgical limb running biomechanics as compared with the non-surgical limb and healthy controls [49]. These running alterations include decreased peak knee flexion angles [49], decreased knee flexion excursion [49], decreased peak knee extensor moment [49] and increased initial impact forces [50]. If a runner has poor kinematics patterns and running form, it will affect the body's ability to absorb external forces and put them at risk of developing overuse injuries [48].

Patients should be exposed to gradual load to allow the clinician to assess the running biomechanics and improve it based on his or her needs. Assessment of running biomechanics with 2D or 3D video on a treadmill has shown to be an accurate way of analysing running style [51]. Assessing the athlete's running gait from the first run gives the opportunity to the clinician to provide feedbacks to the patients to optimize the running technique and decrease joint overload.

Fig. 11.10 Suggested categories of exercise progression to optimize the core function. Progress from working on the local stabilizers to the global stabilizers, and finally introduce load transfer exercises

Table 11.3 Example of application of the progressive overload principle to running. Especially when a new task is introduced, it is suggested to gradually increase the global task demands to allow the body to adapt to the new stimulus

Level	Session	Frequency	Total run (min)
1	5 × (30″ run–30″ walk)	2	5
2	5 × (1′ run–30″ walk)	2	10
3	5 × (1′ run–30″ walk)	3	15
4	4 × (2′ run–1′ walk)	3	24
5	3 × (4′ run–1′ walk)	3	36
6	2 × (8′ run–2′ walk)	3	48
7	2 × (10′ run–2′ walk)	3	60
8	1 × 20′ run	3	60

Patients should resume running from a low load condition (changes of running surface, water running, AlterG treadmill, etc.) (Fig. 11.10) and they should follow the progressive overload principle applied to running (Table 11.3), starting from a foundational level (e.g. walk-run interval training) (Fig. 11.11).

When the patient starts to run for the first time after ACLR it is important to keep in mind that it will not be perfect. The patient might report that he or she feels the injured side different from the other side, that the run is not fluid or that running is "odd", but this has to be considered normal, and part of the rehabilitation process. The first run will set a baseline for the athlete and for the clinician that will assess the running gait of the patient and consequentially adjust the rehabilitation plan (in terms of running feedbacks, treatment goals, exercise selection, etc.) to assist the athlete in re-establishing an optimal running technique.

11.5 Conclusions

Return to running after anterior cruciate ligament reconstruction (ACLR) is not an easy task and lack of guidelines leaves clinicians and patients without a clear route to follow. This chapter guides you through the authors' clinical reasoning utilized before implementing a new task with an athlete after ACLR. From an accurate analysis of the task demands, it is possible to understand the needs that an athlete has to satisfy before resuming that specific task. Assessing the patient's envelope of function gives the clinician information about the patient's readiness in terms of the ability of the athlete to cope with a return

Fig. 11.11 Example of different running surfaces that can be used to modulate the load of the running task. Start from low impact running task demand and depending on the patient response to the task, increase or decrease the task demands

to the running programme. Based on the return to running criteria, the authors suggest practical rehabilitation strategies that can be implemented to expand the patients' envelope of function and assists them on achieving those criteria. Once the athlete achieves all the criteria and is considered ready to resume running by the medical team, the clinicians should implement a gradual return to the running programme. This program should start from low impact running (managing volume, intensity, running surface, etc.) and, applying the principle of overload to running, it should gradually increase the task demands with the goal of achieving the patient's running goals.

References

1. Dragoo JL, Braun HJ, Durham JL, Chen MR, Harris AHS. Incidence and risk factors for injuries to the anterior cruciate ligament in National Collegiate Athletic Association football data from the 2004-2005 through 2008-2009 National Collegiate Athletic Association Injury Surveillance System. Am J Sports Med. 2012;40(5):990–5.
2. Sanders TL, Kremers HM, Bryan AJ, Larson DR, Dahm DL, Levy BA, Stuart MJ, Krych AJ. Incidence of anterior cruciate ligament tears and reconstruction: a 21-year population-based study. Am J Sports Med. 2016;44(6):1502–7.
3. Monk PA, Davies LJ, Hopewell S, Harris K, Beard DJ, Price AJ. Surgical versus conservative interventions for treating anterior cruciate ligament injuries. Cochrane Database Syst Rev. 2016;4(4):CD011166.
4. Buckthorpe M, Della Villa F. Optimising the 'mid-stage' training and testing process after ACL reconstruction. Sports Med. 2019;50(4):657–78.
5. Buckthorpe M, Tamisari A, Della Villa F. A ten task-based progression in rehabilitation after acl reconstruction: from post-surgery to return to play—a clinical commentary. Int J Sports Phys Ther. 2020;15(4):611–23.
6. Dye SF. The knee as a biologic transmission with an envelope of function. Clin Orthop Relat Res. 1996;325:10–8.
7. Cavanagh PR, Lafortune MA. Ground reaction forces in distance running. J Biomech. 1980;13(5):397–406.
8. Moore IS. Is there an economical running technique? A review of modifiable biomechanical factors affecting running economy. Sports Med. 2016;46(6):793–807.
9. Sun X, Lam W-K, Zhang X, Wang J, Fu W. Systematic review of the role of footwear constructions in running biomechanics: implications for running-related injury and performance. J Sports Sci Med. 2020;19(1):20–37.
10. Rambaud AJM, Ardern CL, Thoreux P, Regnaux J-P, Edouard P. Criteria for return to running after anterior cruciate ligament reconstruction: a scoping review. Br J Sports Med. 2018;52(22):1437–44.
11. Della Villa F, Ricci M, Perdisa F, Filardo G, Gamberini J, Caminati D, Della Villa S. Anterior cruciate ligament reconstruction and rehabilitation: predictors of functional outcome. Joints. 2015;3(4):179–85.
12. Della Villa S, Della Villa F, Ricci M, Doral MN, Gasbarro G, Musahl V. ACL risk of reinjury: when is it safe to return (time or criteria). In: Injuries and health problems in football; 2017. pp. 581–592.
13. Jakobsen TL, Christensen M, Christensen SS, Olsen M, Bandholm T. Reliability of knee joint range of motion and circumference measurements after total knee arthroplasty: does tester experience matter? Physiother Res Int. 2009;15:126–34.

14. Sachs RA, Daniel DM, Stone ML, Garfein RF. Patellofemoral problems after anterior cruciate ligament reconstruction. Am J Sports Med. 1989;17(6):760–5.
15. Hébert-Losier K, Wessman C, Alricsson M, Svantesson U. Updated reliability and normative values for the standing heel rise test in healthy adults. Physiotherapy. 2017;103(4):446–52.
16. Freckleton G, Cook J, Pizzari T. The predictive validity of a single leg bridge test for hamstring injuries in Australian rules football players. Br J Sports Med. 2014;48(8):713–7.
17. Crossley KM, Zhang W-J, Schache AG, Bryant A, Cowan SM. Performance on the single-leg squat task indicates hip abductor muscle function. Am J Sports Med. 2011;39(4):866–73.
18. Herrington L, Myer G, Horsley I. Task based rehabilitation protocol for elite athletes following anterior cruciate ligament reconstruction: a clinical commentary. Phys Ther Sport. 2013;14(4):188–98.
19. Shelbourne DK, Wilckens JH, Mollabashy A, DeCarlo M. Arthrofibrosis in acute anterior cruciate ligament reconstruction. The effect of timing of reconstruction and rehabilitation. Am J Sports Med. 1991;19(4):332–6.
20. Logerstedt D, Scalzitti D, Risberg MA, Engebretsen L, Webster KE, Feller J, Snyder-Mackler L, Axe MJ, McDonough CM, Altman RD, Beattie P, DeWitt J, Elliott JM, Ferland A, Fitzgerald GK, Kaplan S, Killoran D, Kvist J, Marx R, Torburn L, Zachazewski J. Knee stability and movement coordination impairments: knee ligament sprain revision 2017. J Orthop Sports Phys Ther. 2017;47(11):A1–A47.
21. Dix J, Marsh S, Dingenen B, Malliaras P. The relationship between hip muscle strength and dynamic knee valgus in asymptomatic females: a systematic review. Phys Ther Sport. 2019;37:197–209.
22. Petersen W, Taheri P, Forkel P, Zantop T. Return to play following ACL reconstruction: a systematic review about strength deficits. Arch Orthop Trauma Surg. 2014;134(10):1417–28.
23. Dorn TW, Schache AG, Pandy MG. Muscular strategy shift in human running: dependence of running speed on hip and ankle muscle performance. J Exp Biol. 2012;215(11):1944–56.
24. Fong C-M, Blackburn JT, Norcross MF, McGrath M, Padua DA. Ankle-dorsiflexion range of motion and landing biomechanics. J Athl Train. 2011;46(1):5–10.
25. Maniar N, Schache AG, Pizzolato C, Opar D. Muscle contributions to tibiofemoral shear forces and valgus and rotational joint moments during single leg drop landing. Scand J Med Sci Sports. 2020;30(9):1664–74.
26. Buckthorpe M, Pirotti E, Della Villa F. Benefits and use of aquatic therapy during rehabilitation after acl reconstruction—a clinical commentary. Int J Sports Phys Ther. 2019;14(6):978–93.
27. Wilk KE, Arrigo CA. Rehabilitation principles of the anterior cruciate ligament reconstructed knee: twelve steps for successful progression and return to play. Clin Sports Med. 2017;36(1):189–232.
28. van Melick N, van Cingel REH, Brooijmans F, Neeter C, van Tienen T, Hullegie W, Nijhuis-van der Sanden MWG. Evidence-based clinical practice update: practice guidelines for anterior cruciate ligament rehabilitation based on a systematic review and multidisciplinary consensus. Br J Sports Med. 2016;50(21):1506–15.
29. Buckthorpe M, La Rosa G, Della Villa F. Restoring knee extensor strength after anterior cruciate ligament reconstruction: a clinical commentary. Int J Sports Phys Ther. 2019;14(1):159–72.
30. Patterson SD, Hughes L, Warmington S, Burr J, Scott BR, Owens J, Abe T, Nielsen JL, Libardi CA, Laurentino G, Rodrigues Neto G, Brandner C, Martin-Hernandez J, Loenneke J. Blood flow restriction: exercise considerations of methodology, application, and safety. Front Physiol. 2019;10:533.
31. Hughes L, Rosenblatt B, Haddad F, Gissane C, McCarthy D, Clarke T, Ferris G, Dawes J, Paton B, Patterson SD. Comparing the effectiveness of blood flow restriction and traditional heavy load resistance training in the post-surgery rehabilitation of anterior cruciate ligament reconstruction patients: a UK National Health Service Randomised Controlled Trial. Sports Med. 2019;49(11):1787–805.
32. Lepley LK, Grooms DR, Burland JP, Davi SM, Mosher JL, Cormier ML, Lepley AS. Eccentric cross-exercise after anterior cruciate ligament reconstruction: novel case series to enhance neuroplasticity. Phys Ther Sport. 2018;34:55–65.
33. Papandreou M, Billis E, Papathanasiou G, Spyropoulos P, Papaioannou N. Cross-exercise on quadriceps deficit after ACL reconstruction. J Knee Surg. 2013;26(1):51–8.
34. Harput G, Ulusoy B, Yildiz TI, Demirci S, Eraslan L, Turhan E, Tunay VB. Cross-education improves quadriceps strength recovery after ACL reconstruction: a randomized controlled trial. Knee Surg Sports Traumatol Arthrosc. 2018;27(1):68–75.
35. Lepley LK, Palmieri-Smith RM. Cross-education strength and activation after eccentric exercise. J Athl Train. 2014;49(5):582–9.
36. Buckthorpe M, Danelon F, La Rosa G, Nanni G, Stride M, Della Villa F. Recommendations for hamstring function recovery after ACL reconstruction. Sports Med. 2020;51(4):607–24.
37. Buckthorpe M, Stride M, Della Villa F. Assessing and treating gluteus maximus weakness—a clinical commentary. Int J Sports Phys Ther. 2019;14(4):655–69.
38. Gokeler A, Dingenen B, Mouton C, Seil R. Clinical course and recommendations for patients after anterior cruciate ligament injury and subsequent reconstruction: a narrative review. EFORT Open Rev. 2017;2(10):410–20.
39. Myer GD, Brent JL, Ford KR, Hewett TE. A pilot study to determine the effect of trunk and hip focused neuromuscular training on hip and knee isokinetic strength. Br J Sports Med. 2008;42(7):614–9.
40. Andrade R, Pereira R, van Cingel R, Staal JB, Espregueira-Mendes J. How should clinicians

rehabilitate patients after ACL reconstruction a systematic review of clinical practice guidelines (CPGs) with a focus on quality appraisal (AGREE II). Br J Sports Med. 2020;54(9):512–9.

41. Hopkins J, Ingersoll CD, Edwards J, Klootwyk TE. Cryotherapy and transcutaneous electric neuromuscular stimulation decrease arthrogenic muscle inhibition of the vastus medialis after knee joint effusion. J Athl Train. 2002;37(1):25–31.

42. Palmieri-Smith RM, Thomas AC, Wojtys EM. Maximizing quadriceps strength after ACL reconstruction. Clin Sports Med. 2008;27(3):405–24.

43. Wellsandt E, Failla MJ, Snyder-Mackler L. Limb symmetry indexes can overestimate knee function after ACL injury. J Orthop Sports Phys Ther. 2017;47(5):334–8.

44. Eitzen I, Holm I, Risberg MA. Preoperative quadriceps strength is a significant predictor of knee function two years after anterior cruciate ligament reconstruction. Br J Sports Med. 2009;43(5):371–6.

45. Kim K-M, Croy T, Hertel J, Saliba S. Effects of neuromuscular electrical stimulation after anterior cruciate ligament reconstruction on quadriceps strength, function, and patient-oriented outcomes: a systematic review. J Orthop Sports Phys Ther. 2010;40(7):383–91.

46. Hewett TE, et al. Biomechanical measures of neuromuscular control and valgus loading of the knee predict anterior cruciate ligament injury risk in female athletes a prospective study. Am J Sports Med. 2005;33(4):492–501.

47. Della Villa F, Buckthorpe M, Grassi A, Nabiuzzi A, Tosarelli F, Zaffagnini S, Della Villa S. Systematic video analysis of ACL injuries in professional male football (soccer): injury mechanisms, situational patterns and biomechanics study on 134 consecutive cases. Br J Sports Med. 2020;54(23):1423–32.

48. Hulme A, Oestergaard Nielsen R, Timpka T, Verhagen E, Finch C. Risk and protective factors for middle- and long-distance running-related injury. Sport Med. 2017;47(5):869–86.

49. Pairot-de-Fontenay B, Willy RW, Elias ARC, Mizner RL, Dubé M. Running biomechanics in individuals with anterior cruciate ligament reconstruction: a systematic review. Sports Med. 2019;49(9):1411–24.

50. Noehren B, Wilson H, Miller C, Lattermann C. Long-term gait deviations in anterior cruciate ligament–reconstructed females. Med Sci Sports Exerc. 2013;45(7):1340–7.

51. Souza RB. An evidence-based videotaped running biomechanics analysis. Phys Med Rehabil Clin N Am. 2015;27(1):217–36.

Part II

Specific Aspects

Athletics Running Disciplines

12

Gian Luigi Canata, Valentina Casale,
Claudio Gaudino, Renato Canova,
and Giacomo Zanon

12.1 Role and Characteristics of Running in Athletics

Athletics is a group of sport events including running, jumping, and throwing. The most common athletics competitions are track and field, road running, racewalking, and cross country running. Track and field running events may be classified as [1]:

- Track events, including sprints (100, 200, and 400 m), middle-distance running (800 and 1500 m), long-distance running (5000 and 10,000 m), hurdling (100 m for women, 110 m for men, and 400 m for both), relays (4 × 100 and 4 × 400 m) and 3000 m steeplechase.
- Road events, including marathon, 20-km race walk for women, 20-km and 50-km race walks for men.
- Combined events, including heptathlon for women and decathlon for men.

G. L. Canata (✉) · V. Casale
Centre of Sports Traumatology, Koelliker Hospital,
Turin, Italy
e-mail: studio@ortosport.it

C. Gaudino
Speed and Hurdles Track and Field Coach and
Football Fitness Coach, Turin, Italy

R. Canova
International Running Coach, Turin, Italy

G. Zanon
Habilita Group, Bergamo, Italy

12.2 Sprint and Short-Distance Running

Also known as sprinting, short-distance running in track and field consists in running over distances no longer than 400 m, at an all-out or nearly all-out burst of speed in a limited period of time.

12.2.1 Technical Aspects

In the Olympic Games and in the National and International Championships, speed races consist in 100 m, 200 m, 400 m, 4 × 100 m relay, and 4 × 400m relay. The 400 m race may be considered as "prolonged speed". The technical aspects of these races concern the starting phase, the acceleration phase, and the maximal speed phase. In relation to the race distance, the impact of these phases on the result is variable. For the 100 m sprinters, but also for the 200 m sprinters, the starting and acceleration phases represent a maximal or almost maximal effort since athletes must aim at quickly reaching 97–99% of their maximal speed. On the other hand, the 400 m sprinters do not exceed 90% of their maximal speed [2].

The correct position to hold on the blocks during the starting phase consists in loading (or compressing) the lower limbs in the interval between the set and the gun shot, combined with a forward imbalance of the body weight, supported by the arms. It is important that both feet are placed in tension against the blocks. The limb placed on the rear block is less loaded than the one placed on the front block, and the knee angle is more open. On the contrary, the limb placed on the front block is more loaded, and the knee flexion is greater. The trunk and head are bent forward, allowing the athlete to unleash his explosive strength to give the maximal impulse in a high-forward direction. At the gun shot, this explosion of force occurs by pushing simultaneously with both feet. Obviously, the back leg (the one with the foot in contact with the rear block) has a minor knee flexion and it ends its push against the block earlier. The forward leg (pushed on the front block) which has a higher knee flexion, completes its action later, thus guaranteeing the maximum continuity to the acceleration.

From the first few steps during the acceleration phase, the arms movement is considerable wide in the sagittal axe to support and balance the lower limbs movement. The trunk gradually tends to reduce its inclination forward and moves vertically. During the acceleration phase, both frequency and step length progressively increase to stabilize at the end of the acceleration phase and maintain the speed (or at least try to). Similarly, flight time tends to increase while contact time and the lower limbs articular angles tend to decrease. In this phase, the importance of the explosive strength first and then the explosive-elastic strength when the athlete reaches high speed is clear.

A peculiar aspect of the acceleration phase is the projection of the center of mass on the ground, which must always be in front of the foot in contact with the ground, to allow the athlete push and not pull. The duration of the acceleration phase and, consequently, the distance covered in this phase vary from athlete to athlete. Only the best sprinters can keep accelerating after 40 m or more. For the 400 m sprinters, the acceleration phase is less intense and, as a consequence, is less energetically demanding [3].

During the maximal speed phase, the percentage of maximal speed reached varies in relation to the race distance: it is 98–99% in the 100 m race;

96–97% in the 200 m race; ~90% in the 400 m race [4]. The aim for the 100 m and 200 m sprinters is to maintain the speed achieved during the acceleration phase, using the elastic component (explosive-elastic strength and plyometric strength/stiffness). The feet are taut, the contact time on the ground is minimal, the trunk is vertical, the shoulders are low and relaxed, the focus is on the rhythm (combination of step frequency and step length) and the arms swing coordinated with the legs. The goal for the 400 m sprinter is to reduce as much as possible the decrease of the speed during the second part of the race (from 200 to 400 m), when the lactic acidosis occurs [5].

12.2.2 Distribution of the Effort

In technical terms, distributing the effort means to express the optimal neuromuscular intensity during the race (in relation to the distance) in order to develop the highest average speed. This is strategically used to reach the best possible result [5]. It is important to underline that the physiological mechanisms of production and transmission of the nervous stimulus and of metabolic energy delivery are unable to sustain a maximum intensity, even for the shortest of races (100 m). Only the 60 m race (official indoor race, not listed in Olympic competitions) can be run at maximal intensity from the start to the end. In this case, the athlete performs a maximal acceleration during 35–40 m followed by a maximal speed phase run at 100% of his maximal speed until the end (60 m) [2, 6, 7].

12.2.2.1   100 m

In this race, the maintenance of velocity in the maximum speed phase is more limited by the production and transmission of the nerve impulse than by metabolic factors. In the last 15–20 m, the step frequency decreases, and the athlete tends to compensate it with an increase in step length.

Therefore, a reduction in speed, albeit limited, occurs in the final part of the race, even in the optimal effort distribution strategies. The best result is achieved by never touching the maxi-

Table 12.1 World athletes race timings (100 m)

	Green	Bailey	Montgomery	Fredericks	Boldon	Lemaitre	Bolt
Total time	9″86	9″84	9″93	9″89	9″92	9″95	9″58
0-50 m	5″55	5″58	5″56	5″55	5″58	5″64	5″47
50-100 m	4″31	4″26	4″37	4″34	4″34	4″31	4″11
Difference	1″24	1″32	1′19	1″21	1″24	1″33	1″36

Table 12.2 U. Bolt 100 m world record details

	Number of steps	Avg step length (m)	Avg frequency (steps/s)	Time per 10 m (s)	Avg speed (m/s)
0–10 m	6.8	1.47	3.59	1″89	5.28
10–20 m	4.2	2.38	4.23	0″99	10.06
20–30 m	4.1	2.44	4.54	0″90	11.06
30–40 m	4.0	2.50	4.63	0″86	11.59
40–50 m	3.9	2.56	4.70	0″83	12.05
50–60 m	3.8	2.63	4.63	0″82	12.20
60–70 m	3.7	2.70	4.57	0″81	12.35
70–80 m	3.6	2.78	4.39	0″82	12.20
80–90 m	3.6	2.78	4.34	0″83	12.05
90–100 m	3.3	3.03	3.98	0″83	12.05
Total 100 m	**41.0**	**2.44**	**4.28**	9″58	**10.44**

mum speed, to avoid a marked final drop. The optimal differential time between the 2 halves of the races (0–50 m and 50–100 m) corresponds to approximately 1.0″–1.1″ (+0.2″ due to the electronic time) in favor of the second part of the race run at higher speed. Tables 12.1 and 12.2 show the times of the two parts of the race of some World specialists (Table 12.1) and the details per each 10 m segment by Usain Bolt in the 100 m World record race.

12.2.2.2 200 m

In the 200 m race, the lactide concentration can even exceed 20 mmol/L. The speed limit to be maintained until almost the end of the race is determined not only by the frequency of the nervous impulse, but also by a massive appeal to the anaerobic lactide mechanism, related to the stiffness capacity. The high levels of lactate [La−] can only be developed by athletes with high percentages of white fibers, and excellent explosive plyometric strength capacity. In this race, the passage time at 100 m, that is half of the race, is normally 0.3″ – 0.4″ higher than the PB (personal best) time of 100 m race. This occurs firstly because the initial 100 m are run in curves and therefore with a relative greater dispersion of energy, and secondly because the speed to be reached and maintained is lower than the one during a 100 m race. Similar to the 100 m race, the inevitable final decrease in speed in the last 50 m is due to a reduction in step frequency, while the step length tends to remain constant. The optimal differential between the two halves of the race (0–100 m and 100–200 m) corresponds to about 0.6″ – 0.8″ in favor of the second part of the race. By dividing the race into four sections of 50 m each, exception made for the initial 50 m, a time difference within 0.3″–0.4″ emerges in the remaining three fractions. Tables 12.3 and 12.4 show the data of some specialists (Table 12.3) and those of the 200 m World record holder (Table 12.4).

The importance of the speed managed by the athlete during the last 50 m is highlighted in Table 12.5. Athletes are listed in order of arrival during a 200 m race. The official final times and those relating to the last quarter of the race are reported. Interestingly, the order of arrival does not change if considering the time relative to the last 50 m.

12.2.2.3 400 m

In the 400 m race, it is important to restrict the range of speed variation, but not for all athletes in the same way. In the 400 m (as well as in the

Table 12.3 World athletes race timings (200 m)

	Deloach J.	Lewis C.	Da Silva R.	Borzov V.	Mennea P.	Bolt U.
Total time	19″75	19″79	20″05	20″00	20″23	19″19
0–50 m	5″81	5″76	5″84			
50–100 m	4″54	4″55	4″57			
100–150 m	4″62	4″66	4″70			4″54
150–200 m	4″78	4″82	4″93			4″74
0–100 m	10″35	10″31	10″41	10″50	10″55	9″91
100–200 m	9″40	9″48	9″63	9″50	9″65	9″28
Difference	0″95	0″83	0″78	1″00	0″90	0″63

Difference is the difference in time between 0–100 m and 100–200 m

Table 12.4 Usain Bolt 200 m World record

	Number of steps	Avg step length (m)	Avg frequency (steps/s)	Time (s)	Avg speed (m/s)
0–100 m	42.0	2.38	4.24	9″91	10.09
100–200 m	38.0	2.63	4.09	9″28	10.78
100–150 m	19.1	2.62	4.21	4″54	11.01
150–200 m	18.9	2.65	3.99	4″74	10.55
Total 200 m	**80.0**	**2.50**	**4.17**	**19″19**	**10.42**

Table 12.5 Total and last quarter times of a 200 m race

Athlete	Total time (s)	Time 150–200 m (s)
Deloach	19″75	4″79
Lewis	19″79	4″82
Da Silva	20″04	4″93
Christie	20″09	5″00
Kiss	21″21	5″22
Semenov	21″22	5″23
Baulch	21″26	5″25
Lack	21″71	5″40

800 m race), the highest concentrations of lactate are reached compared to all other sports disciplines: up to over 25 mmol/L.

The acceleration phase is characterized by a long and smooth acceleration without too high-speed peaks, with the intention of delaying the massive intervention of the glycolytic mechanism, and consequently the premature accumulation of lactate. The speed peak is generally reached around 150 m from the start, and the second fraction (100–200 m) is the fastest one. The effect of acidosis developing during the race and its muscle diffusion reduce the ability to maintain the speed in the third (200–300 m) and especially in the fourth part of the race (300–400 m), as shown in Table 12.6. Consequently, the differential time between the second and the first part of the race (0–200 m and 200–400 m) is in favor of the first part (unlike the 100 and 200 m races) (Table 12.7). Regarding the distribution of effort during the 400 m race, it is possible to compare the passage time at 200 m and the personal best time over the same distance. Essentially, the fastest 400 m sprinters, who set the best times over 200 m and therefore should be less "resistant", show a greater gap between the two times (the personal best of 200 m and the passage time of 200 m in the 400 m races), with the aim of counteracting the speed decrease in the final part of the race.

12.2.3 Speed Training

A significant aspect of speed training in athletics is the strength and in particular the specifications of the strength related to speed:

- Explosive strength and maximal dynamic strength (concentric contractions) correlated with the starting phase

Table 12.6 World athletes total and fractions timings in a 400 m race

	Johnson M.	Reynolds B.	Lewis S.	Everett D.	James K.	Merritt L.	Van Niekerk W.
Total time	43″18	43″29	43″80	44″09	43″76	43″85	43″03
0–100 m	11″10	11″20	11″26	11″03	10″72	10″76	10″72
100–200 m	10″12	10″20	10″15	10″34	9″80	9″64	9″76
200–300 m	10″44	10″70	10″72	10″81	10″66	10″76	10″52
300–400 m	11″52	11″20	11″74	11″91	12″58	12″69	12″03
0–200 m	21″22	21″40	21″41	21″37	20″52	20″40	20″48
200–400 m	21″96	21″90	22″46	22″72	23″24	23″45	22″55
Difference	−0″74	−0″50	−1″05	−1″35	−2″72	−3″05	−2″07

Difference is the difference in time between 0–200 m and 200–400 m

Table 12.7 Van Niekerk W. 400 m world record

	Number of steps	Avg step length (m)	Avg frequency (steps/s)	Time (s)	Avg speed (m/s)
0–100 m	42.8	2.34	3.99	10″72	9.29
100–200 m	38.1	2.62	3.90	9″76	10.24
200–300 m	40.1	2.49	3.81	10″52	9.50
300–400 m	42.0	2.38	3.49	12″03	8.31
Total 400 m	**163.0**	**2.46**	**3.80**	43″03	**9.29**

- Explosive-elastic strength associated with the acceleration phase and the maximum velocity phase
- Plyometric strength mainly connected with the maximum velocity phase.

Each specification of strength considerably differs in relation to the race (100 m, 200 m, 400 m). Strength training is primarily carried out using overloads. However, the increase in strength does not automatically determine an improvement in the performance. The athletes need to go through an adaptation phase during which they perform special exercises for strength (including bounds, uphill sprints, weighted sprints, resisted sprints, etc.) and elastic/technical exercises (including ankle bouncing, bouncing butt kicks, high rebound skips, running rebounds, etc.).

The objectives of the strength training in athletics can be summarized as follows:

1. To create a greater "motor" potential.
2. To use the increased "motor" potential in training and in competition.

The increase of explosive and maximal dynamics strength lead to an improvement of the motor unit's synchronization, and of the inter- and intramuscular coordination (a decisive quality in the relationship between strength and specific gesture/performance). The development of the explosive-elastic and plyometric strength enhances a surplus of power with a lower energy cost, due to the intervention of the serial elastic component (stiffness). Consequently, the athlete can express a higher level of strength, there is a lower inhibition process due to the Golgi receptors, an improvement of neuromuscular sensitivity as well as a reduction of the coupling-time [8, 9].

Overloaded strength exercises:

- Squat
- Explosive half squat
- Half squat continuous
- Half squat jump (starting still)
- Half squat jump continuous

Special strength exercises:

- Bounds
- Uphill sprints
- Resisted sprints

Technical exercises:

- Specific pre-athletism (ankle bouncing, bouncing butt kicks, high rebound skips, running rebounds, etc.)
- Exercises to improve running rhythm (rapid ankling, fast rhythm leg drive, rapid butt kicks, etc.)
- Starting and accelerating phases tests
- Running technique during maximum speed phase.

Besides the strength and technical aspects of speed, a substantial part of training is dedicated to the speed endurance, lactacid capacity and synthesis tests (lactacid power). These three groups of exercises have different purposes according to the races (100 m, 200 m, 400 m).

Speed endurance means the ability to maintain high levels of running speed. In practical terms, these are sets of repetitions of relatively short tests (60 m, 80 m, 100 m). From 2 or 3 to 5 repetitions per set. The running speed is between 92 and 94% of the maximum speed that the athlete can reach on that distance. Both the recovery between repetitions and the recovery between sets are not complete. This way, the athlete accumulates a discrete amount of lactate within the series, which is only partially reduced in the recovery between the series. The goals of this training method are two: on one hand, the stimulation of the production and conduction of the nervous impulse from the central nervous system to the movement effectors and, on the other hand, the increase of biochemical energy reserves.Example of a speed endurance training session for 100 m and 200 m sprinters: 3 sets (4 × 60 m), 2′ rest between repetitions +2 sets (3 × 80m), 3′ rest between repetitions. 6′ rest between sets.

Example of a speed endurance training session for 400 m sprinters: 2 sets (4 × 60m), 2′ rest between repetitions+2 sets (4 × 80m), 3′ rest between repetitions +2 sets (3 × 100 m), 4′ rest between repetitions. 6′ rest between sets.

Lactacid capacity training leads to an increase in its production, and to an increase in the tolerance of high concentrations of lactate in the mus-cles. For this reason, this training would seem suitable for 400 m sprinters, but it is also important for 100 m and 200 m sprinters as they reach quickly high concentrations of lactate.

Example of lactacid capacity training session for 100 m and 200 m sprinters: 100 m, 150 m, 200 m, 250 m, 300 m. Run at 85% of personal maximal speed on each distance. 8′ rest between repetitions.

Example of lactacid capacity training session for 400 m sprinters: 150 m, 200 m, 300 m, 400 m, 500 m. Run at 85% of personal maximal speed on each distance. 8′ rest between repetitions.

Synthesis tests are run on 150 m for 100 m and 200 m sprinters and on 300 m for 400 m sprinters. The athletes run at their personal competition speed, focusing on the distribution of the effort, namely distributing the effort with pre-established intermediate steps based on the performance model. Given the very high intensity of these tests, they are considered to be lactacid power activators.

Example of synthesis tests training session for 100 m and 200 m sprinters: 3 or 4 × 150 m. Rest 15′ between repetition.

Example of synthesis tests training session for 400 m sprinters: 2 or 3 × 300m. Rest 20′ between repetition.

12.2.4 Common and Uncommon Injuries in Sprint

Sprinting requires a huge muscle mass for an explosive lower extremity activity. As a result, the most frequently reported injuries involve the lower limb, in particular hamstring and rectus femoris strains [10–15], as well as Achilles tendon ruptures [11] and/or back injuries [12].

Hamstring injuries are very common among sprinters, especially involving the biceps femoris [16] and are more frequent in males than female athletes [17].

It has been suggested that during both the late swing phase [18] and the early phase of sprinting [19] hamstrings are more prone to develop injury.

Referring to indirect muscle injuries, the eccentric contraction plays a key role in the pathogenesis of muscle strains. Mechanical factors based on overstretching can lead to sarcomeres fibers disruption, vessels lesion, cytoskeletal proteins, and sarcoplasmic reticulum damages [20, 21]. Fast contraction muscle fibers, as well as bi-poliarticular muscles and musculotendinous junctions can be considered as the most susceptible to injuries structures [22].

Achilles tendinopathy is usually correlated with explosive events such as sprints, hurdles, and jumps, but it may develop among middle- and long-distance runners too [23, 24]. It can be considered as the most common overuse injury occurring among sprinters and hurdlers [25]. It has been highlighted how an appropriate loading and adaptation may increase cross-sectional area and tensile strength of the tendons. Conversely, immobilization and inappropriate adaptation may cause tendon degeneration and the development of tendinopathy. Tissue adaptive changes, either physiological or pathological, are in fact influenced by the response of the peripheral nervous system and its messengers to external forces such as mechanical loading to the tendon [26]. Biomechanical causes of the development of Achilles tendinopathy may be intrinsic (such as avascularity, malalignment, hyperpronation, dysbalance of the agonist-antagonist muscle action, poor running style, insufficient warm-up and stretching before sports, age and eccentric loading of the Achilles tendon with a dorsiflexed foot) or extrinsic (ground surfaces, excessive increase in sports activity, repetitive mechanical loadings or old footgear due to changing alignment) [27]. When the inflammation involves the peritendinous sheaths without any pathological changes in the tendon itself, the term *Achilles peritendinopathy* is more appropriate [28, 29].

individual speed characteristics. Therefore, it is correct to address the *specific resistance* of a runner as a fundamental quality connected to the general resistance, at the basis of any kind of endurance discipline.Resistance is the ability to carry out an activity for a prolonged time, without a decrease in its effectiveness. It is the ability to counteract fatigue.

The type of fatigue a runner faces during performance to develop their resistance capacity can be distinguished in three different forms, depending on the muscle groups that participate in the effort:

1. Local fatigue: <1/3 of the overall muscles participate in the work. No noticeable activation of either the respiratory or the cardiovascular systems are seen, and the causes of fatigue are usually related to the neuro-muscular system which directly contributes to the execution of the movements.
2. Regional fatigue: between 1/3 and 2/3 of the overall musculature system participate in the work.
3. General fatigue: >2/3 of the overall muscles participate in the work. The high stresses on the energy exchange systems, such as the respiratory and the cardiovascular systems, highlight the insufficient functional capacities of such systems, with a consequent limitation of the performance.

From a practical point of view, a *central* fatigue can be identified in running, represented by the increase in ventilatory frequency up to an almost labored level; on the other hand, the *peripheral* fatigue is more associated with an increase in muscle lactate and consequent variations on the acidity levels.

12.3 Middle- and Long-Distance Running

12.3.1 Specific Considerations About Endurance and Fatigue

Among the different official running distances in athletics, the main concern is how to extend the

12.3.2 Middle Distances on Track (800 m–1500 m)

Middle distances running on track (i.e., 800 m, 1500 m) require a "Lactate Area" specific training.

In 800 m race, the athlete's approach can be of a fast type, thus similarly to the 400 m runner techniques or, conversely, of a resistant type that is mainly observed in athletes with very high aerobic values.

In this case, 1500 m runners achieve good results also on 800 m, basing the "specific endurance" on their ability to run high volume tests on track with a very short recovery.

For example, David Rudisha (Word Record holder in 800 m with 1′40″91) run 5 times 300 m in 36″ resting 5′ in between (running 400 m near 45″), whereas Steve Cram, with a personal best (PB) in 400 m of middle level (48.05″), was able to run 800 m in 1′42″88, running 10 times 300 m in 39″ during training sessions, with only 30″–45″ of recovery in between. Rudisha is the example of "fast" specialist, with a high percentage of fast fibers, while Cram is an example of "resistant" specialist, considering the Aerobic Power as his main quality.

12.3.3 Long Distances on Track (5000 m–10,000 m–3000 m Steeple)

Athletes endowed with a high Lactate Threshold and a high Aerobic Power are frequently able to run the 5000 m at a competitive level.

In the history of Track and Field, several athletes have set the World Record for both 1500 m and 5000 m at the same event, such as Paavo Nurmi, Gunder Haegg, Sandor Iharos, and recently, Said Aouita.

The following training schemes serve as examples to increase the Aerobic Power (A), and enhance the lactate tolerance in the fibers (B):

(a) 4 × 2000 m + 1 × 1000 m, recovering 4′/5′, a little faster than the speed used by the athlete for running 10,000 m.
(b) 4–5 sets of (600 m + 500 m + 400 m + 300 m + 200 m), resting 2′ between tests and 6′ between sets, starting at the speed used by the athlete for running 3000 m, then increasing the pace to complete the last 200 m at the speed used for running 800 m.

12.3.4 Marathon Training

Specific training already dominates shorter events, such as the 10 km race, however when approaching marathon race, coaches and athletes often accept easier long runs and shorter tempos as sufficient strategy, focusing on the benefits of continued high mileage [30].

Renato Canova trains elite marathon runners differently. He increases the volume and the duration and the single length of every type of interval at a specific speed, to extend the ability to run at that expected speed. His main tenet, in fact, suggests to progressively extend the distance an athlete can run their goal pace, over a period of months and years, to fulfill its potential as marathoner.

From this key concept, some principles derive:

- Pace is more important than distance. Running at paces slower than the marathon race pace does not actually mean training to improve specific marathon endurance. It can be considered a "training for training's sake" instead, the base to develop workouts for specific resistance and aerobic power endurance.
- Extensive workouts require long rest periods and no schedule: especially as the marathon race approaches, a hard long run on Thursday might not be followed by another high-quality session for a week. This notion is quite difficult to accept because the runner must admit that a given physical effort requires a determined rest, and that such an amount of rest is necessary.
- Workouts must get slower, and not faster, as race day approaches. Spending more time near goal pace than building up the support systems might be a mistake.

Furthermore, the main goals on which Renato Canova focuses are:

- Reducing the glycogen consumption at race pace.
- Increasing the speed at which muscle lactate is removed, due to the increase in the cell membrane permeability.

- Improving biomechanical efficiency and consequent performance.
- Preparing body and mind to be able to resist for the time necessary for the race pace.

More specifically, the marathon workout aims at improving the aerobic power levels and raising the anaerobic threshold value [31]. The latter must be used in the full marathon at its higher possible percentage (defined as the coefficient of specific resistance).

It is then fundamental to slightly stress the adaptive mechanisms, by reaching paces from 5 to 15% higher than the runner mean race pace (e.g., from 2′55″ to 2′45″/km. for a 2:08:00 male marathon runner, and from 3′25″ to 3′10″/km for a 2:28:00 female marathon runner). The method includes a progression of fast continuous run, fast progressive run, long speed variations, medium speed variations, short speed variations, mixed speed variations, continuous uphill run, and, finally, competition, as explained in Table 12.8.

Another component to be trained is the aerobic resistance, and a possible scheme is proposed in Table 12.9.

Among all these programs, the periodization plays a fundamental role.

In particular, a well-established scheme for Canova includes three different periods:

1. A general, preliminary, stage (6–8 weeks) to improve muscle efficiency and to increase aerobic resistance, with special attention to muscular extendibility and joint movements.
2. A fundamental preparatory stage (8–10 weeks) which introduces the aerobic power endurance, by increasing the volume of mileage, as well as focusing on the athlete's "internal load". During on-field evaluations, compensatory reactions are encouraged using different

Table 12.8 Training aerobic power

Methods	Quantity	Example (2h 08′ M/2h 28′ W)
Fast continuous run	Time: 20′–40′	10 km (29′/29′20″) Men
	Speed: 104–107% of marathon paced rhythm (M.P.R)	10 km (33′/33′30″) Women
Fast progressive run	Time: 20′–4′	12 km in 35′ (9′ + 8′50″ + 8′40″ + 8′30″)
	Speed: 102–108% of M.P.R	
Long	Distance: 5000/7000 m	3 × 5000 m increasing speed (17′15″/17′/16′15″)
	Volume: 15–21 km	
Intervals	Speed: 103–107% of M.P.R	Recovery (Rec.) 3′
Medium	Distance: 3000/5000 m	5000/4000/3000 m (14′30″/11′25″/8′20″)
	Volume: 10–12 km	
Intervals	Speed: 105–108% of M.P.R	Rec. 3′
Short	Distance: 1000/5000 m	10 × 1000 m at 2′45″ rec. 2′ Men
	Volume: 10-12 km	
Intervals	Speed: 106–110% of M.P.R	5 × 2000 m at 6′30″ rec. 3′ Women
Mixed speed	Distance: 400–3000 m	2000 m (5′25″) rec. 3′ + 10 × 400 m (1′02″) rec. 1′ + 8 km even paced with heart rate
	Total volume: 10–12 km	
Intervals	Speed: 107–112% of M.P.R	
Continuous uphill	Distance: 6–10 km	See the "fast continuous run" scheme
Run	Grades: 3–6%	
Competition	Cross/Road/Track	10.000 m track in 28′15″ Men 5.000 m track in 15′45″ Women

Table 12.9 Training aerobic resistance

Methods	Quantity	Example (2 h 08′ M/2 h 28′ W)
Medium paced Progressive run	Time: 1 h–1 h 30′ Speed: 85–100% of M.P.R	1 h 30′ (30′ at 3′30″/km + 30′ at 3′20″ + 30′ at 3′10″)
Medium-fast Progressive run	Time: 45′–1 h Speed: 95–105% of M.P.R	55′ (20′ at 3′40″/km + 20′ at 3′30″ + 15′ at 3′20″)
Medium even paced Continuous run	Time: 1 h–1 h 30′ Speed: 90% of M.P.R	1 h 30′ at 3′20″/km Men 1 h 30′ at 3′50″/km Women
Long resistance Run	Time: 2 h 15′–3 h Speed: 80% of M.P.R	2 h 45′ at 3′45″/km (44 km) Men 2 h 50′ at 4′15″ (40 km) Women
Long resistance With short variation	Time: 1 h 45′–2 h 15′ Variations: 500–1000 m Base speed: 80% of M.P.R Variation speed: 100% of M.P.R	1 h even paced + 10 × 1′30″ Rec. 1′30″ + 30′ even paced
Long resistance With long variation	Time: 1 h 45′–2 h 15′ Variations: 3–7 km Base speed: 80% of M.P.R Variation speed: 100% of M.P.R	30′ even paced + 7000/5000/3000 m Rec. 10′ slow paced running + 20′/40′ even paced
Continuous rolling Hill run	Distance: 18–30 km Grades: 3–6%	2 h running with 3–4 long and continuous uphill and downhill

stimuli without a strictly modulated training, but always attentive to consistency and continuity of the workload. As a result, the athlete often develops a general state of fatigue, which may prevent muscular freshness. However, this can be considered as a normal passage of this phase which does not necessarily represent a sign of poor preparation or overtraining.

3. A more specific stage ranging 6–8 weeks, during which workouts must be intensified to achieve the specific endurance at marathon pace at a higher level (Table 12.10). The result is the development of aerobic capacity and aerobic resistance. In particular, the main focus will be either extensive aerobic power, if the runner is characterized by a high anaerobic threshold, or intensive aerobic endurance, if the runner shows a high level of general resistance. In this session, the "race pace" gains more importance than during the other phases of training. The athlete must learn how to use the lowest possible quantity of glycogen in order to run for as long as possible at the correct and desired pace.

In Table 12.11, a three-week sample of a specific block of the marathon training program is reported [30].

12.3.5 Common and Uncommon Injuries in Middle-Long-Distance Running

Middle- and long-distance runners are usually leaner, have greater endurance requirements, and are more likely to suffer from chronic injuries than athletes participating in other disciplines [32].

The most frequently diagnosed disorders during long-distance events are in fact medial tibial stress syndrome (MTSS), patellofemoral syndrome (PFS), Achilles tendinopathy, iliotibial band friction syndrome, plantar fasciitis, and stress fractures of the metatarsal bones, the sesamoid bone and the tibia [25, 33–36]. Less common disorders are ankle sprains, hamstring muscle injury

Table 12.10 Training of specific marathon endurance

Method	Quantity	Example (2 h 08′ M/2 h 28′ W)
Marathon paced rhythm	Distance: 18–25 km	Half-marathon race at M.P.R.
Specific extensive	Distance: 19–30 km	4 × 5000, at 15′
	Specific long intervals	Rec. 1000 m at 3′10″/3′15″ Men
Endurance	Repeats 2–7 km (100–102% M.P.R)	3 × 7000 m at 24′30′
	Rec. 1 km (85–95% M.P.R)	Rec. 1000 m at 3′48″/3′55″ Women
Specific intensive	Distance: 15–23 km	8 × 1000 m at 2′55″
	Specific short intervals	Rec. 1000 m at 3′05″ (16 km) Men
Endurance	Repeats 0.5–1 km (103% M.P.R.)	20 × 500 m at 1′42″
	Rec. 0.5–1 km (97% M.P.R)	Rec. 500 m at 1′55″ (20 km) Women
Specific endurance	Distance: 30–35 km	32 km AT 3′06″ (1 h 39′) Men
Long run	Speed: 98–100% M.P.R.	35 km AT 3′40″ (2 h 08′) Women
	Distance: 10 km (85% M.P.R)	10 km at 34′ +
Specific marathon Pace block	+10–15 km (100–103% M.P.R.)	12 km at 36′ (AM + PM) Men
	(morning workout)	10 km at 40′ + 15 km at 52′ (AM + PM) Women
	Repeated in the afternoon	

and tendinopathy, gastrocnemius muscle injury, retrochanteric bursitis, low back pain, tibial posterior and hip adductor tendinopathy, infrapatellar bursitis, and knee sprain [36].

The overall incidence of injuries may vary from 6.8 to 59 injuries per 1000 hours of exposure to running [36]. This large variation is due to differences in the definition of injury, as well as in study populations, type of run, and follow-up periods [37].

Acute injuries are rare among these disciplines, mainly consisting of muscle injuries, sprains, or skin lesions [38]. Eighty percent of running disorders are overuse injuries instead, mostly involving the lower extremity and the knee in particular [33].

The MTSS has been reported as the most frequently diagnosed musculoskeletal injury among middle- and long-distance runners [36]. The main pathomechanical process is related to repetitive contraction of the posterior tibial, soleus, and/or flexor digitorum longus muscles during both the landing and the propulsion phases of running. This repetitive process can generate excessive stress on the tibia, resulting in the development of inflammation at the insertion of the periosteal [39–42]. Another cause of MTSS is an inadequate capacity of bone remodeling, due to the repetitive and constant stress on the tibia after both muscle contraction and vertical ground reaction during the landing phase [40, 43]. Furthermore, several risk factors for the development of MTSS have been recently proposed: female sex, high weight, high navicular drop, previous running injuries, and great hip external rotation with the flexed hip are some examples [44].

12.3.6 Prevention Strategies

According to Alan Hreljac [45], biological tissue adapts to the level of stress placed upon it. Repeated applied stresses below the tensile limit of a structure lead to a positive tissue remodeling if sufficient time is provided between stress applications. On the other hand, inadequate time between stress applications ultimately results in an overuse injury.

Another critical factor affecting the relationship between stress/frequency and injury is the type of stress applied to the structure. One of the most important types of stress, in terms of its effect on the human body, is impact force.

Runners who have developed stride patterns which incorporate relatively low levels of impact forces, and a moderately rapid rate of pronation are at a reduced risk of incurring overuse running injuries.

On a practical level, it may not be possible to teach people to run with a stride which incorpo-

Table 12.11 Example of marathon training schedule

	Week 1	Week 2	Week 3
Mon	Easy run with short variations	Easy run	Easy run with short variations
Tues	Moderate run	Easy run	Fartlek 8 × 2 min hard/1 min easy + 8 × 1 min hard/1 min easy
Wed	Marathon-pace intervals with moderate recovery (e.g., 6 × 3 K w/1 K recovery)	Easy run + strides	Easy run
Thur	Recovery run (extremely easy)	Moderate run	Easy run + 6–8 × 12-s hill sprints (2-min recovery)
Fri	Recovery run + 6–8 × 12-s hill sprints (2-min recovery)	Long intervals on track faster than marathon pace (e.g., 6–8 × 1600 @ 15 K-HM pace w/3-min easy recovery jog)	Continuous intervals (e.g., 6 K–5 K-4 K-3 K-2 K 1 K recovery) start at marathon pace and increase pace a few seconds each interval
Sat	Easy run	Easy run	Recovery run
Sun	Fast long run @ 90–95% goal marathon pace	Moderate run	Easy run + strides

rates lower impact forces and greater rates of pronation, although there are training habits that can reduce impact forces and minimize the effects of these forces on the body. Injured runners, or runners who are at risk of sustaining an injury, should decrease training speed as a means of reducing impact forces which generally increase as speed increases [45].

Most studies evaluating risk factors for preventing musculoskeletal injuries have been conducted in marathon runners and military recruits. Studies have estimated that 60–72% of running injuries are due to training errors, including excessive mileage or sudden change in routine. Factors associated with lower extremity injuries include a prior history of running injuries, training more than 64 km per week, and participation in more than 6 races in the previous 12 months [33, 46–48].

However, it might seem unreasonable to recommend an ultramarathon runner to limit their total running mileage to <64 km per week to prevent injury [33, 46, 48].

Other preventive strategies could include limiting the number of competitions in a race season, adhering to a gradual increase in mileage during training and proper pacing over the course of a race [49].

Unique to marathon and ultramarathons is the type and amount of gear utilized during a race, especially multi-day competitions. Gear recommendations can vary by the type of ultramarathon (continuous vs multi-staged) and the terrain. Appropriate clothing, shoes, backpack, and access to back-up gear, food, and hydration may affect the risk of musculoskeletal injury. Ultramarathon runners should be aware of gear weight, though it is unclear how much this impacts the risk of injury. A runner is less likely to be distracted, affected, or hindered by environmental factors if he has proper gear specific to his race environment. This may allow the runner to focus on avoiding hazards that may cause a fall or twisting injury. Additionally comfortable appropriate gear may allow runners to maintain ideal gait mechanics, thus avoiding abnormal load responses that could lead to acute, insidious onset or overuse injuries. All gear should be utilized and tested prior to the start of any race [46].

Specific characteristics and risk factors of the most common injuries must be always kept in mind: the better you know the problem, the better you can manage it.

For example, hamstring mechanics during sprinting, strength imbalances, athlete's flexibility, effort, age or ethnicity, and severity of previous injuries play an important role in developing hamstring strain [50–53].

Ankle sprains development and gravity depend on previous injuries, as well as proprioceptive or neuromuscular deficits, postural instability, or lack of strength [54–56].

Also thoroughly knowing the biomechanics and metabolic needs of the track and field disciplines with higher injury risks, such as middle- and long-distance running, helps preventing the most common sports-related injuries [17, 57–62].

However, the most essential aspects for an effective prevention can be summarized as: promptly treat acute injuries, especially the first episode, reduce overtraining risks, improve strengthening and recovery schemes [13, 17, 25].

12.4 Conclusions

Athletics is identified with running with a great historical background, covering different distances and speeds. Each running discipline has its own biomechanical and physiological characteristics with specific training modalities, running strategies, and different risks of injuries.

Competitive running athletes are significantly at risk of acute traumatic lesions in short distance disciplines or overuse injuries in the long-distance one, with a relationship between the rate of injuries and training loads. Young, less experienced athletes are more vulnerable to specific injuries. An appropriate training program is the best preventive measure, and an accurate management of acute or overuse injuries is the best way to run safely even in international competitions.

References

1. Rauh MJ, Macera C. Athletics. In: Epidemiology of injury in Olympic sports. Hoboken: Wiley; 2009. p. 26–48.
2. Vittori C, Castrucci P, Nicolini I, Preatoni E. Corse di velocità (Sprint racing). AtleticaStudi. 1983;3(3):1–107.
3. Čoh M, Tomažin K, Juhas I, Čamernik J. Partenza della velocità e accelerazione dai blocchi. Studio di un caso. AtleticaStudi. 2007;3–4:29–38.
4. Vittori C. Le Gare di velocità: la scuola italiana di velocità, 25 anni di esperienze di Carlo Vittori e collaboratori. FIDAL. 1995:190p.
5. Di Mulo F. Analisi ritmica e distribuzione dello sforzo nelle gare di velocità 'i record mondiali dei 100-200 e 400 metri'. AtleticaStudi. 2016;1–2:33–47.
6. Locatelli E. The importance of anaerobic glycolysis and stiffness in the sprints (60, 100 and 200 metres). IAAF New Stud Athl. 1996;11(2–3):121–5.
7. Lacour R. Physiological analysis of qualities required in sprinting. IAAF New Stud Athl. 1996;11(2–3):59–62.
8. Vittori C. L'allenamento della forza nello sprint (Strength training in sprinting). AtleticaStudi. 1990;1–2:3–25.
9. Vittori C. La pratica dell'allenamento. FIDAL; 2003. 80p.
10. Jacobsson J, Timpka T, Kowalski J, Nilsson S, Ekberg J, Renström P. Prevalence of musculoskeletal injuries in Swedish elite track and field athletes. Am J Sports Med. 2012;40(1):163–9.
11. Jacobsson J, Timpka T, Kowalski J, Nilsson S, Ekberg J, Dahlström Ö, et al. Injury patterns in Swedish elite athletics: annual incidence, injury types and risk factors. Br J Sports Med. 2013;47(15):941–52.
12. D'Souza D. Track and field athletics injuries—a one-year survey. Br J Sports Med. 1994;28(3):197–202.
13. Bennell KL, Crossley K. Musculoskeletal injuries in track and field: incidence, distribution and risk factors. Aust J Sci Med Sport. 1996;28(3):69–75.
14. Lysholm J, Wiklander J. Injuries in runners. Am J Sports Med. 1987;15(2):168–71.
15. Pendergraph B, Ko B, Zamora J, Bass E. Medical coverage for track and field events. Curr Sports Med Rep. 2005;4(3):150–3.
16. Malliaropoulos N, Papacostas E, Kiritsi O, Papalada A, Gougoulias N, Maffulli N. Posterior thigh muscle injuries in elite track and field athletes. Am J Sports Med. 2010;38(9):1813–9.
17. Alonso J-M, Edouard P, Fischetto G, Adams B, Depiesse F, Mountjoy M, Determination of future prevention strategies in elite track and field: analysis of Daegu 2011 IAAF championships injuries and illnesses surveillance. Br J Sports Med. 2012;46(7):505–14.
18. Chumanov ES, Schache AG, Heiderscheit BC, Thelen DG. Hamstrings are most susceptible to injury during the late swing phase of sprinting. Br J Sports Med. 2012;46(2):90.
19. Orchard JW. Hamstrings are most susceptible to injury during the early stance phase of sprinting. Br J Sports Med. 2012;46(2):88–9.
20. Morgan DL, Allen DG. Early events in stretch-induced muscle damage. J Appl Physiol. 1999;87(6):2007–15.
21. Proske U, Morgan DL. Muscle damage from eccentric exercise: mechanism, mechanical signs, adaptation and clinical applications. J Physiol. 2001;537(Pt 2):333–45.
22. Potvin JR. Effects of muscle kinematics on surface EMG amplitude and frequency during fatiguing dynamic contractions. J Appl Physiol. 1997;82(1):144–51.

23. Zemper ED. Track and field injuries. Med Sport Sci. 2005;48:138–51.
24. Ahuja A, Ghosh AK. Pre-Asiad '82 injuries in elite Indian athletes. Br J Sports Med. 1985;19(1):24–6.
25. Edouard P, Alonso J-M. Epidemiology of track and field injuries. NSA—New Stud Athl IAAF. 2013;28(1/2):85–92.
26. Ackermann PW, Longo UG, Maffulli N. Aetiology of tendinopathy of the Achilles tendon: mechano-neurobiological interactions. In: Achilles tendinopathy: current concepts. DJO Publications; 2010. p. 15–23.
27. Rosso C, Valderrabano J. Biomechanics of the Achilles tendon. In: Achilles tendinopathy: current concepts. DJO Publications; 2010. p. 25–34.
28. Canata GL, Casale V. Endoscopic Achilles tenolysis. In: EFOST surgical techniques in sports medicine—foot and ankle surgery. JP Medical Limited; 2015. p. 95–100.
29. van Dijk CN, van Sterkenburg MN, Wiegerinck JI, Karlsson J, Maffulli N. Terminology for Achilles tendon related disorders. Knee Surg Sports Traumatol Arthrosc. 2011;19(5):835–41.
30. Canova CR. 101 - how training principles from one of the world's leading marathon coaches can make you a better runner. [Internet] 2021. https://www.runnersworld.com/advanced/a20838198/canova-101/.
31. Canova R. Marathon training principles. Nairobi, bulletin IAAF RDC Kenya. 2016
32. Thing J, Scheer V. Track and field. In: Sports-related fractures, dislocations and trauma: advanced on- and off-field management. Springer Nature; 2020. p. 955–958.
33. van Gent RN, Siem D, van Middelkoop M, van Os AG, Bierma-Zeinstra SMA, Koes BW. Incidence and determinants of lower extremity running injuries in long distance runners: a systematic review. Br J Sports Med. 2007;41(8):469–80; discussion 480.
34. Wen DY. Risk factors for overuse injuries in runners. Curr Sports Med Rep. 2007;6(5):307–13.
35. Taunton JE, Ryan MB, Clement DB, McKenzie DC, Lloyd-Smith DR, Zumbo BD. A retrospective case-control analysis of 2002 running injuries. Br J Sports Med. 2002;36(2):95–101.
36. Lopes AD, Hespanhol Júnior LC, Yeung SS, Costa LOP. What are the main running-related musculoskeletal injuries? A systematic review. Sports Med. 2012;42(10):891–905.
37. Hoeberigs JH. Factors related to the incidence of running injuries. A review. Sports Med. 1992;13(6):408–22.
38. van der Worp MP, ten Haaf DSM, van Cingel R, de Wijer A, Nijhuis-van der Sanden MWG, Staal JB. Injuries in runners; a systematic review on risk factors and sex differences. PLoS One. 2015;10(2):e0114937.
39. Beck BR, Osternig LR. Medial tibial stress syndrome. The location of muscles in the leg in relation to symptoms. J Bone Joint Surg Am. 1994;76(7):1057–61.
40. Craig DI. Medial tibial stress syndrome: evidence-based prevention. J Athl Train. 2008;43(3):316–8.
41. Michael RH, Holder LE. The soleus syndrome. A cause of medial tibial stress (shin splints). Am J Sports Med. 1985;13(2):87–94.
42. Moen MH, Tol JL, Weir A, Steunebrink M, De Winter TC. Medial tibial stress syndrome: a critical review. Sports Med. 2009;39(7):523–46.
43. Mubarak SJ, Gould RN, Lee YF, Schmidt DA, Hargens AR. The medial tibial stress syndrome. A cause of shin splints. Am J Sports Med. 1982;10(4):201–5.
44. Reinking MF, Austin TM, Richter RR, Krieger MM. Medial tibial stress syndrome in active individuals: a systematic review and meta-analysis of risk factors. Sports Health. 2017;9(3):252–61.
45. Hreljac A. Impact and overuse injuries in runners. Med Sci Sports Exerc. 2004;36(5):845–9.
46. Krabak BJ, Waite B, Lipman G. Injury and illnesses prevention for ultramarathoners. Curr Sports Med Rep. 2013;12(3):183–9.
47. Fields KB, Sykes JC, Walker KM, Jackson JC. Prevention of running injuries. Curr Sports Med Rep. 2010;9(3):176–82.
48. Van Middelkoop M, Kolkman J, Van Ochten J, Bierma-Zeinstra SMA, Koes BW. Risk factors for lower extremity injuries among male marathon runners. Scand J Med Sci Sports. 2008;18(6):691–7.
49. Krabak BJ, Waite B, Schiff MA. Study of injury and illness rates in multiday ultramarathon runners. Med Sci Sports Exerc. 2011;43(12):2314–20.
50. Schache AG, Dorn TW, Blanch PD, Brown NAT, Pandy MG. Mechanics of the human hamstring muscles during sprinting. Med Sci Sports Exerc. 2012;44(4):647–58.
51. Yeung SS, Suen AMY, Yeung EW. A prospective cohort study of hamstring injuries in competitive sprinters: preseason muscle imbalance as a possible risk factor. Br J Sports Med. 2009;43(8):589–94.
52. Malliaropoulos N, Isinkaye T, Tsitas K, Maffulli N. Reinjury after acute posterior thigh muscle injuries in elite track and field athletes. Am J Sports Med. 2011;39(2):304–10.
53. Opar DA, Williams MD, Shield AJ. Hamstring strain injuries: factors that lead to injury and re-injury. Sports Med. 2012;42(3):209–26.
54. Malliaropoulos N, Ntessalen M, Papacostas E, Longo UG, Maffulli N. Reinjury after acute lateral ankle sprains in elite track and field athletes. Am J Sports Med. 2009;37(9):1755–61.
55. Holmes A, Delahunt E. Treatment of common deficits associated with chronic ankle instability. Sports Med. 2009;39(3):207–24.
56. Edouard P, Chatard J-C, Fourchet F, Collado H, Degache F, Leclair A, et al. Invertor and evertor strength in track and field athletes with functional ankle instability. Isokinet Exerc Sci. 2011;19:91–6.

57. Edouard P, Morel N. Suivi prospectif des blessures en athlétisme. Étude pilote sur deux clubs durant une saison. Sci Sports. 2010;25(5):272–6.
58. Alonso JM, Junge A, Renström P, Engebretsen L, Mountjoy M, Dvorak J. Sports injuries surveillance during the 2007 IAAF world athletics championships. Clin J Sport Med. 2009 Jan;19(1):26–32.
59. Watson MD, DiMartino PP. Incidence of injuries in high school track and field athletes and its relation to performance ability. Am J Sports Med. 1987;15(3):251–4.
60. Alonso J-M, Tscholl PM, Engebretsen L, Mountjoy M, Dvorak J, Junge A. Occurrence of injuries and illnesses during the 2009 IAAF world athletics championships. Br J Sports Med. 2010;44(15):1100–5.
61. Edouard P, Samozino P, Escudier G, Baldini A, Morin J-B. Injuries in youth and National Combined Events Championships. Int J Sports Med. 2012;33(10):824–8.
62. Rebella GS, Edwards JO, Greene JJ, Husen MT, Brousseau DC. A prospective study of injury patterns in high school pole vaulters. Am J Sports Med. 2008;36(5):913–20.

Running in Football

13

Lorenz Huber, Henrique Jones, Paolo Gaudino, Claudio Gaudino, and Werner Krutsch

13.1 Role and Characteristics of Running in Football

13.1.1 History of Football

Football as we know it today arose in the nineteenth century. In 1863 the Football Association (FA) was founded in England as the first association in the world. But the beginnings of the game go back to the second century BC. The first ball sport in which the ball was kicked with the foot is called *cuju* and it is first documented in China 200 BC. The players had a ball which was made of leather and filled with feathers or fur. A similar game on the same continent was practised in Japan under the name *kemari*. Many other cultures all over the world had similar ball games. One of the most famous is *Calcio*, which was played in Florence in the sixteenth century and included plenty of people playing in huge places in town. These rude game forms often caused damage on towns and severe injuries to players. That is why the game has been banned in many countries. The first attempts to develop and establish consistent rules began in Cambridge in 1844 [1]. The current numerically largest organisation is the Chinese Football Association with over 26 million male and female members. In particular, women's football is enjoying increasing popularity in the twenty-first century. Over the past decade, the number of female players has grown by over 50 per cent [2]. Football is played and eagerly pursued in all age groups, through all social classes and all over the world. The medical and scientific interest in football has increased strongly over the past decades [3].

13.1.2 Physiological Demands on Players

Different running and playing styles as well as various levels of play cause different physiological demands on the athletes. During a football match, players spend time above, and below, the anaerobic threshold speed: going considerably above it during the high-intensity periods and staying below it during the low-intensity phases of the match. The average heart rate for a football player during a match is between 80 and 95% of

L. Huber (✉) · W. Krutsch
Department of Trauma Surgery, University Medical Centre Regensburg, FIFA Medical Centre of Excellence, Regensburg, Germany
e-mail: lorenz1.huber@ukr.de; werner.krutsch@ukr.de

H. Jones
Orthopaedic and Sports Medicine Clinic, Montijo, Portugal

Lusofona University, Lisbon, Portugal

P. Gaudino
Manchester United, Manchester, UK

C. Gaudino
Turin, Italy

his maximal heart frequency. As a consequence, is it clear that the training sessions should focus on these physical match demands. Investigation showed that elite level players had mean blood lactate concentrations of 5.1 mmol/L (SD ±0.5) after the first half and 2.7 mmol/L (SD ±0.4) after second half, with peaks up to 12 mmol/L. The lower concentration following second half has been assigned to a reduced distance and physical workload [4].

The following figure shows some physical requirements which define a good runner in football (Fig. 13.1).

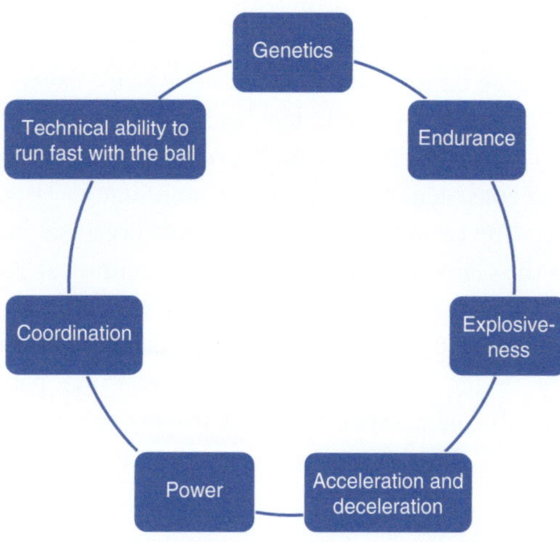

Fig. 13.1 Some qualities that define a good runner during a football game

13.1.3 Run Performance Analysis with GPS

Football is a highly complex team sport with changing dynamics and multistructural movements [5]. Running involves accelerations, decelerations, very high-speed running, maximal speed running and moderate intensity running. Therefore, performance analysis is crucial in the evaluation of players' achievements [6]. The global popularity of football has led to the implementation of scientific and technological knowledge in everyday use, and this is particularly evident within the field of performance analysis. Through the use of Electronic Performance and Tracking System (EPTS) devices, teams can track player's movements on the pitch and collect huge amounts of data on their performance, such as running speed, distance, position on the pitch, heart rate and body's work rate.

One of the important aspects of performance analysis is termed "running performance", which is nowadays mostly evidenced by global positioning software systems (GPS) [7]. Specifically, GPS (Fig. 13.2) allows collecting data about players' running performance regarding individual performance and position on the field correlation. These match analysis systems have been crucial in determining the external load of the football players during official matches. Thank to these technologies it has been possible to know specific total distance run, distances run at differ-

Fig. 13.2 GPS device

ent speed categories, maximal speed, number of accelerations and decelerations, maximal acceleration and maximal deceleration and, as a consequence, be able to estimate the metabolic load match performance indicators as well as another set of variables used to improve individual performance [8, 9]. Basically, match performance indicators are defined as a "selection and combination of variables that define some aspect of performance and that help achieve athletic success" [10]. Although match performance indicators and running performance are often investigated separately, to the best of our knowledge, there is no study that simultaneously observed both groups of performance variables during official soccer matches. Additionally, there is no information about the relationship that may exist between these 2 groups of variables.

13.1.4 Gait Analysis

When we conduct a gait analysis, the feet are only one small piece of the body's biomechanical puzzle. What happens to the feet is merely part of a complete, whole body, integrated movement pattern. Running, like most other whole-body activities (such as swimming or many field sports), is essentially a unique way of moving. When an athlete is analysed statically, dynamically, and then running on the treadmill during a gait analysis, it serves to provide a unique, personal movement design. That complete movement design reveals the programming of everything happening within the body—from kinaesthetic awareness and habit, to individual levels of mobility, stability, flexibility, and functional strength. The analysis of all these different elements taken together is what creates a complete picture of a person's gait. In essence, it is far more than just gait analysis. It is a true "movement" analysis. Some research shows that altering running style can prevent injury and improve running performance, but evidence is low [11]. However, in football there are some differences to other running sports. The running technique is particular due to the fact that players run while looking at the ball, a teammate (for example, when they wait to receive a pass) or the opponent (at defending when they try to get the ball back). Another characteristic of running in football is that it can be unilateral when players dribble with the ball on one foot.

Most players have a dominant right foot which they use for dribbling and kicking the ball. It turned out that this asymmetry also affects the gait. The athletes tend to jump off with their non-dominant foot and have a longer step length and a higher late swing phase percentage on the dominant side. This imbalance leads to asymmetrical muscles, which can cause injuries [12]. Especially for hamstring injuries, it was found that there is an association between the increased occurrence of injury and anterior pelvic tilting and thoracic side bending during swing phases of sprinting. It is therefore important that coaches and medical staff ensure an upright and bilaterally symmetrical running technique [13].

13.1.5 Statistical Data of Running in Football

In 90 min, an average professional female athlete covers 9–10.5 km [4], while professional male players cover distances of 10–12 km, with peaks of over 14 km [8]. Midfielders run the longest ranges (over 11 km), followed by defenders and attackers. But the running distances also depend on tactical formations: teams, which use formations with 3 defenders rather than 4, cover greater distances in all positions [14]. More than 75% of the total distance is covered at walking or jogging speed ($\leq$15 km/h) [10]. Only 200–300 m, depending on the literature, are run at a high speed (over 25 km/h) [15] (Table 13.1). Furthermore, observation of match running performance in Brazilian professional football players indicated that winning teams and home playing teams had greater total distance covered, but there are different findings in the literature whether the "stronger" or "weaker" teams, based on the league standings, cover greater distances [16, 17].

 Distances in different sex categories and pace [4, 8, 14]

Total distance male players	10.1–10.6 km
Total distance female players	9.1–10.4 km
Walking/jogging (≤15 km/h)	7726–8005 m (76.2%)
Running (15.1–24.9 km/h)	2173–2411 m (21.5%)
Sprinting (≥25 km/h)	211-238 m (2.3%)

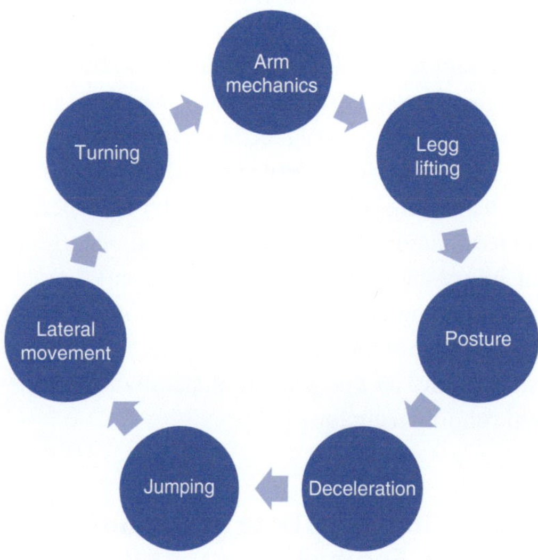

Fig. 13.3 Movement mechanics to improve

13.2 Training Strategies

Running in football is not just the players chasing the ball. By different running paths you can determine the tactics and influence your opponent's style of play. In football, the speed in the first few meters is particularly important [18]. Therefore, it is not only important to improve the physical endurance of the players, but also to improve the gait and explosiveness. Movement mechanics can be improved in training session for a more economic and explosive running style (Fig. 13.3).

In addition to the improvement of the running style, it was shown that improved core strength had positive effects on both agility and explosiveness. In the season preparation period and during return-to-competition, it should be implemented in training besides maximal strength exercises [19]. Training loads and their effect on injuries are frequently discussed and investigated in several studies. It was shown that workload during training can either have positive or negative effects on injuries and illness. But there is a consensus to avoid acute peaks. Coaches and medical staff who work with athletes should respect adequate resting periods and recurring spikes in training load [20]. Another important part of the training session are also the tactical exercises. These can be considered specific resistance (specific endurance) according to the player's movements on the pitch where running is training according to specific position on the pitch.

13.3 Prevention Strategies

Workload during training represents a modifiable risk factor for injuries. In addition to the physically and mentally demanding training, professional soccer players only have short recovery phases, as they sometimes play in competition every 3 days [21]. Football-related injuries are associated with non-modifiable factors, such as sex and age, and modifiable factors. These can be improved by training programs that influence strength, balance, and flexibility, which is especially important at young age. The general aspects of injury prevention in football include several strategies: proper preparation and health care, knowledge about injury prevention and prevention programs, neuromuscular training, adequate rest and sleep, adequate dietary and energetic supplementation, adequate warm up and cool down, adequate post-match recover strategies, appropriate equipment, safe environment, and, after a possible injury, safe and appropriately timed return to play [22]. Specific exercises are required to develop strength and include stretching to cope with eccentric forces during rapid movements. Nordic hamstring exercise for example reduces muscle injuries by 50% [23]. Problem is that especially in amateur categories without supporting medical staff, the exercises are not carried out correctly, which reduces the preventive and rehabilitative part and even increases the risk of re-injury [23]. Therefore, the FIFA 11+ injury prevention program was developed under the leadership of the FIFA Medical Assessment and

Research Centre and in collaboration with the Oslo Sports Trauma Research Centre and the Santa Monica Orthopaedic and Sports Medicine Centre in 2006. The program offers a complete warm-up procedure to reduce injuries in soccer players. It includes 15 structured exercises; it is available as printed material or online and can be executed easily. The exercises include core stabilization, eccentric thigh muscle training, proprioceptive training, dynamic stabilization, and plyometric exercises, all performed with proper postural alignment. The program's effectiveness was confirmed by various studies involving female and male players that revealed significant decreases in the incidence of non-contact injuries [24]. In 2016 the warm-up programme "FIFA 11+ Kids" was launched with the intention of preventing and reducing the number and severity of football-related injuries by enhancing children's fundamental and sport-specific movement skills through a range of evidence-based exercises.

The external factors like equipment, boots, the ball, shin and mouth guards and the field surfaces are always updated in investigation. There are some studies which show that injuries can also be related to the equipment used and the playing surfaces. In conclusion, it can be said that injury prevention depends on many factors. Most important are things like correct training load, adequate rest periods, sport-specific prevention programs such as FIFA 11+, the ideal medical support and diagnosis (Fig. 13.4) as well as timely return to play.

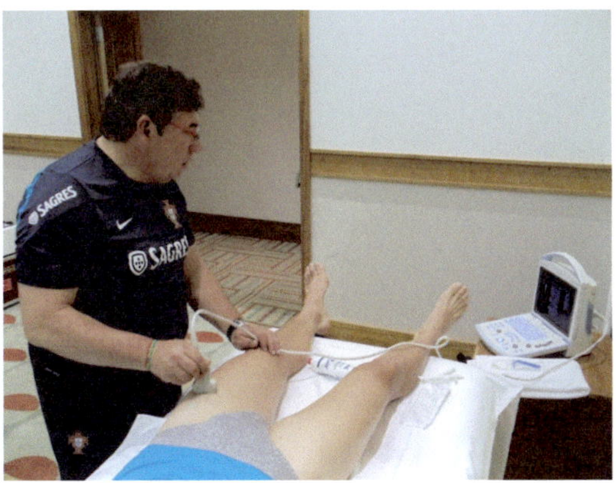

Fig. 13.4 Team doctor during sonography examination

> **Fact Box**
> - Running is one of the major skills in football.
> - Professional female athletes cover 9–10.5 km per game, professional male players cover 10–11 km. Seventy-five percent of the distance are at walking or jogging speed, 200–300 m per game are sprinted.
> - Training strategies should include improvement of gait, endurance, explosiveness, and core strength.
> - Injury prevention depends on peak loads in training and competition, adequate rest periods, football-specific prevention programs, ideal medical support, and timely return to sports.

References

1. Goldblatt D. The ball is round: a global history of soccer. London: Penguin; 2007.
2. Kunz M. Big count 2006. FIFA Magazine. 2007;7:10–5.
3. Dvořák J. Give hippocrates a jersey: promoting health through football/sport. Br J Sports Med. 2009;43(5):317–22.
4. Datson N, Hulton A, Andersson H, Lewis T, Weston M, Drust B, et al. Applied physiology of female soccer: an update. Sports Med. 2014;44(9):1225–40.
5. Kubayi A, Toriola A. Physical performance analysis of elite soccer players during the extra-time periods of the 2016 UEFA Euro Championship. S Afr J Sports Med. 2018;30(1):1–3.
6. Carling C, Williams AM, Reilly T. Handbook of soccer match analysis: a systematic approach to improving performance. Reprinted. London: Routledge; 2007.
7. Ehrmann FE, Duncan CS, Sindhusake D, Franzsen WN, Greene DA. GPS and injury prevention in professional soccer. J Strength Cond Res. 2016;30(2):360–7.
8. Vigne G, Gaudino C, Rogowski I, Alloatti G, Hautier C. Activity profile in elite Italian soccer team. Int J Sports Med. 2010;31(5):304–10.
9. Gaudino P, Iaia F, Alberti G, Strudwick A, Atkinson G, Gregson W. Monitoring training in elite soccer players: systematic bias between running speed and metabolic power data. Int J Sports Med. 2013;34(11):963–8.

10. Modric T, Versic S, Sekulic D, Liposek S. Analysis of the association between running performance and game performance indicators in professional soccer players. Int J Environ Res Public Health. 2019;16(20):4032.

11. Moore IS, Jones AM, Dixon SJ. Mechanisms for improved running economy in beginner runners. Med Sci Sports Exerc. 2012;44(9):1756–63.

12. Leroy D, Polin D, Tourny-Chollet C, Weber J. Spatial and temporal gait variable differences between basketball, swimming and soccer players. Int J Sports Med. 2000;21(3):158–62.

13. Schuermans J, van Tiggelen D, Palmans T, Danneels L, Witvrouw E. Deviating running kinematics and hamstring injury susceptibility in male soccer players: cause or consequence? Gait Posture. 2017;57:270–7.

14. Modric T, Versic S, Sekulic D. Position specific running performances in professional football (soccer): influence of different tactical formations. Sports (Basel). 2020;8(12):161.

15. Asian Clemente JA, Requena B, Jukic I, Nayler J, Hernández AS, Carling C. Is physical performance a differentiating element between more or less successful football teams? Sports (Basel). 2019;7(10):216.

16. Aquino R, Munhoz Martins GH, Palucci Vieira LH, Menezes RP. Influence of match location, quality of opponents, and match status on movement patterns in Brazilian professional football players. J Strength Cond Res. 2017;31(8):2155–61.

17. Aquino R, Carling C, Palucci Vieira LH, Martins G, Jabor G, Machado J, et al. Influence of situational variables, team formation, and playing position on match running performance and social network analysis in Brazilian professional soccer players. J Strength Cond Res. 2020;34(3):808–17.

18. Jones A, Jones G, Greig N, Bower P, Brown J, Hind K, et al. Epidemiology of injury in English professional football players: a cohort study. Phys Ther Sport. 2019;35:18–22.

19. Akif AY, Olcay M, Abdurrahman B. The effects of core trainings on speed and agility skills of soccer players. Int J Sports Sci. 2017;7(6):239–44.

20. Drew MK, Finch CF. The relationship between training load and injury, illness and soreness: a systematic and literature review. Sports Med. 2016;46(6):861–83.

21. Malone S, Owen A, Mendes B, Hughes B, Collins K, Gabbett TJ. High-speed running and sprinting as an injury risk factor in soccer: can well-developed physical qualities reduce the risk? J Sci Med Sport. 2018;21(3):257–62.

22. Arnason A, Sigurdsson SB, Gudmundsson A, Holme I, Engebretsen L, Bahr R. Risk factors for injuries in football. Am J Sports Med. 2004;32(1 Suppl):5S–16S.

23. Al Attar WSA, Soomro N, Sinclair PJ, Pappas E, Sanders RH. Effect of injury prevention programs that include the Nordic Hamstring exercise on Hamstring injury rates in soccer players: a systematic review and meta-analysis. Sports Med. 2017;47(5):907–16.

24. Sadigursky D, Braid JA, de Lira DNL, Machado BAB, Carneiro RJF, Colavolpe PO. The FIFA 11+ injury prevention program for soccer players: a systematic review. BMC Sports Sci Med Rehabil. 2017;9:18.

Running in Rugby

14

Michael R. Carmont, Fraser Morgan,
and Keji Fakoya

14.1 Characteristics of the Sport

Rugby is a popular contact collision sport played worldwide. In 2014, there were more than six million people playing, of whom 2.36 million were registered players. In 1845, the first rugby football laws were written after when William Webb Ellis picked up a football during a game and run with it. The sport of rugby became distinct from football when Blackheath Club left the Football Association in 1863. The sport was always considered amateur; however, players in some professions suffered financial hardship by having to take unpaid time off from work to play. Discussions regarding payment-in-lieu for players occurred at the George Hotel in Huddersfield, England, on the 29 August 1895 leading to the formation of the Northern Rugby Football Union. Over the next 15 years many clubs left the Rugby Football Union to form what became known as the Rugby Football League. Since then the sport has progressed in two codes known as Rugby Union and Rugby League.

The first world cup in rugby was the Rugby League World Cup held in 1954. Since then the competition has been held every 2–8 years. In the Union code the Rugby World Cup was first held in 1987 and now consists of a tournament of 20 nations. A seven a side union game is contested at the Commonwealth Games and was first contested in the 1900 Olympic Games in Paris.

Both codes have similar rules in that the oval-shaped ball must be touched on the ground on or over the try line to score a try. A subsequent place kick, level with the place at which the try was scored, allows a conversion for further points if the ball goes over the horizontal crossbar of the posts. The ball may be carried forwards or kicked forwards but may only be passed backwards. Only the ball carrier may be tackled, or stopped by wrapping the tackler's arms around the body or legs of the ball carrier. There is no limit to the amount of force that may be applied but a tackled player may not be lifted up and forced into the ground (spear tackle). Once forward progression has been stopped, in Union the ball must be released backwards to maintain possession. The opposing team may be able to reach over the tackled player to retrieve a loose ball but their feet must not leave the ground, without kneeling.

In League, six tackles are permitted before possession of the ball is exchanged. Both codes have scrums to permit restart after minor infringe-

M. R. Carmont (✉)
Department of Trauma and Orthopaedic Surgery, Princess Royal Hospital, Shrewsbury and Telford Hospital NHS Trust, Shrewsbury, England

Great Britain Winter Universiade, Telford, UK

F. Morgan · K. Fakoya
Department of Trauma and Orthopaedic Surgery, Princess Royal Hospital, Shrewsbury and Telford Hospital NHS Trust, Shrewsbury, England

© The Author(s), under exclusive license to Springer-Verlag GmbH, DE, part of Springer Nature 2022
G. L. Canata et al. (eds.), *The Running Athlete*, https://doi.org/10.1007/978-3-662-65064-6_14

ments. In a scrum the forwards are bound together and link with the opposition. In League, scrums remove forwards from open play giving more room for the unoccupied field of play. In Union, the scrum is contested and both packs push against each other aiming to move forwards, after the ball has been put in between them. In open play the ball may be kicked over the crossbar of the posts to score a drop goal. If a penalty is awarded further points can be scored by the success of the ball from a place kick over the crossbar.

14.2 Match, Field of Play and Team Characteristics

Matches last 80 min and are comprised of halves of 40 min of play separated by a half time period of 15 min. Upon completion of 80 min play will continue until the ball leaves the field of play or "goes dead".

The pitch or field of play, between the try lines, is no more than 100 m long and 70 m wide. Each in goal area is no longer than 22 m the distance from the in-goal line. The dead ball line is not less than 10 m.

In Rugby Union the playing team is comprised of 15 players with 8 substitutions being permitted per game. Positions 1–8 are termed forwards, and numbers 9–15 the backs. To restart play, the forwards may be bound together to form a scrum. In Union those in the scrum push against the opposing team to win the ball and gain ground on the field of play. The Hooker is in the middle of the front row and aims to retrieve the ball by hooking it back. The Hooker is supported on either side by a Prop. The two players in the second row of the scrum link the front row, and they in turn are each supported by a Flanker with the Number 8 in between them. The backs consist of the scrum half and the fly half, an inside and an outside centres and two wings on each side of the field. The remaining player is termed the full back.

In Rugby Union players may be selected for specific play for example locks tend to be taller for contesting lineouts and front row tend to be heavy.

In Rugby League each playing team has 13 players, 2 less players than Union. There are 17 per team and 12 interchanges are allowed between these. This means that managers may exchange players for tactical benefit. The 13-a-side game creates more open space on the pitch and potentially more running play. Following a restart each team is allowed to be tackled 6 times before the ball is passed to the opposition. Scrums are formed but usually are not contested. This binds the more physical players together in the scrum creating more space for further play. In terms of player size the lack of contested scrums and line outs mean that players have less of a variation for specific positions although forwards tend to be heavier and backs leaner and faster.

Although rugby pitches are of similar length to a 100 m-sprint track, the process of running is far more complicated. Play may be sustained over a number of phases progressing towards the try line. Forwards may drive the ball forwards over short distances frequently pushing against or even carrying members of the opposite team. Backs however particularly centres or those on the wing may break through the opposite defensive line running sprints of commonly 30 m. It is not uncommon however to run the full length of the pitch. At full sprint whilst carrying the ball, players may draw defenders towards them, creating space for another player to receive the ball, to give a clear run to the try line.

14.3 Role and Characteristics of Running in Rugby

As with other sports computer technology and the use of Global Positioning Systems (GPS) has allowed increased understanding of the demands of rugby in all its codes. The technical and financial demands of this analysis understandably mean that most data has been obtained at the elite level of the game. The literature will be presented according to the two codes of the game. Additionally, running patterns can be analysed and compared between different positions, phases of play and halves of the game.

14.3.1 Rugby Union

14.3.1.1 Running Patterns Between Positions

Players will typically cover 6950 m during a game of 83 min with 2800 m spent walking (37%), walking and standing, 1900 m 27% jogging, cruising 700 m (10%), 990 m 14% striding, 320 m 5% high-intensity running and 420 m 6% sprinting. Greater distances are spent running in the second half of the game. Backs once again had a greater number of sprints than forwards (34 vs. 19) but spent less time standing and walking (66.7 vs. 77.8%). Average distances of 15.3 m for backs and 17.3 m for forwards were recorded for speeds >20 km/h respectively suggesting that forwards ran greater distances in earlier studies [1].

Conversely in Elite English rugby union play backs covered greater distances than forwards (6127 m vs. 5581 m) and greater distances in walking and high intensity running (448 m vs. 298 m). Whereas forwards were engaged in a greater number of activities with static exertion, backs had greater activities in high intensity running. Players notably travelled greater distances in the first 10 min compared with the final 10 min of the game [2]. This confirms the distinction of positional play with forwards repeatedly tackling and retrieving the ball moving in mauls.

Ninety-eight elite players from 8 English Premiership clubs were tracked during 44 competitive matches throughout the 2010/2011 season. In this study the scrum half covered the greatest total distance during a match 7098 m and the front row the least 5158 and the back row covered the greatest distance at sprinting speeds particularly the number 8 position (77 m) [3]. This contrasts with the perceived opinion that the backs are involved in more of the running game and forwards more stationary.

French data from international play has been tracked using the Amisco Pro during 5 international games. Exercise periods typically lasted 4 s. Back row players had the highest mean acceleration values over the game. No significant difference was seen between the halves except for back rows who showed a decrease in mean accel-

eration. This suggests the specific activity of back row players [4].

Forwards have greater initial sprint momentum and maximal sprint momentum compared with backs, which have greater initial sprint velocity and maximal sprint velocity [5]. This finding is predictable due to the greater weight of forwards over backs.

Video analysis has shown that forwards will achieve sprints of 90% of maximal velocity 5 times during a match whereas for backs this is much higher at 9 times. Forwards would start from a standing position in 40% of sprint whereas backs would already be moving either walking, jogging or striding in over 60% of sprints [6].

When game movement demands from a professional club and senior international rugby union players were compared from 188 players from 4 professional teams and the affiliated international team during the 2014–2015 season. Significant effects between club and international were found for repeated high-intensity locomotion efforts, in all 6 positioning groups with a higher number on RHILE in international vs. club games. Significantly greater total distance and meterage were also shown in international compared to club for Outside Backs position. This allows performance staff to concentrate on training protocols to enhance quality [7].

Fornasier-Santos et al. have analysed the influence of competition level on running patterns for five playing position in the most successful 2014–2015 European rugby union team. During European Champions Cup games, front row forwards performed a higher number of repeated high-intensity efforts compared to national championship games (5.8 vs. 3.6, +61%) and back row forwards travelled greater distance both at high-speed movements (3.4 vs. 2.4, +41.7%) and after high-intensity accelerations 78 vs. 68, +14.8%. In backs, scrum-halves carried out more high-intensity accelerations (24 vs. 14.8, +66.3%), whereas outside backs completed a higher number of high-speed movements (63 vs. 48, +29.8%) and repeated high-intensity efforts (13.5 vs. 9.7, +39%). This highlighted that the competition level affected the high-intensity activity differently among the five playing positions.

Consequently, training programs in elite rugby should be tailored taking into account both the level of competition and the high-intensity running pattern of each playing position [8].

14.3.1.2 Running Patterns within Phases of Play

Outside backs, i.e., the outside centre and wings, tend to be involved in a greater aspect of running play. Fitness testing and training should therefore be tailored specifically to positional groups rather than between forwards and backs [9].

Fitness testing compared with game performance when sprint times over 10, 20 and 30 m were analysed in training there was a moderate to small negative correlations with metres advanced line breaks, tackle breaks and tries scored [10].

When the speed of running, phase of play and success of the move for attacking the 22-yard line were compared, greater success occurred when forwards had greater intensity running speeds of 3.6 m/min compared with 1.8 m/min for unsuccessful attacking play [11].

In terms of phases of play there were similar distances for attacking play for forwards 112 m/min compared with backs 114 m/min but greater in forwards 114 vs. 109 m/min compared with backs in defensive play and greater in backs during ball out of play periods of the game [12].

GPS data from 22 players at 8 international test matches were collected during 2016, based upon ball-in-play periods. Mean and max ball-in-play high-speed running were significantly higher for backs vs. forwards, whereas collisions per minute were significantly higher for forwards in plays lasting 61 s or greater. Max ball-in-play and high-speed running and collisions per minute were all time-dependent showing that movement metrics and collision demands differ as the length of play continues [13].

In team sports fatigue is manifested by a self-regulated decrease in movement distance and intensity. GPS data were collected from 46 professional match participants to determine the temporal effects on movement patterns. The total relative distance was decreased in the second half for both forwards and backs. A larger reduction in high-intensity running distance in the second half was observed for forwards than backs. Similar patterns were observed for sprint frequency and accelerations. There were different pacing strategies for forwards and backs. Forwards display a "slow-positive" pacing strategy while the pacing strategy of backs is flat. Backs were able to maintain performance intensity [14].

A prospective exploratory analysis has indicated that subsequently injured athletes demonstrated a tendency for greater thoracic lateral flexion, greater hip extension moments and greater knee power absorption, compared with uninjured athletes. All variables demonstrated an ability to descriptive differentiate between injured and uninjured athletes at approximately 60% of the late swing phase [15].

Technical substitutions are allowed per match as well as substitutions for head and blood injuries. Forwards entering the game as substitutes demonstrate significantly greater high-speed running distance and acceleration frequency than whole game players. For backs starters displayed greater high-speed running distance than whole game players. Forwards displayed "slow-positive" pacing strategies, while backs displayed "flat" pacing strategies. Forwards demonstrated greater performance decrements over the course of the match [16].

14.3.1.3 Running Patterns of Officials

In terms of officials, the movement demands of 9 elite officials across 12 super rugby matches were calculated in 2018. The total distance covered was 8030 m with a relative distance of 83 m/min with no differences observed between halves. Most game time was spent at lower movement speeds 76% with a larger effect for time spent >7 m/s between halves. Most game time was spent between 81% HRmax and 90% HRmax with no observable differences between halves. Distances covered at >5.1 m/s were highest during the first 10 min of a match while a distance at speeds 3.7–5 m/s decreased during the final 10 min of play [17].

The On-Field movements, heart rate responses and perceived exertion of lead referees in the 2019 Rugby World Cup matches were analysed.

Referees covered on average 6674 m with 586 m in high-speed running. A large reduction of high-speed running distance and moderate reductions in average speed over 1 and 2-day epochs were found in knockout compared with pool matches. This may be explained by contextual factors of the game [18].

14.3.2 Rugby League

14.3.2.1 Running Patterns Between Positions

In Rugby League distances covered were slightly shorter with backs covering 5573 m and forwards 4982 m per game. Backs also achieved greater maximum running speeds, completed a greater number of sprints, had less time between sprints and achieved a greater duration total duration of sprinting and covered a greater distance in sprinting 321 m vs. 153 m than forwards [19].

The overall distances covered in Rugby League during the 2010 season were 5964 m for forwards and 7628 m for backs with the backs also undertaking a greater number of high intensity running episodes 35 compared with 23 whereas the maximum work rate per 10 min block was 115 m/min and 120 m/min for forwards and backs respectively [20].

Thirty-seven elite players were evaluated during 104 National Rugby League (NRL) appearances. The majority 67.5% of sprint efforts were across distances of <20 m. The commonest distance for forwards was 6–10 m (46.3%), whereas outside backs had a greater proportion 33.7% of sprint efforts over distances of >21 m. The proportion of sprint efforts over 40 m was 9.7% for outside backs. Of sprints 48% involve contact and 58% were preceded by forward movement. For forwards the majority of sprints came from a standing start. The majority of sprint efforts were performed without the ball 78.7% and 67.5% were followed by a long recovery ≥5 min. Sprint training should be tailored to meet the individual demands of specific playing positions [21].

A comparison between NRL and European Super League (SL) matches in 192 performances. The relative distance covered in SL matches 95.8 m/min was greater than in NRL matches 90.2 m/min. Low-speed activity and moderate speed running were highest in the SL matches and relatively high-speed distance was greater during NRL matches 7.8 m/min vs. 6.1 m/min. This has been suggested to imply an NRL provides a greater standard or rugby league competition than the SL [22].

The physical demands of defence are considerably greater than attack for distance covered 109 m/min vs. 82 m/min and repeated high-intensity effort bouts 1 per 4.9 min vs. 1 every 9.4 min. These effort bouts were greater when defending in the opposition's 30 m zone, when defending their own try line and when attacking the opposition's try line [23].

Competition running intensities varied by both position and moving average duration. Hookers exhibited the greatest metabolic power of all positions due to a high involvement in both attack and defence. Full backs also reached a high Pmet due to a higher volume of absolute running [24].

There were significant differences between 10 m and 40 m speed and maximum speed principal components between U18 academy and youth international backs in 11 English RL academies. Multiple measures are recommended when assessing physical performance [25].

In women's rugby league backs and adjustable players demonstrated greater running and average acceleration/deceleration demands than forwards across all durations. This suggests specific training programmes for players according to their position [26].

Comparisons of elite rugby union and rugby league athletes, rugby union athletes typically displayed greater short-distance sprint performance, which may be linked to an ability to generate high level of horizontal force and power. Rugby union forwards had a higher absolute force despite having 12% more bodyweight than rugby league forwards [27].

14.4 Sevens

In "Sevens" competitions games are seven-a-side and are played on a full size pitch, however game duration is much shorter with 7 min halves. Team still comprise forwards and backs although the increased space tends to offer greater running.

Competitions involve 6 games across 2–3 days in a tournament format. The Sevens season involves a dozen competitive tournaments over less than 9 months. The sport also involves substantial international long-haul travel and repetitive chronic load demands [28].

International rugby sevens competition is more intense than domestic matches with substantially greater distance covered at high velocity 27% at $\geq$6 ms^{-1} and 4–39% more accelerations and decelerations. There was a small reduction in velocity from the first to the second half [29].

The relative distances covered by players throughout sevens matches ($n = 23$) was 102 m/min with 10% covered at low intensity, 14% running at both low and medium intensity, 4.6% at high intensity and 9.5% 9.7 m/min sprinting. For the backs a substantial decrease in total distance and distance covered in low, medium and high intensity was observed in the second half. Backs covered more distance at medium and sprinting speeds than forwards [30].

Conversely, Ross et al. [31] have studied the match demands of an international sevens tournament and found almost trivial differences between forwards and backs and similar trivial differences in pool and cup matches. Running intensity changes between and within halves [32].

In a study of 12 international male sevens players during international competitive matches the relative distance covered was 112.1 m/min, 35% was covered at medium speed and 17.1% 19.2 m/min high speed. A decrease in the distance covered at >14 and >18 km/h, the number of accelerations and repeated sprint acceleration sequences was observed in the second half compared with the first half. An increase in the mean heart rate, maximal heart rate (HRmax), %age of time at >80% and >90% HRmax was observed in the second half compared with the first half [33].

Twenty international female rugby sevens players were studied using GPS in 5 World Cup Sevens series events. Total, moderate speed and high-speed running distances were greater in the first half, during losses and against top 4 opponents. Additionally, total distance increased significantly from day 1 to day 2 of tournaments and very high-speed running distance increased during losses. The maximum time spent between 90 and 100% max heart rate and player load was significantly greater during the second half compared with the first. No difference in activity profiles was observed between forwards and backs [34].

With many tournaments occurring in hot and humid countries exertion during the heat is a major concern. Core temperatures collected in Fiji from 11 elite men's sevens athletes with 8 athletes reaching temperatures of >39 °C with several athletes reaching >39 °C during warm up. The post-game temperature was related to playing minutes, total running distance and high-speed distance and post-game activity. The temperatures obtained frequently exceeded those known to be detrimental to repeated sprint performance. Warm up temperature represents the easiest predictor of post-game temperature via time/intensity modulation and the use of pre and per cooling strategies. Warm ups should be modulated based on likely conditions in hot/humid conditions [35].

14.5 Training and Prevention Strategies

Running in rugby has been identified as the second most common causative activity during injury (12%) after tackling (43%) in the Professional Rugby Injury Surveillance Project from 2002/2003 to 2018/2019 [36]. Over time, however, there was a significant decrease in the incidence of running-related injuries (β −0.3195% CI −0.57 to −0.05, $p = 0.02$).

The incidence of hamstring muscle injuries was found to be 0.27 per 1000 player training hours and 5.6 per 1000 player hours. Second row forwards sustained the fewest (2.4 injuries/1000 player hours). Running activities accounted for 68% of hamstring exercises in addition to conventional stretching and strengthening exercises had lower incidences and severities of injuries during training and competition [37].

The risk of injury was 2.7 (95% CI 1.2–6.5) times higher when very high-velocity running, i.e., sprinting exceeded 9 m per session. Greater distances covered in mild moderate and maximum accelerations and low and very low-intensity movement velocities were associated with a reduced risk of injury [38].

The aetiology of hamstring strain injuries is multifactorial with playing position previous injuries leg imbalances, lack of readiness to return to play and running actions identified as contributing factors across levels. Combining strategies to prevent hamstring injuries and recurrences and to inform return to play is likely worthwhile and should include Nordic strength assessment and Nordic exercises [39].

Playing at senior competition level, a history of calf cramping and a history of lower back pain resulted in missed field minutes due to calf cramping. Preseason screening is recommended to identify those at risk of calf cramping and the development of prevention strategies [40].

14.6 Summary

The sport of rugby in all its codes includes considerable running activity. Although backs may be perceived to do more high-speed running and sprinting, forwards often undertake considerable sprinting more frequently from a standing start and with greater weight leading to higher momentum. High-intensity and sprint training must be specific for players of all positions.

References

1. Cunniffe B, Proctor W, Baker JS, Davies B. An evaluation of the physiological demands of elite rugby union using global positioning system tracking software. J Strength Cond Res. 2009;23(4):1195–203.

2. Roberts SP, Trewartha G, Higgitt RJ, El-Abd J, Stokes KA. The physical demands of elite English rugby union. J Sports Sci. 2008;26(8):825–33.

3. Cahill N, Lamb K, Worsfold P, Headey R, Murray S. The movement characteristics of English Premiership rugby union players. J Sports Sci. 2013;31(3):229–37.

4. Lacome M, Piscione J, Hager J-P, Bourdin M. A new approach to quantifying physical demand in rugby union. J Sports Sci. 2014;32(3):290–300.

5. Barr MJ, Sheppard JM, Gabbett TJ, Newton RU. Long-term training-induced changes in sprinting speed and sprint momentum in Elite rugby union players. J Strength Cond Res. 2014;28(10):2724–31.

6. Duthie GM, Pyne DB, Marsh DJ, Hooper SL. Sprint patterns in rugby union players during competition. J Strength Cond Res. 2006;20(1):208–14.

7. Beard A, Chambers R, Millet GP, Brocherie F. Comparison of game movement positional profiles between professional club and Senior International Rugby Union Players. Int J Sports Med. 2019;40(6):385–9.

8. Fornasier-Santos C, Millet GP, Stridgeon P, Brocherie F, Girard O, Nottin S. How does playing position affect fatigue-induced changes in high-intensity locomotor and micro-movements patterns during professional rugby union games? Eur J Sport Sci. 2021;21(10):1364–74.

9. Deutsch MU, Kearney GA, Rehrer NJ. Time–motion analysis of professional rugby union players during match-play. J Sports Sci. 2007;25(4):461–72.

10. Smart D, Hopkins WG, Quarrie KL, Gill N. The relationship between physical fitness and game behaviours in rugby union players. Eur J Sport Sci. 2014;14(sup1):S8–S17.

11. Tierney P, Tobin DP, Blake C, Delahunt E. Attacking 22 entries in rugby union: running demands and differences between successful and unsuccessful entries. Scand J Med Sci Sports. 2017;27(12):1934–41.

12. Read DB, Jones B, Williams S, Phibbs PJ, Darrall-Jones JD, Roe GAB, et al. The physical characteristics of specific phases of play during Rugby union match play. Int J Sports Physiol Perform. 2018;13(10)

13. Pollard BT, Turner AN, Eager R, Cunningham DJ, Cook CJ, Hogben P, et al. The ball in play demands of international rugby union. J Sci Med Sport. 2018;21(10):1090–4.

14. Tee JC, Lambert MI, Coopoo Y. Impact of fatigue on positional movements during professional rugby union match play. Int J Sports Physiol Perform. 2017;12(4):554–61.

15. Kenneally-Dabrowski C, Brown NAT, Warmenhoven J, Serpell BG, Perriman D, Lai AKM, et al. Late swing running mechanics influence hamstring injury susceptibility in elite rugby athletes: a prospective exploratory analysis. J Biomechan. 2019;92:112–9.

16. Tee JC, Coopoo Y, Lambert M. Pacing characteristics of whole and part-game players in professional rugby union. Eur J Sport Sci. 2020;20(6):722–33.

17. Blair MR, Elsworthy N, Rehrer NJ, Button C, Gill ND. Physical and physiological demands of elite rugby union officials. Int J Sports Physiol Perform. 2018;13(9):1199–207.

18. Elsworthy N, Blair MR, Lastella M. On-field movements, heart rate responses and perceived exertion of lead referees in Rugby World Cup matches, 2019. J Sci Med Sport. 2021;24(4):386–90.

19. McLellan CP, Lovell DI, Gass GC. Performance analysis of elite rugby league match play using global positioning systems. J Strength Cond Res. 2011;25(6):1703–10.

20. Austin DJ, Kelly SJ. Positional differences in professional rugby league match play through the use of global positioning systems. J Strength Cond Res. 2013;27(1):1703–10.

21. Gabbett TJ. Sprinting patterns of national rugby league competition. J Strength Cond Res. 2012;26(1):1703–10.

22. Twist C, Highton J, Waldron M, Edwards E, Austin D, Gabbett TJ. Movement demands of elite rugby league players during Australian National Rugby League and European Super League Matches. Int J Sports Physiol Perform. 2014;9(6):925–30.

23. Gabbett TJ, Polley C, Dwyer DB, Kearney S, Corvo A. Influence of field position and phase of play on the physical demands of match-play in professional rugby league forwards. J Sci Med Sport. 2014;17(5):556–61.

24. Delaney JA, Duthie GM, Thornton HR, Scott TJ, Gay D, Dascombe BJ. Acceleration-based running intensities of professional rugby league match play. Int J Sports Physiol Perform. 2016;11(6):802–9.

25. McCormack S, Jones B, Elliott D, Rotheram D, Till K. Coaches' assessment of players physical performance: subjective and objective measures are needed when profiling players. Eur J Sport Sci. 2021;

26. Cummins C, Charlton G, Paul D, Shorter K, Buxton S, Caia J, et al. Women's Rugby league: positional groups and peak locomotor demands. Front Sports Act Living. 2021;3:648126.

27. Cross MR, Brughelli M, Brown SR, Samozino P, Gill ND, Cronin JB, et al. Mechanical properties of sprinting in elite rugby union and rugby league. Int J Sports Physiol Perform. 2015;10(6):695–702.

28. Schuster J, Howells D, Robineau J, Couderc A, Natera A, Lumley N, et al. Physical-preparation recommendations for Elite Rugby Sevens performance. Int J Sports Physiol Perform. 2018;13(3):255–67.

29. Higham DG, Pyne DB, Anson JM, Eddy A. Movement patterns in rugby sevens: effects of tournament level, fatigue and substitute players. J Sci Med Sport. 2012;15(3):277–82.

30. Suarez-Arrones L, Portillo J, Pareja-Blanco F, Sáez de Villareal E, Sánchez-Medina L, Munguía-Izquierdo D. Match-play activity profile in Elite women's rugby union players. J Strength Cond Res. 2014;28(2):452–8.

31. Ross A, Gill N, Cronin J. The match demands of international rugby sevens. J Sports Sci. 2015;33(10):1035–41.

32. Furlan N, Waldron M, Shorter K, Gabbett TJ, Mitchell J, Fitzgerald E, et al. Running-intensity fluctuations in elite Rugby Sevens performance. Int J Sports Physiol Perform. 2015;10(6):802–7.

33. Suarez-Arrones L, Núñez J, de Villareal ES, Gálvez J, Suarez-Sanchez G, Munguía-Izquierdo D. Repeated-high-intensity-running activity and internal training load of Elite Rugby Sevens players during international matches: a comparison between halves. Int J Sports Physiol Perform. 2016;11(4):495–9.

34. Goodale TL, Gabbett TJ, Tsai M-C, Stellingwerff T, Sheppard J. The effect of contextual factors on physiological and activity profiles in International Women's Rugby Sevens. Int J Sports Physiol Perform. 2017;12(3):370–6.

35. Fenemor SP, Gill ND, Driller MW, Mills B, Casadio JR, Beaven CM. The relationship between physiological and performance variables during a hot/humid international rugby sevens tournament. Eur J Sport Sci. 2021:1–9.

36. West SW, Williams S, Kemp SPT, Eager R, Cross MJ, Stokes KA. Training load, injury burden, and team success in professional Rugby Union: risk versus reward. J Athlet Train. 2020;55(9):960–6.

37. Brooks JHM, Fuller CW, Kemp SPT, Reddin DB. Incidence, risk, and prevention of hamstring muscle injuries in professional Rugby Union. Am J Sports Med. 2006;34(8):1297–306.

38. Gabbett TJ, Ullah S, Finch CF. Identifying risk factors for contact injury in professional rugby league players—Application of a frailty model for recurrent injury. J Sci Med Sport. 2012;15(6):496–504.

39. Chavarro-Nieto C, Beaven M, Gill N, Hébert-Losier K. Hamstrings injury incidence, risk factors, and prevention in Rugby Union players: a systematic review. Phys Sportsmed. 2021:1–19.

40. Summers KM, Snodgrass SJ, Callister R. Predictors of calf cramping in Rugby League. J Strength Cond Res. 2014;28(3):774–83.

Running in Gymnastics

15

Angelina Lukaszenko, Natalia Chelnokova,
Dmitry Sidorkin, Vadim Stepanov,
Artem Sergienko, and Mikhail Novikov

Gymnastics was born in ancient Greece, where only men were eligible to train. Later on, this type of sport found a place at the Olympic Games. We can define 7 different types of gymnastic variants. We will focus only on the 2 main ones: artistic gymnastics and rhythmic gymnastics, based on injury incidence.

15.1 Role and Characteristics of Running in Gymnastics

In gymnastics, there is no differentiate "running" as an additional component of the performance. Usually running, or jumping elements are part of the general composition sustaining pirouettes,

evolutions, and handstands. Athletes mainly use running as a transition element between the specific figures on the arena. Nevertheless, during those short periods of dynamic movements, muscles are exposed to a big tension. Pointed feet and permanent stand on the toes can lead to an increased tension over gastrocnemius muscle (predisposing to Achilles tendinopathy) or to increase forces over patello-femoral joint predisposing to patella external pression or instability.

Proper diagnosis and training can create a special preventive methodology to avoid problems.

Gymnastics, and particularly artistic and rhythmic gymnastics, are one of the most difficult sports requiring excellent coordination, balance, and movement skills to perform at a competitive level.

Running represents a small percentage of the whole training and is a transitional element which allows the athlete to coordinate the various elements of its performance (Fig. 15.1a, b). All the movements on the gymnastic floor should be done with artistic inclination, dancing charm, coordinating head, and upper limbs movements. Different amplitude and dynamic is always required.

For each phase of running the most difficult skill is to coordinate all kinematic chains of the body. If a coordination problem occurs in one of these chains, the possibility to get an injury is highly increased.

A. Lukaszenko
Physiotherapy Department, International Knee and Joint Center, Abu Dhabi, United Arab Emirates

N. Chelnokova (✉) · D. Sidorkin
Jauza Clinical Hospital, Moskow, Russia
e-mail: Info@yamed.ru

V. Stepanov
Major Clinic, Moskow, Russia
e-mail: info@major-clinic.ru

A. Sergienko
Burnazyan Federal Medical Biophysical Center, Moscow, Russia
e-mail: fmbc-fmba@bk.ru

M. Novikov
Fit-Studio, St-Petersburg, Russia
e-mail: info@ftstudio.ru

Fig. 15.1 Different phases of running in gymnastics: during performance (**a**) and training (**b**) (Konstantin © Santalov, All rights reserved)

15.2 Running Phases in Gymnastics

15.2.1 Amortization Phase (Stabilization): Initial Contact (Stance Phase)

At the beginning of this stage, the biggest force impact is taken by the quadriceps muscle.

In this phase, foot and lower limb biomechanics are important. Valgus foot and weak muscles can create an inappropriate amortization and abnormal forces distribution along the whole kinematic chain, which can lead to Achilles' tendon and ankle joint ligaments injuries.

Specific training should be implemented to improve muscle balance and prevent overuse lesions.

15.2.2 Mid Stance Phase

In this phase, the most important muscles are iliopsoas, sartorius, and rectus femoris muscles, engaged in hip flexion and rotation. Decreased elasticity of these muscles, especially iliopsoas, may be the cause of pain while performing.

Idiopathic scoliosis in gymnastic athletes may cause imbalance in body muscles and lead to a lateral flexion of the body.

15.2.3 Toe Off Phase

In this phase, gastrocnemius and soleus muscle and additionally all the group of gluteus muscles are engaged. In gymnastics, performing exercises on tiptoes is the aesthetic requirement of the sport itself. Running performed on the toes is responsible for overuse injuries of the gastrocnemius and soleus muscles. Long training sessions and frequent performances may also cause shortening and elasticity reduction of these muscles, with consequent Achilles tendon injuries. Furthermore, running on the toes eliminates amortization typically given by landing on the whole foot, thus forces are not distributed properly.

15.3 Characteristics of Main Injuries

15.3.1 Achilles Tendinopathy

Classification and the terminology of the Achilles tendinopathy has been updated by C. Niek Van Dijk in 2011. Terminology was based on the clinical assessment, anatomy and pathology [1]. Within professional gymnasts, Achilles tendinopathy appear usually as a cause of the ankle joint instability, hypermobility syndrome and as a typical consequence of "point" feet while performing or training. Gastrocnemius and soleus muscles are always in tension during the flexion movement and eccentric workload when biggest pressure is putt on the place of the distal attachment of the Achilles tendon. This situation can exacerbate when running or during jumping activities in absence of proper training floor, adequate shoes or correct warm-up procedures.

Shoes selection is very important for avoiding pain at the distal part of the Achilles' tendon [2] and magnetic resonance imaging (MRI) is the key to confirm diagnosis.

Achilles tendinopathy [3] could have several types of treatment:

- Conservative methods such as taping, dry cupping, and exercising.
- Invasive treatments [4], i.e., injections [5–8], dry needling or surgery [9].

Conservative non-invasive treatment may require short timing immobilization of the joint or more frequently may rely on kinesio taping of muscles [10] (Fig. 15.2). Customized shoes insoles are used to better stimulate the foot and stabilize the joint. In case of professional athletes no running and jumping until full regression of symptoms is advisable. Proper physiotherapy plan is based on muscles relaxation, balance, and strength.

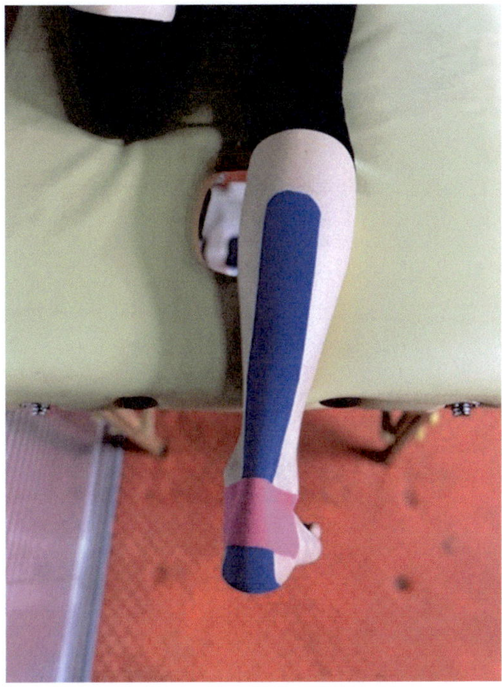

Fig. 15.2 Kinesio taping of Achilles tendon (Konstantin © Santalov, All rights reserved)

15.3.2 Patello-Femoral Syndrome (PFS)

Patello-femoral syndrome often occurs among professional gymnasts as a result of an unstable patella. Instability of the patella usually happens in this sport due to hypermobility syndrome, hip dysplasia, patella alta, hip and knee muscles imbalance. Pain and discomfort can be felt in the area of the knee joint but the true location may be somewhere else (Fig. 15.3).

Flat foot and cavus foot can be a cause of pain as well. Foot pronation can lead to an internal rotation of the hip which can lead to changes in the patella-femoral biomechanics (Figs. 15.4 and 15.5).

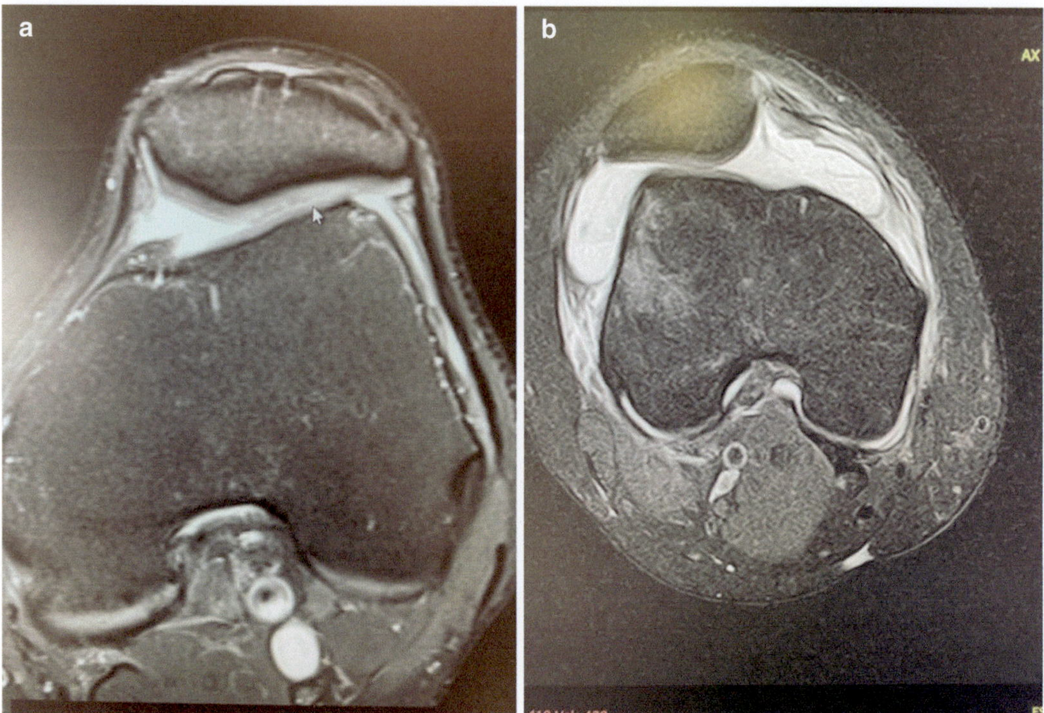

Fig. 15.3 Lateralization (**a**) and subluxation of the patella (**b**)

Fig. 15.4 Training with ball roller

Running or jumping can lead to increased forces over lateral patella border and femur bone, which lately is a main cause of developing a patello-femoral arthritis, patellar tendon tendinopathy [3] or overuse injuries of the quadriceps muscle.

In case of patella lateralization physiotherapy, treatments with kinesio taping method (Fig. 15.6), dry cupping or in case of patella tendon tendinopathy eccentric exercises and injections [11–13] are used. Athletes should stay away from running or jumping activities until full symptoms disappear.

15.4 Training Strategies

During training and physiotherapy, the main focus should be on exercises balancing muscle function (Fig. 15.7) improving lower limb alignment. Stabilization exercises are implemented in each stage of training.

Fig. 15.5 Foot ball/foot roller to decrease the tension of the plantar muscles

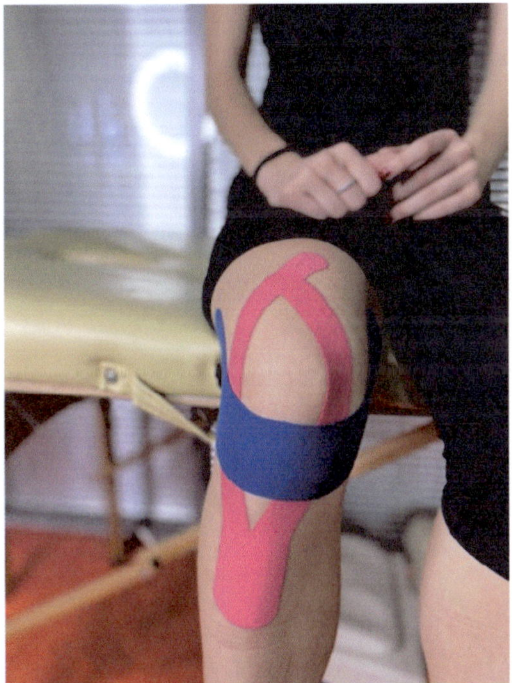

Fig. 15.6 Kinesio taping of the knee joint

Fig. 15.7 Prevention exercises: resistance training for gluteus muscles (**a**, **b**), squats with proper knee alignment (**c**), and balance training (**d**)

References

1. Calder J, Karlsson J, Maffulli N, Thermann H, van Dijk CN. Disorders of the Achilles tendon insertion: current concepts. London: DJO Publications; 2012.
2. Smigielski R, Zdanowicz U. Achilles tendon pathology. In: van Dijk CN, Dejour D, Denti M, Randelli P, Seil R, editors. Arthroscopy: basic to advanced. Berlin: Springer Berlin Heidelberg; 2016. p. 1115–24.
3. Papalia R, Moro L, Franceschi F, Albo E, D'Adamio S, Di Martino A, Vadalà G, Faldini C, Denaro V. Endothelial dysfunction and tendinopathy: how far have we come? Musculoskelet Surg. 2013;97(3):199–209. https://doi.org/10.1007/s12306-013-0295-7.
4. Erroi D, Sigona M, Suarez T, Trischitta D, Pavan A, Vulpiani MC, Vetrano M. Conservative treatment for insertional Achilles tendinopathy: platelet-rich plasma and focused shock waves. A retrospective study. Muscles Ligaments Tendons J. 2017;7(1):98–106. https://doi.org/10.11138/mltj/2017.7.1.098.
5. Abate M, Di Carlo L, Belluati A, Salini V. Factors associated with positive outcomes of platelet-rich plasma therapy in Achilles tendinopathy. Eur J Orthop Surg Traumatol. 2020;30(5):859–67. https://doi.org/10.1007/s00590-020-02642-1.

6. Abate M, Di Carlo L, Salini V. Platelet rich plasma compared to dry needling in the treatment of non-insertional Achilles tendinopathy. Phys Sportsmed. 2019;47(2):232–7. https://doi.org/10.1080/00913847.2018.1548886.

7. Abate M, Di Carlo L, Salini V. Platelet-rich plasma diffusion in Achilles tendon: relationship with therapeutic outcomes. Med Princ Pract. 2019;28(4):367–72. https://doi.org/10.1159/000499528.

8. Boesen AP, Hansen R, Boesen MI, Malliaras P, Langberg H. Effect of high-volume injection, platelet-rich plasma, and sham treatment in chronic midportion Achilles tendinopathy: a randomized double-blinded prospective study. Am J Sports Med. 2017;45(9):2034–43. https://doi.org/10.1177/0363546517702862.

9. Alfredson H. Traditional treatment vs biologics for Achilles tendinopathy. In: Roi GS, Della Villa S, editors. Football medicine outcomes: are we winning? XXVII isokinetic medical group conference abstract book. Torgiano: Calzetti Mariucci Editori; 2018. p. 166.

10. Lee JH, Yoo WG. Treatment of chronic Achilles tendon pain by Kinesio taping in an amateur badminton player. Phys Ther Sport. 2012;13(2):115–9. https://doi.org/10.1016/j.ptsp.2011.07.002.

11. Abate M, Di Carlo L, Verna S, Di Gregorio P, Schiavone C, Salini V. Synergistic activity of platelet rich plasma and high volume image guided injection for patellar tendinopathy. Knee Surg Sports Traumatol Arthrosc. 2018;26(12):3645–51. https://doi.org/10.1007/s00167-018-4930-6.

12. Filardo G, Kon E, Della Villa S, Vincentelli F, Fornasari PM, Marcacci M. Use of platelet-rich plasma for the treatment of refractory jumper's knee. Int Orthop. 2010;34(6):909–15. https://doi.org/10.1007/s00264-009-0845-7.

13. Kon E, Filardo G, Delcogliano M, Presti ML, Russo A, Bondi A, Di Martino A, Cenacchi A, Fornasari PM, Marcacci M. Platelet-rich plasma: new clinical application: a pilot study for treatment of jumper's knee. Injury. 2009;40(6):598–603. https://doi.org/10.1016/j.injury.2008.11.026.

Running in Tumbling

16

Henrique Jones and José Martinez

16.1 History of Tumbling [1]

The activity dates back to ancient China, Egypt, and Greece. Tumbling was performed by traveling bands of entertainers in the European Middle Ages and later by circus and stage performers.

While the origins of tumbling are unknown, ancient records have shown acts of tumbling in many parts of the world including China, India, Japan, Egypt, and Iran. Tumbling became part of the educational system of ancient Greece, from which early Romans borrowed the exercise for use in military training [2, 3]. During the Middle Ages, minstrels incorporated tumbling into their performances, and multiple records show tumblers performed for royal courts for entertainment [4]. It is at the end of this period in 1303 that the verb *tumble* is first attested in this sense in English [5]. There was renewed interest in formalized physical education during the Renaissance, and shortly thereafter gymnastics began to be introduced into some physical education programs, such as in Prussia as early as 1776

[3, 4]. The Federation of International Gymnastics (FIG) was officially formed in 1881, then known as the European Gymnastics Federation [4]. Tumbling, however, was not governed by the FIG until 1999. Before this time, the International Trampoline Federation governed the sport since its founding in 1964 [3]. National federations have even longer histories, such as the Amateur Athletic Union of the United States which included tumbling in events as early as 1886 [4]. Tumbling has only been included as an official event in one Olympic games, the 1932 Summer Olympics, and was exclusively a men's event. It was around this time that the floor exercise, which includes many elements of tumbling, became an individual event at the Olympics [4]. Tumbling has been an event at the World Games since the event's founding in 1980, first appearing at the 1981 World Games. Gymnastics has a long history founded in entertainment and demonstration of athleticism. Over the course of several centuries, it has evolved into the current sport we know today. The rapid growth in popularity of gymnastics created the need for regulation. Currently the FIG is the governing body for the sport with eight unique disciplines including women's artistic gymnastics, men's artistic gymnastics, acrobatic gymnastics, tumbling and trampoline, rhythmic gymnastics, aerobic gymnastics, Gymnastics for All, and parkour. Each discipline has an individual code of conduct, a code of points, eligibility criteria for participa-

H. Jones (✉)
Orthopaedic and Sports Medicine Clinic, Montijo, Portugal

Lusofona University, Lisbon, Portugal

J. Martinez
Setubal, Portugal

Clube Naval Setubalense, Tumbling, Trampolining and Gymnastics coach, Portugal

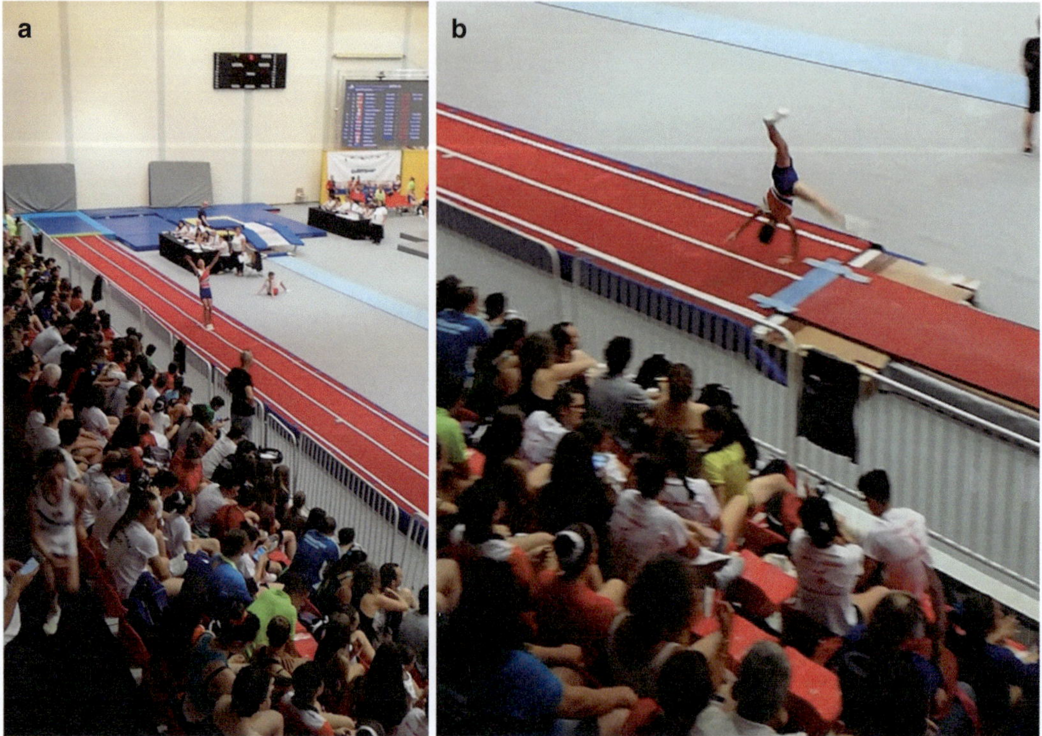

Fig. 16.1 (a, b) Tumbling competition

tion, and varying requirements of strength, power, and flexibility (Fig. 16.1). Recently, and being a discipline practiced mostly on Europe, team gym also includes tumbling.

16.2 Role and Characteristics of Running in Tumbling

Tumbling was brought to the United States in 1830. Tumbling is a characterized by the complex, swift, and rhythmical succession of acrobatic bounding from hands to feet, feet to hands, or even feet directly back onto feet. A tumbling pass may be over in a matter of seconds and is performed on a tumbling track that is 25 m in length. In tumbling, a gymnast performs a tumbling pass which sees the gymnast gain speed and power by running along a track and performing a series of somersaults and twists.

World-class tumblers perform no less than two double somersaults in one run, the best of them three, with twisting elements in addition. A typical tumbling competition will include the gymnast completing three tumbling passes. The first is called a straight pass (composed of somersaults); the second is called the twisting pass (twists) with the third and final pass composed of both somersaults and twists. Tumbling is the perfect sport for those looking for a fast-paced and daring challenge and involves gravity-defying stunts, power, and grace in a very coordinated running specificity. No doubt that tumbling is the fastest sport in the world!

A pass consists of eight skills which can be done in under 8 s; that's less than a second a skill! Competitors perform two passes during the qualification rounds and one or two passes for finals depending on the level. Single, double, and triple somersaults are all commonplace with twisting adding to the difficulty.

The action takes place on a 25-m rod floor usually alongside trampoline and double mini trampoline competitions.

16.3 Training Strategies

Whenever a fast running is needed, although it is not a maximal speed, the running training is included in the planning strategy as one more component of serial choreography to be included in the training planning. Running training is necessary to increase controlled running speed. Very important in training planning is also upper and lower limbs strengthening exercises, according with annual goals, in association with resistance, maximal, and explosive strength exercises and technical-specific gests in the apparatus. Coordination motor skills, movement variations, and alternative floor skills are also fundamental.

Tumbling requires power, fast reflexes, and courage to really excel at this fast-paced sport. Running is necessary to achieve speed for somersaults and twists series and is always part of training sessions.

Nassib et al. [5] study, 2020, findings revealed that the strength (static strength, speed strength, and endurance strength), power, and flexibility seem to be important and essential for good performance. Another characteristic that emerged from the results of the physical domain is coordination. This motor skill may seem relevant to gymnastics, they are applicable to the gymnasts' ability to perform all apparatus, and, more generally, they relate to the ability to accurately perform whole-body skills supported by the leg on the floor, balance beam, and vault. Therefore, athletic performance can be boosted using a combination of several characteristics that seems to be important for an elite gymnast. This fact reinforces the view that a systematic approach for the development and multidimensional profile seems promising. On the evolution planning, the gymnast begins the learning of a new technical element on facilitator apparatus, like the trampoline, fast track, or inflatable tumbling. This current practice intents to facilitate the new skills, avoiding traumatic injuries.

16.4 Types of Tumbling Moves

There are several specific and recognized tumbling movements; however the sport and the moves are always improving with new choreographies, interesting running, balance, skills, and ability to perform in an apparatus.

16.4.1 Flick

Flick is the most basic and fundamental skill in gymnastics and is a jump backward, with the support of the hands, alternating with the landing of the feet.

16.4.2 Whip Back

It is a movement similar to the flick but without the support of the hands, being made from feet to feet, increasing horizontal speed.

16.4.3 Mid Routine Doubles

Mid routine doubles are performed in a way to finish with a significant unbalance, so it is possible to resume the horizontal speed lost while performing the skill.

16.4.4 Final Skill

It's mandatory to finish with a somersault, usually presented as a double or triple with or without twists.

The tumble tracks are evolved in a way to facilitate harder skill and reduce injuries induced by the vibration of the apparatus itself.

The tumbling used to be more eclectic in a sense that it needed to have forward and back routines, short and long routines. In this discipline it's possible to perform forward skills that are not mandatory, and the backward exercises allow for a higher increase of speed making it easier to connect mid routines with doubles that increase difficulty of the pass. In team gym, backward and forward routines are mandatory.

16.5 Basic Skill to Begin Gymnastics as an Activity (Basic Tumbling Moves)

16.5.1 Roll

A roll is the most basic and fundamental skill in gymnastics. There are many variations to the skill. Rolls are similar to flips in the fact that it is a complete rotation of the body. However, the rotation of the roll is usually executed on the ground, while a flip is executed in midair with hips passing over the head, without hands touching the floor.

16.5.2 Pencil and Log Rolls

This can be started lying down on the back and front with the body outstretched. The gymnast then rolls onto their side and does a complete rotation of the body. The pencil roll is with the hands stretched above the head. The log roll is a sideways roll with the hands next to the waist.

16.5.3 Forward Roll

The forward roll is one of the most basic elements in gymnastics. The forward roll is started from a standing position, and then the gymnast then crouches down and places their hands shoulder width apart and next to the ears. They tuck their chin to their chest and place their hands onto the floor slightly in front of the knees. Then they place the back of their head onto the floor. They then push off of the floor with their legs and

rotate over their head onto their back. The gymnast then presses their feet onto the floor and whips the arms forward to stand up.

16.5.4 Backward Roll

The backward roll is similar to the forward roll but in reverse. The gymnast starts in a standing position and bends to a squat/sitting position with their arms in front. They then lower and lean back slightly until their bottom reaches the floor. They then continue this momentum and roll over their back onto their shoulders. They should then place their hands next to their shoulders and tuck their head into their chin. The hands then push the floor strongly and straighten their arms and continue to rotate their body over their head. The feet are then placed on the floor and the gymnast stands.

16.5.5 Cartwheel

A classic cartwheel performed with proper gymnastics form always starts with a lunge, lead leg (stronger leg) in the front of the lunge and the weaker leg in the back. During the lunge the gymnast will have their arms high in the air and straight, and the hips are square facing the forward direction. The gymnast will then push off of their back leg of the lunge, followed by placing their hands side by side on the ground in front of them. As they do this, they will begin to kick their legs up and over their torso and head as the body becomes inverted (upside down). During the rotation the legs stay apart in a large, wide straddle (as far apart as the gymnast can get them); the legs are straight; and the toes and feet always stay pointed. Finally the gymnast will set their first foot on the ground, followed by the second foot, landing in a lunge with the weaker leg in the front of the lunge, and the lead leg will be in the back. The gymnast will land facing the opposite direction than they started in. Their hands and arms will be perfectly straight, pointed high in the sky.

16.5.6 Round Off

It is a gymnastic technique that turns horizontal speed into vertical speed (to jump higher); it is also used effectively to turn forward momentum from a run into backward momentum, giving speed and power to backward moves such as flips and somersaults.

16.5.7 Bridge

Lie on your back, bend your knees up and keep your feet flat on the ground, and lace your hands by your ears with palms facing the ground push up on your arms and legs, bending your back and straightening your arms and legs and lifting your head off of the ground.

16.6 Other Move Options

16.6.1 Straddle Jump

Jump with legs straight out to the side. Tuck is a move where the knees are brought up to the chest. Turn is used in many sports including dance, artistic gymnastics, rhythmic gymnastics, and ice skating where it is called a spin. A turn is usually a complete rotation of the body, although a quarter (90°) and half turn (180°) is possible. A turn less than 360° is often called a pivot. Multiple rotations are possible, and these are named by the number of complete rotations, e.g., double turn or triple turn. In every turn the head must spot.

16.6.2 Front Pike Roll (Somersault)

A forward somersault is performed with knees kept straight. Front split is a split where one leg is forward and one is in the back. Straddle is a position in which the gymnast's legs are far apart at each side. Straddle split is a split where legs are out at each side. The move is used in all of the four women's events.

16.6.3 Gymnastic Injuries

The most serious problem faced by contemporary gymnasts is injury. The majority of injuries in tumbling occurs in acrobatic exercises more than in running time or in relationship with running. Overuse injuries, mostly tendon injuries, are quite frequent. Achilles tendon ruptures (Fig. 16.2), ACL injuries, and muscles injuries are rare but can occur during jumping exercises [6].

A review of the literature, in 2005 [7], reveals a reasonably consistent picture of pediatric gym-

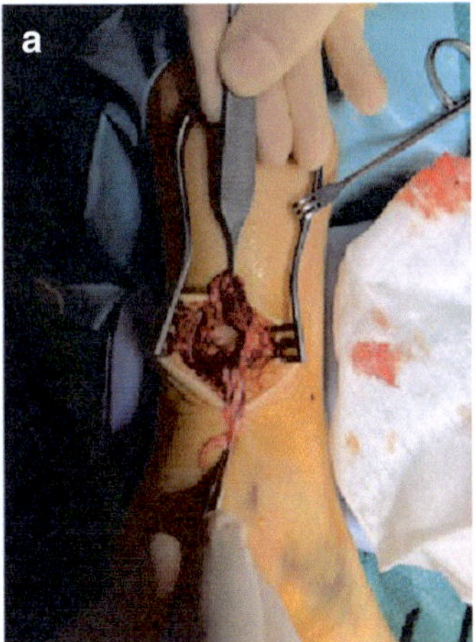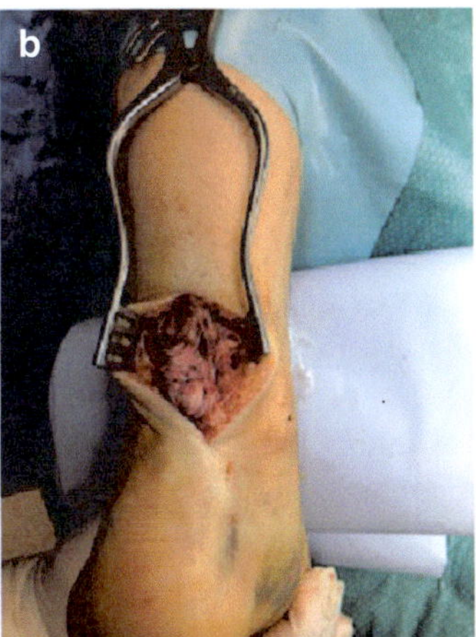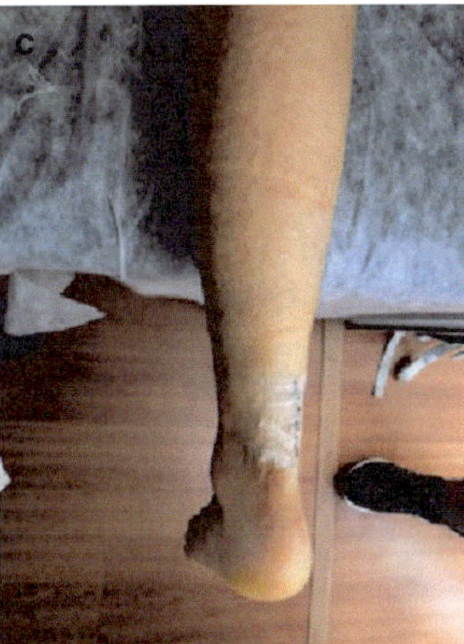

Fig. 16.2 (**a–c**) Sixteen-year-old tumbling athlete with a complete rupture of the Achilles tendon

nastics injuries. The incidence and severity of injuries is relatively high, particularly among advanced-level female gymnasts. Body parts particularly affected by injury vary by gender and include the ankle, knee, wrist, elbow, lower back, and shoulder. Ankle sprains are a particular concern. Overuse and nonspecific pain conditions, particularly the wrist and low back, occur frequently among advanced-level female gymnasts. Factors associated with an increased injury risk among female gymnasts include greater body size and body fat, periods of rapid growth, and increased life stress. Above all, this overview of the gymnastics injury literature underscores the need to establish large-scale injury surveillance systems designed to provide current and reliable data on injury trends in both boys' and girls' gymnastics and to be used as a basis for analyzing injury risk factors and identifying dependable injury preventive measures.

16.7 Prevention Strategies

Given that prevention is superior to treatment, can the gymnastics community and the scientific and medical community do a better job at injury prevention? Most research [8] in gymnastics has been descriptive in nature. Injury prevention ultimately requires that one can predict the outcome of certain activities and their injurious nature. Making such predictions requires a knowledge of the scientific and medical aspects of injury, but more than that, one must have an intimate knowledge of the sport. Injury prevention efforts must be firmly grounded in science and medicine while making pragmatic linkages to gymnastics as it exists and is practiced. The results of a study conducted by Sand [9] indicated that gymnasts were most fearful of injuries because of the difficulty in returning from an injury and being unable to participate in practices and competitions while injured. Gymnasts described aspects of their past performance experience, such as success, consistency, and communication with

significant others, as important sources of self-efficacy. Some examples of psychological strategies [6] used to overcome their fear of injury were mental preparation (e.g., imagery, relaxation), just "going for a skill," and the coaches' influence.

> **Fact Box**
> - Tumbling is the perfect sport for those looking for a fast-paced and daring challenge and involves gravity-defying stunts, power, and grace in a very coordinated running specificity.
> - Running pretends the generation of horizontal speed that transforms in vertical speed for double intermediate exercises or final jumping.
> - Also it is important to generate energy for transversal rotation according with body alignment in contact with tumbling apparatus.
> - It is a controlled running, rarely achieving maximal speed.

References

1. Coleen W. What is tumbling? 2017. cupdf.com › document › what-is-tumbling-tumbling-is-…
2. Carter E, Orlofsky F. History. In: Beginning tumbling and floor exercise. Belmont: Wadsworth Publishing Company; 1971. isbn: 9780534006464.
3. Britannica, The Editors of Encyclopedia. Tumbling. Encyclopedia Britannica; 2015. https://www.britannica.com/sports/tumbling-acrobatics. Accessed 24 Feb 2021.
4. Kilijanek K, Sanchez K. History and overview of gymnastics disciplines. In: Sweeney E, editor. Gymnastics medicine. Cham: Springer; 2020. https://doi.org/10.1007/978-3-030-26288-4_1.
5. Nassib SH, Mkaouer B, Riahi SH, Wali SM, Nassib S. Prediction of gymnastics physical profile through an international program evaluation in women artistic gymnastics. J Strength Cond Res. 2020;34(2):577–86.
6. Chase MA, Michelle Magyar T, Drake BM. Fear of injury in gymnastics: self-efficacy and psychological strategies to keep on tumbling. J

Sports Sci. 2005;23(5):465–75. https://doi.org/10.1080/02640410400021427.

7. Caine DJ, Maffulli N (eds). Epidemiology of pediatric sports injuries. Individual sports. Med Sport Sci. Basel: Karger; 2005, vol 48. pp. 18–58.

8. Claire C, Fabienne AL, Fournier JF, Soulard A. Competitive strategies among elite female gymnasts: an exploration of the relative influence of psychological skills training and natural learning experiences. Int J Sport Exerc Psychol. 2003;1(4):327–52.

9. Sands WA. Injury prevention in women's gymnastics. Sports Med. 2000;30:359–73. https://doi.org/10.2165/00007256-200030050-00004.

Running in Basketball 17

Thomas Geoffroy, Claudio Gaudino,
Daniele Mozzone, Luigi Talamanca,
François Tassery, and Patricia Thoreux

17.1 Roles and Characteristics of Running in Basketball

17.1.1 Movements in Basketball

Basketball is characterized by continuous change of movement. This sport is physically demanding, with many defensive and offensive actions, requiring players to repeatedly engage in spells of intense activities: sprinting, shuffling, jumping, etc. [1]. The performance model of the players varies significantly between playing positions and roles. There are several types of runs or movements in basketball, of which each of them have a specific purpose:

- **Acceleration and deceleration**: Basketball is characterized by rapid changes in direction involving complex movements that require a large number of transitions from jogging to sprinting and jumping, as well as accelerating and decelerating. Accelerations which can be maximal or submaximal and can start from a still position or from a certain speed are usually followed by decelerations. Decelerations can lead to a stop or just a decrease in speed and can include changes of direction at different angles. Accelerations and decelerations together represent the most important part of the running activity that characterize the game. Even if very high, during the first 2–3 steps, the acceleration does not allow the player to reach his personal maximal speed due to the limited court size. Both accelerations and decelerations movements are crucial for the players who carries the ball, the one

T. Geoffroy (✉)
CIMS (Centre d'Investigations en Médecine du Sport (CIMS)—APHP, Hôpital Hôtel Dieu, Paris, France

Département médical de l'INSEP (Institut National du Sport, de l'Expertise et de la Performance), Paris, France
e-mail: thomas.geoffroy@aphp.fr

C. Gaudino
Turin, Italy

D. Mozzone
Istituto di Medicina dello Sport di Torino, FMSI, Torino FC, Turin, Italy

L. Talamanca
Pallacanestro Biella, Biella, Italy

F. Tassery
Fédération Française de Basket—FFBB (Medical Commission), Paris, France

P. Thoreux
CIMS (Centre d'Investigations en Médecine du Sport (CIMS)—APHP, Hôpital Hôtel Dieu, Paris, France

Département médical de l'INSEP (Institut National du Sport, de l'Expertise et de la Performance), Paris, France

Institut de Biomécanique Humaine Georges Charpak—Arts et Métiers Paris Tech—Université Sorbonne Paris Nord, Paris, France
e-mail: patricia.thoreux@aphp.fr

© The Author(s), under exclusive license to Springer-Verlag GmbH, DE, part of Springer Nature 2022
G. L. Canata et al. (eds.), *The Running Athlete*, https://doi.org/10.1007/978-3-662-65064-6_17

who receives it, and the one who is moving without it. Accelerations, decelerations, and consequent changes of directions have been demonstrated to have a higher energy expenditure if compared to constant speed running [2]. The energy cost of an acceleration is three to four times higher than the deceleration one.

- **Runs:** They are peculiar in basketball during the transition phases. In these moments of the game (or specific training), the player runs without the ball, with the chest or the head turned on one side in order to follow the game and be ready to intercept the ball (if defending) or receive the ball (if attacking).
- **Shuffling:** The shuffle step describes a defensive form used to stay in front of the ball handler. Shuffling or moving side to side quickly is a lateral displacement. The goal is to contain the dribble as best as possible. To be effective this movement needs to be done quickly. These movements can be considered runs as they require a short flight phase between the steps.
- **Cutting**: An offensive move is when an offensive player tries to get past a defender by moving quickly toward the *basket* or to get open for a pass.
- **Jumping:** The purpose of jumping is offensive or defensive, and it is one of the most frequent moves in Basketball. Offensively, the jump is used to score with a jump shooting, a dunk, or by a layup most of the time. Another characteristic run-in basketball is the preparation for the takeoff in the layup. This includes a more or less marked loading phase on the last two steps similar to what normally occurs in all the run-up to jump. Defensively, a jump's goal is to regain possession of the ball (rebound) or make a shot blocking.

Between these high-intensity movements, there are also periods of walking and slight jogging, enabling players to recover before starting again on a high-intensity action.

Many technical gestures, fast and explosive, are incorporated in the various movements of basketball such as rebounding, driving, jump shooting, etc. The most important are presented here:

- **Rebounding**: It refers to catching the ball after a missed shot and before it hits the ground.
- **Driving**: *"Drive"* means to get the ball from outside the three-point line inside the *basket*. Most of the time, it is done by dribble penetration from the attacking team. To drive to the hoop effectively in basketball, it's necessary to have a good ball handling and dribbling skills as well as being able to move quickly on the court with the ball.
- **Layup:** A layup is a two-point attempt made by leaping from the ground, releasing the ball with one hand up near the basket, and using one hand to tip the ball over the rim and into the basket (layin) or banking it off the backboard and into the basket (layup). The motion and one-handed reach distinguish it from a jump shot. The layup is considered the most basic shot in basketball.
- **Jump shot**: A player may attempt to score a basket by leaping straight into the air. The elbow of the shooting hand is cocked, ball in hand above the head, and lancing the ball in a high arc toward the basket for a jump shot.
- **Shot blocking**: A block or blocked shot occurs when a defensive player legally deflects a shot from an offensive player to prevent a score. The defender is not allowed to make contact with the offensive player's hand (unless the defender is also in contact with the ball), or a foul is called. In order to be legal, the block must occur while the shot is traveling upward or at its apex.
- **Dribbling**: Bouncing the ball continuously with one hand at a time without holding the ball. Dribbling is necessary in order to take steps while possessing the ball. Once a player picks up their dribble (stops it by holding the ball), they may not dribble again until they pass, shoot, or otherwise lose possession of the ball and the ball touches a different player or the rim or backboard.

Roles and movements on the field vary according to each player's position. Guards, the smallest players, are in charge of the game organization. They travel the most distance at the fastest speed on the field because they have to bring the ball up the court and set up the plays. They are excellent ball handlers with high dribbling skills. The guards and forwards have great coverage of the field, staying on top of their opponents who try to create space with cutting and dribbling. Forward players, who are taller than guards but smaller than centers, alternatively assist the guards or centers during offence or defense. They are very often positioned close to the outer lines of the field. The center is usually the tallest one on the team, around 185–190 cm for women and more than 200 cm for men, and uses his size to score and defend from a position close to the basket [3]. The is only one center per team.

17.1.2 Activity and Physiological Demand in Basketball

17.1.2.1 Generalities

Basketball is a high-intensity intermittent sport: players execute repeated intense actions separated by short periods of active recovery (walking, jogging) or unactive recovery (on the bench, time-outs). As a result, playing basketball at the highest level requires a well-developed physical aerobic and anaerobic fitness [4].

Anaerobic power is often considered the most important. Many intensive skills and movements are involved, such as acceleration and jumping, for which the contribution of the anaerobic system is predominant. Indeed, the anaerobic component is very much in demand for vertical jumps, agility, and speed. However, a basketball game lasts 14 min, and so an efficient aerobic metabolism is also fundamental. Aerobic metabolism allows, during the active and passive recovery phases, to resynthesize creatine phosphate and to decrease lactate level in skeletal muscle. Aerobic capacity enables recovery from high-intensity phases to sustain high-intensity efforts. Moreover, some studies have shown that players with the best aerobic metabolism were also the best in terms of anaerobic performance; both are correlated [5].

Different field tests are used to evaluate these two metabolisms. For anaerobic metabolism, it's the lower body explosive strength (vertical Jump), which is very important in basketball, and strongly correlated to the result of the Wingate test, the standard test to evaluate anaerobic metabolism [6]. For aerobic metabolism, the *Yo-Yo Intermittent Recovery Test* (YYIR1) is used [4–6]. This test evaluates the ability to engage in repeated high-intensity exercise. Explosive strength (anaerobic metabolism), interval endurance capacity (aerobic metabolism), and VO_2 max are significant predictors of the repeated sprint ability, a key element of performance [6].

17.1.2.2 Physical Attributes

Considering their position (guard, forward, or center), players do not have the same morphology and physical composition and develop different physical attributes [3].

Centers are the tallest and heaviest players on the field (over 2 m and 100 kg). Centers and power forwards, who play in low and middle positions which involve lots of contacts, need to have a good mass to resist the opponent physical challenge. Moreover, centers have the best muscular strength, vertical jump, and absolute power [3, 7]. They use their body mass, strength, and power to be efficient in their tasks during the game: body opposing, winning ball possession, score points close to the hoop, and block opponent's shot [7, 8]. Having a larger body size has consequences: these players are involved in less high-intensity actions and are the slowest players on the team; their VO_2 (41.7 ± 1.1 mL/min/kg) max is lower than forwards and guards (45.5 ± 0.7 and 54 ± 1.6 mL/min/kg) [3, 7].

Guards are the shortest and lightest players of the team (about 185 cm and 85 kg). Guards and small forwards have less fat mass than centers. These physical attributes are necessary for them to fulfil their role as game organizers. Their role requires high sprint frequency and the greatest total distance covered of any player [7]. Guards run faster and have a better ball handling and better agility than other players, particularly centers. Moreover, they have better aerobic and anaerobic metabolism, so they are more endurant [4].

These morphological characteristics combined with well-developed aerobic and anaerobic metabolism enable the guards to have a shorter recovery period between high-intensity offensive or defensive tasks.

17.1.2.3　Distance Covered

The distance covered during a basketball game varies depending on the playing level and the competition. It is usually between 5 and 6 km according to a recent systematic review (Fig. 17.1) [9]. The frequent changes of intensity in basketball make it also interesting to know the distance covered in performing various types of activities. Usually a large part of the total distance covered is by jogging and running and not by sprinting. Stojanovic et al. [9] define the different activities like multidirectional movements performed at various velocity: 0–1 m.s^{-1} for stand/walk, between 1.1 and 3 m.s^{-1} for jogging, between 3 and 5 to 7 m.s^{-1} (depending on the author) for running, and a velocity greater than 5 m.s^{-1} or 6.66 m.s^{-1} or 7 m.s^{-1} for sprinting.

The relative distance, distance travelled per minute, is also a relevant variable because of different playing times between players. Male and female players travel in average 110 m/min during total game play (including stoppages of play such as free throws and exits) and 130 m/min during live gameplay.

Depending on game quarter, the distance covered is different; fatigue mechanisms and tactical issues can potentially induce these changes (Fig. 17.2).

Each player has position-specific activity demands: guards (6635 ± 448 m) cover significantly more distance than centers (5225 ± 659 m). Overall, backcourt players (5288–6390 m) travel more distance compared to frontcourt players (4976–6230 m) [1].

17.1.2.4　Number and Frequency of Actions

Each player performs close to 1000 different movements per game, such as jumping, dribbling, or sprinting. Change in direction and

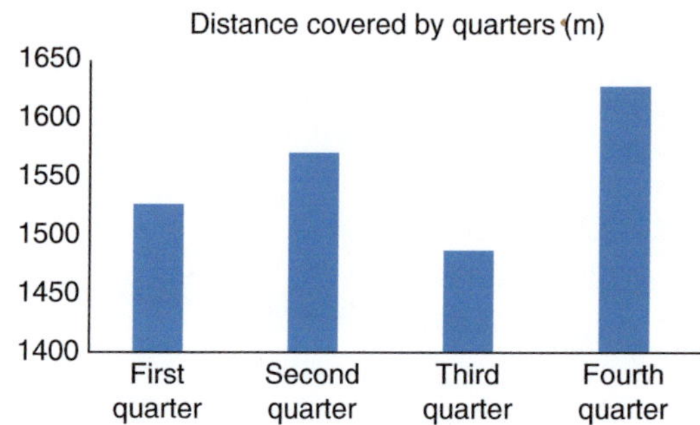

Fig. 17.2 Distance covered relative to playing period. (From Milanović Z. et al., (2020) Activity and Physiological Demands During Basketball Game Play [1])

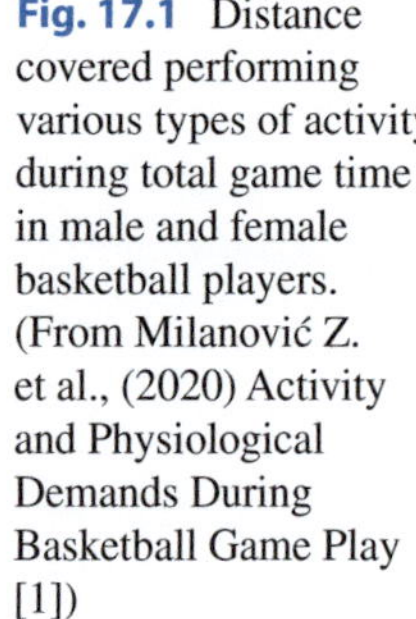

Fig. 17.1 Distance covered performing various types of activity during total game time in male and female basketball players. (From Milanović Z. et al., (2020) Activity and Physiological Demands During Basketball Game Play [1])

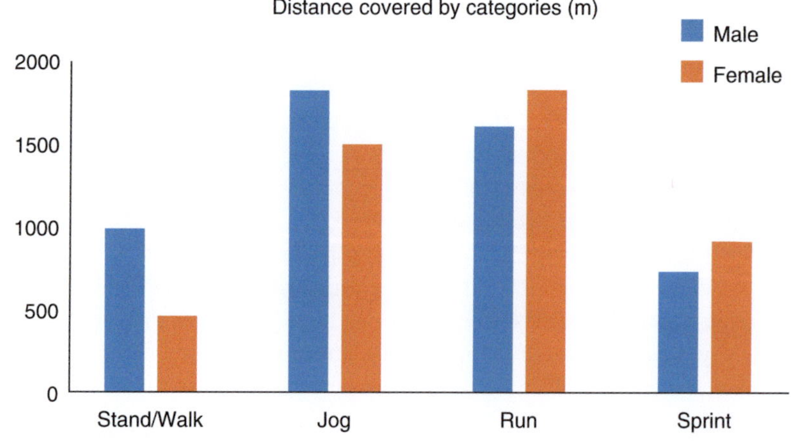

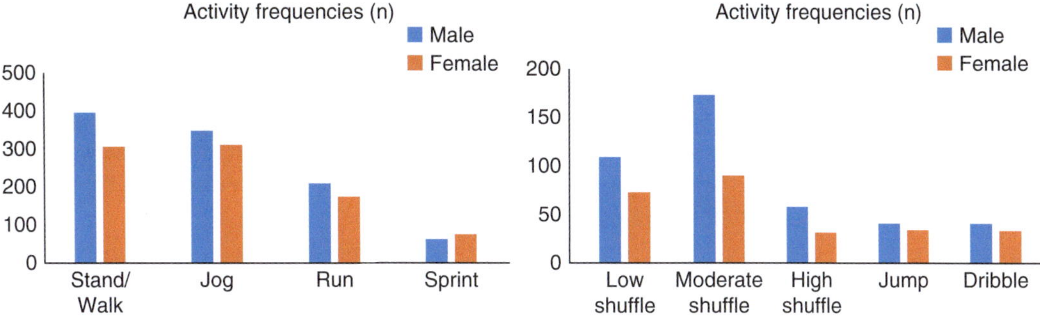

Fig. 17.3 Activity frequencies for various types of activities according to sex during total game time in basketball. (From Milanović Z. et al., (2020) Activity and Physiological Demands During Basketball Game Play [1])

rhythm occurs every 1–3 s, with a relative rate of change of 21–57 movements per minute (Fig. 17.3) [1].

Backcourt players (guards) make more dribbles during games than frontcourt players (center and forwards) (66 vs 21). Overall, guards perform more skills than other players.

Playing at a higher level requires higher-intensity activities like sprints and cutting.

17.1.2.5 Activity Duration

During a basketball game, players spend a certain amount of time performing each type of activity. Low- and moderate-intensity activities account for a large proportion of playing time: stand/walk, 29%; jog, 18%; low and moderate shuffle, 23%; dribbling, 5%; and static actions, 2% (average in male basketball) [1, 9], while all high-intensity activities represent a small proportion of play time: run, 12%; sprint, 4%; high shuffle, 5%; and jump, 2%. Male players spend more time dribbling (5 vs 4% of play time). There is similar proportions of playing time for standing, walking, and running between male and female basketball players. Guards spend more time engaging in high-intensity activities (sprint, 5.9%; high-intense shuffle, 9.3%) when compared with frontcourt players (sprint: forwards, 5.4%; centers, 4.5%; high intense shuffle: forwards, 9.2%; centers 7.9%). Near the end of a game, fewer high-intensity actions are performed by players. It is therefore essential for coaches to manage substitutions and time-outs, to allow players to recover and increase their ability to perform at high-intensity level during the money time [1, 9].

17.1.2.6 Heart Rate (HR)

Heart rate is a very useful variable to monitor during a game. It is used to understand the physiological stress induced by basketball. Then it is used to set training parameters, identify progress, and detect overtraining situations. In addition, monitoring the HR during games can help with tactical decisions such as changes, time-outs, and intentional fouls, with the aim of providing recovery time for the team. It has been reported that elite basketball players spend 75% of their playing time with a heart rate greater than 85% of its maximum value [1, 9]. This value is used as the threshold to identify high-intensity activities. The average heart rate during a game is between 82 and 95% of the maximum HR, highlighting the intensity of this sport [1, 9]. Usually, the mean heart rate is significantly lower in the last quarter, showing a reduction in the intensity of the game near the end. Higher HR are observed in backcourt players. Centers and forwards have equivalent HR even though forwards spend more time performing high-intensity actions. This may be due to a greater demand for static movements for centers, screening, blocking, and physically resisting their opponent [1, 9].

17.1.3 Specificities of 3v3 Basketball

The last few years, three-player basketball (3v3) has increased in popularity and will be an Olympic sport for the first time at the Olympic Games in Tokyo (2021). This discipline, a variant of basketball, opposes two teams of three players instead of five. The game is played on a half-court of 15 × 11 m with only one basket. 3v3 basketball is played over a single 10-min period or as soon as a team reaches 21 points, with a 12-s shot clock. If the defending team recovers the ball, they must get out of the one-point zone (two-point zone for regular basketball) before attempting a shot. These few different rules from the original basketball make this sport otherwise physically and physiologically demanding.

Less distance is covered in 3v3 basketball for both male and female, respectively, 876 ± 220.8 m and 856.7 ± 220.8 m, much less than the 5000–6000 m for the 5v5 Basketball [10, 11]. Regarding the number of meters per minute between both sports, 5v5 competitions report similar values with 3v3 [10]. One explanation could be that running speed during 5v5 transitions is often relatively slow, decreasing the distance-to-time ratio. Regarding physiological demand with player Load, heart rate, and RPE (rating perceived exertion), very few studies have compared 5v5 full-court and 3v3 half-court [10–12]. The available data does not allow us to conclude if one practice is more demanding than the other. Indeed, measurements in 3v3 are similar to 5v5 (very discreetly lower for PRE and mean heart rate). 3v3 players may not produce high values for whole body demands expressed as PlayerLoad™ or cover a large amount of distance, but their high-speed actions intensity and changes in pace and direction appear to be more frequent. One thing is certain, 3v3 basketball players engage in more player's ball contacts, and therefore each player participates more often in an offensive attack [10–12].

17.2 Training Strategies

Since physical demands, physical fitness, and morphology are very different depending on the position played, a lot of variables must be considered to develop personalized and specific training programs.

By analyzing the specificity of each player movements during the game, it is possible to determine the most efficient training strategies and methods. The training goals would be to improve the specific physical capacities of a player and his fitness.

Basketball includes frequent changes of sprint intensity, with variations between moderate to high intensity (every 21 s on average) [1]. Therefore, the running speed capacity over a short period of time is important, as well as the repeated sprint ability (RSA). These two variables are accepted to be performance indicator. In sports that require sprinting and jumping, peak power during the activity is typically the most important variable associated with success. Peak power (product of force and velocity) is the highest instantaneous value achieved during a movement. Counter movement jump (CMJ) is widely used to assess athletes' lower limb explosive power and is a predictor of RSA in elite basketball players [13]. Several studies investigated the use of repeated sprint ability training in basketball practice. It was observed that this type of training was correlated with maximal jump performance and with a potential role in the development of repeated sprint ability. Other studies found relationships between sprint times and jump height. A potential explanation for the relationship between sprint velocity and jump height could be that at the start of the sprint, the athlete applies maximal muscle effort action. The maximal speed obtained is determined by the different forces (multiplied by the contact time) applied on the ground in the acceleration phase. Power is thus essential for the jumps but also for the speed of the sprints [6, 13].

The development of the lower limbs' explosive and explosive-elastic strength is crucial for the accelerations and decelerations phases, changes of direction, and run-up to jumps. For this reason, overload exercises performed from a still position (explosive) and with the counter-movement (explosive-elastic) are crucial [14, 15]. Other exercises called "of connection" (vertical and horizontal jumps, uphill sprints, sled sprints) act as a conjunction between overload strength and short sprints (accelerations) without changes of direction [15].

The use of soft surfaces such as sand, especially when performing various types of jumps, accelerations, decelerations, and changes of direction, can help to avoid excessive stress on the muscle-tendon protecting both ligaments and joints. Running on sand surface entails longer contact times which can facilitate the explosive strength component more than the explosive-elastic one. In the advanced warming up phase, pre-athletic sprint exercises, including those focused on elasticity and those targeting speed, can have a certain importance in training sessions dedicated to the abovementioned phases (accelerations, decelerations, changes of direction) [16].

Regarding lateral displacements, sliding's and more generally any movements that doesn't require a loading emphasize of the lower limb but fast contact times. The improvement of the explosive-elastic-reflex strength must be pursued with exercises including consecutive rebounds using light loads or bodyweight in order to develop maximal feet tension and reaction.

During a basketball game, the movements mentioned above generically considered as anaerobic alactic can be repeated several times with incomplete recoveries. Therefore, it's involving an intervention of the lactic mechanism [17, 18]. Technical specific drills suitably modulated in regard of durations and methods of execution (speed, tactical focus, etc.) can be considered an appropriate specific resistance training and therefore balance out the metabolic need [16, 19].

The development of the aerobic component (average HR during the match: 160–180 b/min) is considered a mix of aerobic power and lactic capacity. It generally allows greater resistance to fatigue and reduce the need of the anaerobic mechanism when the intensity of the activity is lower [14, 18, 19]. On top of that, the aerobic component is also important as it allows a faster recovery of the PC (phosphocreatine) reserve linked to mitochondrial activity [17].

Nowadays it is common to incorporate aerobic training in the team technical exercises. However, aerobic power training using intermittent runs should be included especially in the pre-season period [14].

Core stability in general and all the twisting and turning movements included in the functional training framework play a decisive role in injury prevention and allow all the movements (including running, in all its different modalities) to be efficient [19].

Regarding physical output in different playing positions in basketball, a substantial difference has been established between external and internal players—with the external players performing a greater number of high-intensity actions leading to a higher external-load [17, 20].

Finally, it has been reported that elite basketball players have a better ability to sustain high-intensity resistance-specific work and to recover quicker than lower-level basketball players [20, 21].

17.3 Prevention Strategies

There is a high injury rate in basketball similarly to other team sports, and it is a major issue. A rate of ten injuries per 1000 h of play has been reported [22]. Injuries have a detrimental effect on performance and a significant number of games missed at the end of the season. Understanding the mechanism of these injuries and prevent them is therefore essential. In the injury prevention field, different values of external load in training or competition are used to determine if the athlete is at risk. As seen previously, external load variables, that can be determined using different electronic devices and systems, are related to the run: total distance, distance covered above specific speeds and metabolic power, total number of acceleration/deceleration, or the number of high-intensity efforts performed. The same way that an excessive training load imposed on players increases the risk of injuries, a diminished training loads can have similar consequences. A proper load management is essential; adequate levels of training might have protective effect and decrease the injury risk. Development of a minimum chronic quantitative workload (distance) and acute qualitative workload (accelerations and decelerations) seems to have an important factor to prevent injuries [22].

Basketball is a contact sport that involves large and heavy players, combined with the rhythm and the efforts developed by the players which make basketball a high-risk sport for injuries. Two types of injuries are distinguished: non-contact and contact injuries.

Ankle sprains are recognized to be the most common injury. During the reception of jumps or quick changes of direction, the ankles are extremely solicited, as well as the knees. The solicitation of those regions is even greater on wet and slippery surfaces. The strongest risk factor for an ankle sprain is having suffered from a previous ankle sprain, along with other factors such as poor balance control and limited range of motion. For this reason, primary prevention is fundamental. The strategies that granted a significant protection against this kind of injury are the implementation of a neuromuscular training program (proprioception and balance) and the use of taping or ankle braces. The stability of the foot inside the shoe appears to be more important than the shoe height in the prevention strategy. High-top shoes were not found to be safer than low-top shoes.

Muscle lesion is a very common injury among professional basketball players. In recent years, the epidemiology has drastically changed, so much that recent studies suggest that muscles

strain incidences might have surpassed ankle sprain among professional players [23]. The increase in muscle injuries in professional basketball players could be explained by an increase in game frequency and therefore an increase in physical demands. Stretching and muscular eccentric exercise may be very useful to increase the elasticity and the strength of muscles and tendons. The goal is to increase the body tolerance of specific athletic capacities that are necessary for this sport and to reduce the risk of muscular injury. Muscle injuries in basketball affect mainly lower limbs (90% of reported muscle injuries), with hamstrings and triceps surae being the most significant in terms of time loss, followed by the adductors, the obliquus abdominis, and finally the quadriceps [23]. Some of the most significant risk factors for these kinds of lesions are, as for ankle injuries, a previous strain, muscle weakness, muscle fatigue, limited range of motion, and biomechanical abnormalities increasing the internal load on the particular muscle district [24–26]. More specifically hamstring lesions and triceps surae lesions usually occur during high-intensity sprints, while adductor strains occur more frequently after sudden changes of direction with rapid adduction of the hip. Muscle injuries are often underestimated by the athlete because of the pain disappearing well before muscle recovery, and an early return to play may lead to reinjury. The recurrence of thigh lesions may add up to 15% in some cases [27]. Ultrasound assessment is fundamental for diagnosis, along with MRI for major lesions, and for follow-up. Before granting clearance to practice again and participating in games, it's mandatory to perform a series of functional tests to evaluate possible strength, proprioception, or range of motion impairments. A poor rehabilitation of muscular strains may lead to fibrosis and muscular scars that will possibly decrease muscle elasticity and alter the general biomechanics, thus increasing the risk of injury recurrence.

Patellofemoral tendinopathy is the most significant injury regarding lost competition days. This pathology is due to the repetition of jumps and landings, requiring plyometric and eccentric work. Young players are the most affected. The treatment should be preventive for the first episode including reinforcement of the torso, stability and alignment of the lower limbs, neurosensitivomotor reprogramming exercises, and stretching. Curative treatment includes rest from sports, physiotherapy care (physiotherapy, radial shockwaves, etc.), and adapted rehabilitation.

These three different types of injuries that make the lower limbs the most affected region with two cases out of three.

Since basketball is a high-speed contact sport, concussion and dental traumas are also frequent. Concussions has become one of the most challenging traumas during the last 15 years. Major advances have been made in the prevention, the diagnosis, and the management of concussions. A concussion protocol has been implemented in the NBA since the 2011–2012 season. Respecting the adequate duration of post-concussion rest is the best way to prevent long-term sequelae or early career termination. A study in 2017 interrogated 358 French elite basketball players; 10.4% reported having suffered from concussion in their career. In addition, basketball has a high incidence of oral trauma. The best prevention is to use intraoral custom-made protection (mouthguard), made by a dental technician under the supervision of a dental surgeon [23, 28].

References

1. Milanović Z, Stojanović E, Scanlan AT. Activity and physiological demands during basketball game play. In: Basketball sports medicine and science. Berlin: Springer; 2020. p. 13–23.
2. Di Prampero PE, Fusi S, Sepulcri L, et al. Sprint running: a new energetic approach. J Exp Biol. 2005;208(14):2809–16.
3. Ostojic SM, Mazic S, Dikic N. Profiling in basketball: physical and physiological characteristics of elite players. J Strength Cond Res. 2006;20(4):740.
4. De Araujo GG, De Barros Manchado-Gobatto F, Papoti M, et al. Anaerobic and aerobic performances in elite basketball players. J Hum Kinet. 2014;42(1):137–47.
5. Hoffman JR, Epstein S, Einbinder M, Weinstein Y. The influence of aerobic capacity on anaerobic performance and recovery indices in basketball players. J Strength Cond Res. 1999;13(4):407–11.
6. Te Wierike S, De Jong MC, Tromp EJY, et al. Development of repeated sprint ability in talented youth basketball players. J Strength Cond Res. 2014;28(4):928–34.
7. Abdelkrim N-B, Chaouachi A, Chamari K, et al. Positional role and competitive-level differences in elite-level men's basketball players. The. J Strength Cond Res. 2010;24(5):1346–55.
8. Pojskić H, Šeparović V, Užičanin E, et al. Positional role differences in the aerobic and anaerobic power of elite basketball players. J Hum Kinet. 2015;49(1):219–27.
9. Stojanović E, Stojiljković N, Scanlan AT, et al. The activity demands and physiological responses encountered during basketball match-play: a systematic review. Sports Med. 2018;48(1):111–35.
10. Montgomery PG, et Maloney BD. Three-by-three basketball: inertial movement and physiological demands during elite games. Int J Sports Physiol Perform. 2018;13(9):1169–74.
11. McCormick BT, Hannon JC, Newton M, et al. Comparison of physical activity in small-sided basketball games versus full-sided games. Int J Sports Sci Coach. 2012;7(4):689–97.
12. Schelling X, et Torres L. Accelerometer load profiles for basketball-specific drills in elite players. J Sports Sci Med. 2016;15(4):585.
13. Maggioni MA, Bonato M, Stahn A, et al. Effects of ball drills and repeated-sprint-ability training in basketball players. Int J Sports Physiol Perform. 2019;14(6):757–64.
14. Egger JP. Stratégie de préparation physique pour les J.O. de Sydney 2000, Equipe de France masculine de basketball.
15. Vittori C. Le gare di velocità: la scuola italiana di velocità: 25 anni di esperienze di Carlo Vittori e collaboratori. FIDAL. 1995;
16. Trachelio C. La preparazione fisica agli sport di squadra con riferimenti specifici per il basket. Libreria dello sport. 1997;
17. Benelli P. Physiological load analysis related to technical commitment in the basketball player. Primo Master per Formatori, CAN SIPI; 2011.
18. McInnes SE, Carlson JS, Jones CJ, et al. The physiological load imposed on basketball players during competition. J Sports Sci. 1995;13(5):387–97.
19. Barnaba SD, Corso O. Acceleration and deceleration in basketball. From the movement analysis to the methodological tests. C.N.A. Senigallia; 2011.
20. Petway AJ, Freitas TT, Calleja-González J, et al. Training load and match-play demands in basketball based on competition level: a systematic review. PLoS One. 2020;15(3):e0229212.
21. Berkelmans DM, Dalbo VJ, Kean CO, et al. Heart rate monitoring in basketball: applications, player responses, and practical recommendations. J Strength Cond Res. 2018;32(8):2383–99.
22. Caparrós T, Casals M, Solana Á, et al. Low external workloads are related to higher injury risk in pro-

fessional male basketball games. J Sports Sci Med. 2018;17(2):289.

23. Rodas G, Bove T, Caparrós T, et al. Ankle sprain versus muscle strain injury in professional men's basketball: a 9-year prospective follow-up study. Orthop J Sports Med. 2019;7(6):2325967119849035.

24. Mosler AB, Weir A, Serner A, et al. Musculoskeletal screening tests and bony hip morphology cannot identify male professional soccer players at risk of groin injuries: a 2-year prospective cohort study. Am J Sports Med. 2018;46(6):1294–305.

25. Mosenthal W, Kim M, Holzshu R, et al. Common ice hockey injuries and treatment: a current concepts review. Curr Sports Med Rep. 2017;16(5):357–62.

26. Tokutake G, Kuramochi R, Murata Y, et al. The risk factors of hamstring strain injury induced by high-speed running. J Sports Sci Med. 2018;17(4):650.

27. Malliaropoulos N, Isinkaye T, Tsitas K, et al. Reinjury after acute posterior thigh muscle injuries in elite track and field athletes. Am J Sports Med. 2011;39(2):304–10.

28. Foschia C, Tassery F, Cavelier V, et al. Les blessures liées à la pratique du basketball: revue systématique des études épidémiologiques. Journal de Traumatologie du Sport. 2019;36(4):242–60.

Running in Tennis

18

Claudio Zimaglia, Philippe Afriat, Dalibor Sirola, and Pino Carnovale

18.1 Role and Characteristics of Running in Tennis

During a tennis match, the player moves quickly on the court with different multidirectional combinations in order to hit back in time the ball thrown by the opponent. The rhythmic way to perform the individual steps must combine a quick move with the need to hit the ball with power and precision, through an optimal acceleration of the racket.

Although the running of a tennis player is different from that of a sprinter in athletics, the ability to run at high speed with decontraction and with a rapid and smooth transition from amplitude to frequency in order to hit the ball is nevertheless achieved through perfect contact with the ground and with a powerful propulsive impulse; a

correct running technique can certainly help the tennis player to better move on the court.

Tennis requires a sophisticated interaction of physical components. In order to be competitive and successful, tennis players need a mixture of speed, agility and power combined with medium to high aerobic capacities. Consequently, they require training increasingly focused on power, explosive strength and plyometrics [1, 2].

Players also need to show high reactive, anticipatory and decision-making skills, to possess the mental rigour to cope with the resulting fatigue and pressures of decisive points and significant extrinsic rewards (e.g. ranking, money sponsorships) [3] as well as being able to perform dynamic and continuous on-court movements under constant pressure [4, 5].

Furthermore, tennis is a sport whose matches do not have a predetermined time limit; it is therefore difficult to quantify the length of time of a match [6]. In addition to the high aerobic and anaerobic requirements, tennis is a sport that submits the player to constant changes of direction and speed during the course of a match, with multidirectional stopping and restarting movements that can result in on-court injuries or chronic injuries and disorders: depending on the studies, the probability of injury is between 0.04 and 3 per 1000 h of play [7].

It should also be taken into account that tennis is played on various playing surfaces, ranging from concrete or acrylic courts (hard courts), red

C. Zimaglia (✉)
Piatti Tennis Center, Bordighera, IM, Italy

P. Afriat
International Medical Center of Monaco, Montecarlo, Monaco
e-mail: afriat-cmim@monaco.mc

D. Sirola
Strength and Conditioning Coach, VEMA SPORT
d.o.o, Rijeka, Primorsko Goranska Zupanija, Croatia
e-mail: info@sirolatrainingsystem.com

P. Carnovale
Italian National Tennis Team, Torino, Italy

© The Author(s), under exclusive license to Springer-Verlag GmbH, DE, part of Springer Nature 2022
G. L. Canata et al. (eds.), *The Running Athlete*, https://doi.org/10.1007/978-3-662-65064-6_18

Fig. 18.1 Movements on red clay courts

clay courts, natural grass and artificial grass, which expose the athlete to different strains. Changing playing surfaces can be a risk factor for tennis injuries, especially for high-level professional players. Studies in this regard have shown a higher rate of overuse injuries in players who played on different surfaces compared to players who played primarily on a single surface [8]. The different friction coefficients and the different shock absorption coefficients between the various playing surfaces can be predisposing factors to joint and muscle disorders. The different speed of the ball, the different ability to slide on the ground and the different accelerations and decelerations on the field can predispose the athlete to abnormal musculoskeletal strains [9]. Research has shown that the speed of the ball bouncing on concrete courts is higher than on clay courts or other surfaces and this results in an increase in the reaction forces of the athlete's lower limbs and shoulder girdle [10].

Tennis is also a discontinuous sport, with a mixed metabolic pattern where the actions of the game alternate with moments of pauses established by the rules. Most of all, it is a 'situation' sport, where the outcome of the action depends on technical gesture, physical performance and the ability to find the right adaptation responses to different stimuli [6].

The average duration of the game actions ranges from 4 to 8 s with high variability depending on the surface. There are usually no consecutive actions lasting more than 20–30 s. Speed, explosiveness and quickness are the main characteristics to be considered. A study carried out on clay courts showed that tennis player's movements on the court are performed in limited spaces: in 80% of cases, the tennis player moves between 0 and 3 m; in 10%, between 0 and 6 m and in the remaining 10%, between 0 and 12 m [11].

During a tennis match on clay courts, about 80% of shots are played in less than 2.5 m, about 10% are executed with movements ranging between 2.5 and 4.5 m, and only 5% are executed with more than 4.5 m. The ability to move correctly with a high level of dynamism in all directions and, above all, backwards and forwards enables the player to better coordinate and to hit effectively. The athlete's movements on the court in a slow surface are distributed as follows: forward movements 20%, movements to the right 33%, movements to the left 35% and backward movements 12% (Fig. 18.1).

The breakdown of movements in different directions on fast surfaces is as follows: forward movements 53%, movements to the right 15%, movements to the left 19% and backward movements 13% (Fig. 18.2).

Research has shown that tennis players move laterally on the court for about 71.8% of their movements, forward for about 20% and less than 8% backward [12]. Each change of direction is performed on average every 4 m [13]. Tennis players execute about 80% of their shots within the so-called comfort zone, i.e. with movements up to a distance of 3 m [14].

The tennis player moves in an area that requires a continuous development in terms of design, movement schemes and speed changes depending on the specific game situations [15]. Professional tennis players make an average of four directional

Fig. 18.2 Movements on hard courts

changes per point; the number of dribbles varies from 1 to 15 in longer changes [16].

Furthermore, the most recent statistical studies have shown that athletes move around the court within 10″ 70% of the time with an average of 4/8 changes, between 10″ and 20″ 20% of the time with 9/16 changes and the remaining 10% of the time over 20″ with more than 16 changes. According to the latest ITF regulations, recovery breaks last 20 s between points, 90 s at changeovers and 120″ between sets [17].

18.2 Training Strategies: Technical Aspect

The tennis player must be able to produce energy in the shortest possible time; coordination and the ability to accelerate are the most influential aspects of the tennis player's game. The physical training of a tennis player is very complex; this is why adequate training for the athlete according to his characteristics and short- and long-term goals is essential [18].

18.2.1 Starting Position (Split Step)

During a match, the player must adopt a position that enables him to react quickly to his opponent's shots. The player takes this starting position when hitting from the baseline, returning a serve or following a drop at the net, and is commonly known as the **split step** (Fig. 18.3). This movement prepares the body to move towards the incoming ball and provides the necessary balance to react to any change in angle, pace and trajectory. It is a preliminary movement executed by a player just before the opponent's shot [15]. The split step enables a more dynamic neuromuscular reaction, a quicker start to the movement and to a change of direction [19]. It involves a wide jump with spread legs, which the player executes automatically.

In tennis, the split step is one of the most important footwork and movement preparation techniques. This movement is an integral part of preparation for hitting the ball on the volley, a return serve or a groundstroke. It prepares the

Fig. 18.3 Split step

quadriceps muscle, enabling the storage of energy and its following release to improve rapid movement in preparation for running, sprinting and moving around the court (Elliott 2006). Roetert et al. defined the split step as a tennis-specific movement pattern in which eccentric-concentric muscle action occurs [20].

If the split step is crucial to prepare for a change of position and to begin the movement, no less important is the 'first step,' which is useful to continue the movement by starting the push forward, increasing strength and power at the push-off and improving sprint performance over short distances [21].

Jacobs and Van Ingen Schenau pointed out that the aim of the motor system in the first phase of sprinting is to control the transformation of joint angular accelerations into an increase in starting speed in the horizontal direction. The ability to quickly and explosively execute the first step after the split step is critical to achieving a coordinated and effective movement.

Research by Lamond et al. has shown that players who must respond to an opponent's serve use a lateral step (i.e. jab step) with the intent of increasing their base of support and becoming more stable before performing the change of position or before hitting the ball [22].

The starting movement is often used by tennis players on the baseline when executing shots in the comfort zone via a lateral reaction phase [15].

This step enables tennis players to accelerate faster after the split step. This type of starting movement is most effective when tennis players move forward or backward by taking the first step in the opposite direction of the movement [21]. According to Parsons and Jones, a quick lateral movement is crucial when a tennis player has to react immediately and change direction. Komi pointed out that this type of movement is more effective than a pure concentric muscle action [23].

The split step, as an essential part of the preparation for the next shot, positively influences the efficiency of the tennis player's movement both from a neuromuscular and a proprioceptive point of views. It changes according to the different game situations, and it is closely related to the characteristics of the opponent's ball.

Fig. 18.4 Lateral shuffle to the forehand

Professional tennis players who prepare the split step with greater speed are able to execute a faster eccentric-concentric muscular movement. Tennis players use a lateral reaction phase for sprints over distances of 2.5–6 m. Two to three fast and explosive steps are performed after the split step for movements that force the athlete to sprint and accelerate to reach a short ball towards the net or a side ball [15].

18.2.2 Lateral Movement Towards the Ball Over Short Distances

In modern tennis footwork is becoming increasingly important. A movement included in the tennis player's running and court movement training is the *lateral shuffle or multidirectional shuffle*. The lateral step and small adjustment passes enable a tennis player to execute shots with good balance in both offensive and defensive situations [24].

The lateral shuffle is used when the athlete moves over lateral distances of about 2 m maximum; those movements enable the athlete to regain the best position to resist against the opponent's response (Figs. 18.4 and 18.5).

The lateral movement from the baseline is carried out by pivoting on the foot of the leg closest to the ball and subsequently through the powerful and complete extension of the same leg and is performed after the split step. This will enable the player to overcome the state of inertia and gain speed. The ensuing steps will be adjustments and will enable the player to decelerate and get into position to execute the shot.

18.2.3 Lateral Movement Towards Angled Balls

During a match less than 5% of shots are played at distances greater than 4.5 m, and at the given moments in the match, the player must be able to reach angled balls (Figs. 18.6 and 18.7). When the player returns from a previous move towards the centre and needs to run to the other side of the court, he draws the foot nearer the ball towards the centre of gravity at a closed angle on his knee. This will enable the trunk to be unbalanced out of the base of support, thus facilitating the speed of the movement in the desired direction. The supporting leg, on the other hand, pushes hard by extending the hip, knee and ankle joints. In the subsequent steps, the player will rapidly develop speed, bringing the feet in contact with the ground in a position 'below and slightly behind' the hips, until the moment when he strikes.

Fig. 18.5 Lateral shuffle to the backhand

Fig. 18.6 Backhand lateral movement towards angled balls

As tennis is a dynamic, fast and unpredictable game, the efficiency of tennis players' movement is affected by spatial and temporal factors. These are primarily determined by direction, depth, height, rotation and speed. Furthermore, movement, stroke movement and execution are influenced by a player's position in relation to an opponent's stroke, the distance and expected position of the next stroke, the desired direction of the executed stroke and, from a tactical point of view, the opponent's movement and positioning while executing the move [25].

Fig. 18.7 Forehand lateral movement towards angled balls

Some recent scientific research has tried to analyse the physical performance of athletes during a match [17] in order to obtain information useful in particular for physical trainers and thus to be able to develop focused training programmes. In modern tennis, the athlete has to perform rapid multidirectional movements with continuous acceleration and deceleration and continuous changes of direction. Then, he has to return to the starting position with short, high-intensity efforts followed by brief rest periods regulated by the international federation. This is why tennis is called an *intermittent* sport [26].

18.2.4 The Drop Shot and the Lob: Recovering a Ball by Sprinting Forward or Backward

The pure form of acceleration in tennis is characterised by an increase in the initial speed when the player must run forward to recover a drop shot or backward to recover a lob.

In the first case, the player, after the split step, will perform a forward propulsion by a marked extension of the hip, knee and ankle joints of the supporting limb, thus rapidly increasing his speed until he recovers the ball (Fig. 18.8).

In the second case, the player will run backwards to catch a lob, i.e. the shot used in tennis to bypass an attacker who is running towards the net to perform a volley. To chase the lob, the player must decelerate his run towards the net until he stops and immediately reverses direction (Fig. 18.9).

18.2.5 Inside-out Forehand: Movement Towards the Ball

The inside-out forehand is a shot played from the angle of the backhand (Fig. 18.10). The current trend in the game of top-level players is to increase the dynamics and pace of the game from the backhand. This requires remarkable coordination and speed both to move and hit the ball and to quickly regain the correct position on the court. After the split step, the player moves the foot nearer to the ball by crossing back and sideways to the opposite foot; the following short and crossed steps will enable the player to find the optimal distance to hit the ball.

Fig. 18.8 Recovering a ball by sprinting forward

Fig. 18.9 Recovering a ball by sprinting backward

18.2.6 Forward and Backward Movement Towards the Ball within 3 m

An important aspect of the game of tennis is the ability to move quickly forwards and backwards. Most of the movements from the baseline forwards, backwards or diagonally, over a distance of about 2 or 3 m, are performed with a movement technique characterised by a combination of sliding or crossing steps, before hitting the ball.

In particular, the 'double step' represents a forward movement that is in preparation for the shot and enables the player to keep the trunk upright, with the head perfectly aligned, so that a

Fig. 18.10 Inside-out forehand

linear movement is conveyed forward and a high speed of the racket head can be obtained.

18.2.7 Recovery of the Position

This is facilitated by a 'recovery' step that helps the athlete to decelerate and at the same time regain balance in order to resume rapidly the position. This movement pattern varies according to the surface of the court, the tactical situation and the type of ball played. The running technique to return and regain the ideal position will depend on the extent of the movement. If the player is chasing angled balls that take him far away from the court, the repositioning will take place by means of a crossover step in front of the body, while the subsequent repositioning steps consist of sliding or crossing steps depending on the size of the displacement. If the player, after the recovery step, does not need to recover the position quickly, he moves towards the ideal position through sliding steps (Fig. 18.11).

18.3 Prevention Strategies

Tennis is regarded as a high-intensity sport [26]. If, for instance, we compare football players with tennis players, we can find that the former reach their maximum speed between 7 and 10 m/s, while the latter spend 80% of the game running at a speed between 0 and 2 m/s, 16% of the time between 2 and 3.5 m/s and only 4% between 3.5 and 7 m/s [18].

The most desirable conditional skills are those related to maximum acceleration and the ability to sprint over short distances [11]. For a tennis player, speed represents both the speed of movement of the body, the speed of reaction between the recognition of a stimulus and the relevant motor response, as well as the segmentary or angular speed possible in moving parts of the body over itself.

In tennis, it is not easy to talk about movement speed because of the size of the court. As the spaces are very limited, and the player has to find a good balance in order to position himself

Fig. 18.11 Recovery of position towards the centre of the ground with a cross step

appropriately to hit the ball, it is more accurate to speak of acceleration and deceleration capability, rather than speed as generally defined. A 2016 paper on ITF future tournaments showed that players spend around 80% of the match running at speeds between 0 and 2 m/s, 16% between 2 and 3.5 m/s and only 4% between 3.5 and 7 m/s.

Assessment and training must be specific for tennis in order to achieve maximum efficiency. Traditional linear sprint tests (20–40 m) have shown that this is not an accurate indicator of the player's running speed because it differs too much from the actual playing conditions [27]. One of the requirements of the professional tennis player is to be able to repeat and perform continuous movements on the court through motor reactions linked to sensory stimuli (visual and acoustic) in order to position himself appropriately to hit the ball.

Measuring movement speed during a tennis match is different from measuring speed by means of general movement tests performed outside a tennis court. More precisely, in order to understand how to train a tennis player, speed has to be measured during the game or after the split step to detect whether there are differences in deceleration or acceleration in various types of strokes. To reach the highest efficiency, assessment and training must be specific for this sport. Furthermore, the speed and accuracy of the serve has also an impact on the style of movement, acceleration and execution of the return response. This indirectly affects the efficiency of the player's movements and his movements on the court.

We usually try to train acceleration by diversifying the technical training activities, in which we aim to focus the training on explosiveness and acceleration tests or continuous changes of direction over short distances (5–10 m). This type of training is called the *ability to endure repeated sprints*. The ability to develop movement patterns outside a tennis court is required, where the focus should be on the explosiveness and speed of the tennis player's movement on the court by extensively working on the starting phase of the move on the court, but particularly on the deceleration phase by training the capacity for eccentric muscular contraction of the lower limbs (especially the quadriceps) and to prepare the correct body balance before executing the shot in order to immediately adjust to unpredictable game situations. A tennis player's movements on the court take place in an area that requires continuous development in terms of planning movement patterns, in terms of speed control and in terms of the choices of playing situations that a tennis coach or physical trainer must know. The improvement of all these skills must always and inevitably be related to the performance model.

For adequate training of the tennis player, especially in relation to running, work cycles are developed so that they can be divided into micro and macro cycles, containing components that are to be trained in a specific period of the season, in relation to competitive programming. Observing a tennis player, it is noticeable how he tries to get to the ball as quickly as possible while keeping his centre of gravity low. This is necessary in order to increase the effect of the loading by performing a rapid series of steps to find the correct balance and equilibrium when executing the shot. In tennis, high levels of speed are achieved by moving in very limited spaces, so speed will always refer to very short movements in which maximum static/dynamic balance is required.

In conclusion, we must remember the relevance of working on the recovery and maintenance of accurate and functional joint mobility. This is essential for the correct execution of technical gestures, efficiency in movement and the possibility of using the joints to their maximum capacity for movement. Furthermore, good joint biomechanics helps prevent the onset of any muscular and joint damage.

References

1. Fernandez-Fernandez J, Saez de Villarreal E, Sanz-Rivas D, Moya M. The effects of 8-week plyometric training on physical performance in young tennis players. Pediatr Exerc Sci. 2016;28(1):77–86. https://pubmed.ncbi.nlm.nih.gov/26252503/. Accessed 5 Mar 2021
2. Loturco I, Pereira LA, Kobal R, Zanetti V, Kitamura K, Abad CC, Nakamura FY. Transference effect of vertical and horizontal plyometrics on sprint performance of high-level U-20 soccer players. J Sports Sci. 2015;33(20):2182–91. https://pubmed.ncbi.nlm.nih.gov/26390150/. Accessed 3 Mar 2021
3. Hornery DJ, Farrow D, Mujika I, Young W. Fatigue in tennis: mechanisms of fatigue and effect on performance. Sports Med. 2007;37(3):199–212. https://pubmed.ncbi.nlm.nih.gov/17326696/. Accessed 25 Feb 2021
4. Nieminen MJ, Piirainen JM, Salmi JA, Linnamo V. Effects of neuromuscular function and split step on reaction speed in simulated tennis response. Eur J Sport Sci. 2014;14(4):318–26. https://pubmed.ncbi.nlm.nih.gov/23600926/. Accessed 22 Feb 2021
5. O'Donoghue P, Ingram B. A notational analysis of elite tennis strategy J Sports Sci. 2001;19(2):107–13. https://pubmed.ncbi.nlm.nih.gov/11217009/. Accessed 5 Mar 2021
6. Kovacs MS. Applied physiology of tennis performance. Br J Sports Med. 2006;40(5):381–5. https://www.ncbi.nlm.nih.gov/pmc/articles/PMC2653871/. Accessed 3 Mar 2021
7. Pluim BM, Staal JB, Windler GE, Jayanthi N. Tennis injuries: occurrence, aetiology, and prevention. Br J Sports Med. 2006;40(5):415–23. https://www.ncbi.nlm.nih.gov/pmc/articles/PMC2577485/. Accessed 15 Feb 2021
8. Fu MC, Ellenbecker TS, Renstrom PA, Windler GS, Dines DM. Epidemiology of injuries in tennis players. Curr Rev Musculoskelet Med. 2018;11(1):1–5. https://www.ncbi.nlm.nih.gov/pmc/articles/PMC5825333/. Accessed 24 Feb 2021
9. Pluim BM, Clarsen B, Verhagen E. Injury rates in recreational tennis players do not differ between different playing surfaces. Br J Sports Med. 2018;52(9):611–5. https://pubmed.ncbi.nlm.nih.gov/28209569/. Accessed 11 Mar 2021

10. Dines JS, Bedi A, Williams PN, Dodson CC, Ellenbecker TS, Altchek DW, Windler G, Dines DM. Tennis injuries: epidemiology, pathophysiology, and treatment. J Am Acad Orthop Surg. 2015;23(3):181–9. https://pubmed.ncbi.nlm.nih.gov/25667400/. Accessed 23 Feb 2021

11. Adriano Pereira L, Freitas V, Arruda Moura F, Saldanha Aoki M, Loturco I, Yuzo NF. The activity profile of Young tennis athletes playing on clay and hard courts: preliminary data. J Hum Kinet. 2016;50:211–8. https://www.ncbi.nlm.nih.gov/pmc/articles/PMC5260656/. Accessed 7 Mar 2021

12. Weber K, Pieper S, Exler T. Characteristics and significance of running speed at the Australian open 2006 for training and injury prevention. Med Sci Tennis. 2007;12(1):14–7. https://fis.dshs-koeln.de/portal/de/publications/characteristics-and-significance-of-running-speed-at-the-australian-open-2006-for-training-and-injury-prevention(2a547d51-156d-4-316-9723-27ba35b450bd)/export.html. Accessed 24 Feb 2021.

13. Pieper S, Exler T, Weber K. Running speed loads on clay and hard courts in world class tennis. J Med Sci Tennis. 2007;12(2):14–7. https://fis.dshs-koeln.de/portal/en/publications/running-speed-loads-on-clay--and-hard-courts-in-world-class-tennis(5a7b0c8a-a5b8-4181-99b8-94235687b0da).html. Accessed 7 Feb 2021

14. Over S, O'Donoghue PG. What's the point? Tennis analysis and why. CSSR 2008. 2016;15:19–21. http://en.coaching.itftennis.com/media/113918/113918.pdf

15. Filipčič A, Leskošek B, Munivrana G, Ochiana G, Filipčič T. Differences in movement speed before and after a split-step between professional and junior tennis players. J Hum Kinet. 2017;55:117–25. https://www.ncbi.nlm.nih.gov/pmc/articles/PMC5304280/. Accessed 15 Feb 2021

16. Kovacs MS. CSCS movement for tennis: the importance of lateral training. Strength Condit J. 2009;31(4):77–85. https://journals.lww.com/nsca-scj/Fulltext/2009/08000/Movement_for_Tennis__The_Importance_of_Lateral.9.aspx. Accessed 24 Feb 2021.

17. Kovacs MS. Tennis physiology: training the competitive athlete. Sports Med. 2007;37(3):189–98. https://pubmed.ncbi.nlm.nih.gov/17326695/. Accessed 3 Mar 2021.

18. Kutzdub M. The truth about speed and acceleration training for tennis. www.mattspoint.com/blog. 2018. https://www.mattspoint.com/blog/speed-acceleration-training-tennis. Accessed 16 Feb 2021.

19. Salonikidis K, Zafeiridis A. The effects of plyometric, tennis-drills, and combined training on reaction, lateral and linear speed, power, and strength in novice tennis players. J Strength Cond Res. 2008;22(1):182–91. https://pubmed.ncbi.nlm.nih.gov/18296973/. Accessed 22 Feb 2021

20. Roetert P, Ellenbecker T, Chu D. ITF strength and conditioning for tennis. Movement mechanics. 2003.

21. Frost DM, Cronin JB. Stepping back to improve sprint performance: a kinetic analysis of the first step forwards. J Strength Cond Res. 2011;25(10):2721–8. https://pubmed.ncbi.nlm.nih.gov/21912339/. Accessed 23 Feb 2021

22. Lamond F, Lowdon B, Davis K. Determination of the quickest footwork for teaching return of the tennis serve. In: Healthy Lifestyles Journal. 1996;43(4):5–8. https://search.informit.org/doi/abs/10.3316/aeipt.73800. Accessed 23 Feb 2021

23. Komi PV. Stretch-shortening cycle: a powerful model to study normal and fatigued muscle. J Biomech. 2000;33(10):1197–206. https://pubmed.ncbi.nlm.nih.gov/10899328/. Accessed 15 Feb 2021

24. Reid M, et al. ITF strength and conditioning for tennis. London: International Tennis Federation Ltd.; 2003.

25. Elliott B. Biomechanics and tennis. Br J Sports Med. 2006;40(5):392–6. https://www.ncbi.nlm.nih.gov/pmc/articles/PMC2577481/. Accessed 3 Mar 2021

26. Hoppe MW, Baumgart C, Bornefeld J, Sperlich B, Freiwald J, Holmberg HC. Running activity profile of adolescent tennis players during match play. Pediatr Exerc Sci. 2014;26(3):281–90. https://pubmed.ncbi.nlm.nih.gov/25111161/. Accessed 18 Feb 2021

27. Leone M, Alain S, Comtois FT, Léger L. Specificity of running speed and agility in competitive junior tennis players. Med Sci Tennis. 2006;11:10–1.

Running in Paddle

19

Manuel Duarte Marques Virgolino

19.1 Characteristics of Running in Paddle

The paddle is played in a smaller court, which means a faster game with short displacement [1]. Changes in direction are frequent while running, pivoting, and cutting maneuvers. These requires a mixture of speed, agility, and quick bursts of energy [2–4], which makes athletes more susceptible to injury. Many players are middle-aged "weekend warriors" who don't strengthen or stretch their muscles and ligaments in between games or practices with injury consequences.

As features, paddle uses motor anticipation, force explosive and reactive, speed (acceleration reaction and braking deceleration), and resistance [3, 4].

One of the most important factors that the players possess is the motor anticipation. Without anticipation it is impossible to play. The reading of the movement, of the stroke, of the action, and of the opponent is the richest and purest and most important skill information that the player obtains and, the one that, in addition, allows him to make the motor adjustment prior to the arrival of the ball [5]. Different types of force are present in the game. Explosive force, as the ability to generate high levels of force in a short period of time, and reactive force, as the ability of the muscle to produce a cycle of stretching – shortening, producing high amount of strength [3, 4]. These manifestations of the force are used in each one of the blows, in each jump, in each of the sprints, and in the different actions that allow to reach a faster ball and better positioned to technically perform the indicated stroke.

Another quality is speed, defined as the ability to achieve, based on cognitive processes, the maximum volitional force and functionality of the neuromuscular system, a maximum speed of reaction and movement under certain established conditions. The specificity of the sport is reaction speed (i.e., the ability to react in the shortest time to a stimulus) and acceleration and deceleration speed.

Resistance, defined as the muscular ability to maintain a task for a prolonged period of time, is also important [3–5]. Flexibility and agility are two more essential qualities to play paddle.

Racquet sports, such as paddle, can be considered a well-proven static and dynamic form of exercise with many health benefits including improved aerobic fitness, reduced body fat percentage, the development of a more favorable

M. D. M. Virgolino (✉)
Clínica Ortopedica do Montijo, Hospital das Forças Armadas - PL, Hospital da Luz Setúbal, Lisbon, Portugal

lipid profile, reduced risk of cardiovascular diseases, and improved bone health [6].

19.2   Training Strategies

It may frequently happens that an athlete does not see any change or improvement of the physical training.

Both cardiovascular aspects (e.g., intervals or work/rest ratio fitness) and musculoskeletal functions (e.g., strength, mobility, stability) should be taken into account during workouts. Cardiovascular fitness improves faster for a given activity before your musculoskeletal system (tendon, bone, muscle, etc.) has adapted for that activity, improving the upper limb, trunk, and core strength and stability in order to reduce the likelihood of upper limb injury [7, 8].

Plyometrics exercises are those in which the muscle is loaded with an eccentric contraction (stretching) immediately followed by a concentric contraction (shortening). In paddle these exercises are designed to improve the ability of harmonizing and coordinating the speed and force, i.e., "the power player." Training of coordination, speed and strength, improves the quick changes of direction in the field, increases the acceleration of the movements, makes the movements faster and more explosive, being able to hit the ball higher [3]. The plyometric training can increase the explosive reaction of the muscle contractions as a result of rapid eccentric contractions.

Exercises to train reflexes and improve peripheral vision are also essential. It is important to gradually increase training volume (sessions/week) over 3–6 months to allow musculoskeletal adaptation. This will permit a gradual exposure to the stability and strength required for the game.

Paddle can be important to health, but the body must be prepared in physical, psychological, and nutritional ways [6]. Along with training, and proper rest, fluid intake and energetic support are fundamental to prevent injuries and keep athletes healthy.

19.3   Most Frequent Injuries

Paddle is a demanding sport for the joints, tendon, and muscles. The small court size in paddle, as well as the quicker pace of play than tennis, may also lead to faster and more intense gameplay, which may contribute to injury risk in paddle players. Such a gameplay may also explain improper technique and fatigue as common mechanisms of injury [5–8]. Fatigue may affect the kinematics and kinetics of running. Also, given that paddle gameplay includes a high frequency of jumps, fatigue may exacerbate the ability to control landings. The high intensity of running and jumping may consequently be associated with noncontact injury. In addition, excessive training or exercise load may also be associated with injury risk [3–9]. Focusing on proper technique, rest, and recovery is essential.

Injuries can be classified as below:

- Accidental injuries: the most common and frequent (walking on a ball, being shot in the body, hitting yourself against the glass or the grill).
- Overload injury: focused on muscles, joints, or ligaments, this injuries appears during intense matches or during an accumulation of matches over a short time.
- Injury due to poor technique: performing repetitive and incorrect movements can lead to injuries.
- Injury due to poor heating: this is very frequent. A good warm-up is important in order to avoid muscle lesions or serious injuries from your first acceleration.
- Climate injury: cold, humidity, and wet field surface or the field's bad condition could contribute to injuries in relationship with fatigue or increased ball weight.

According to some studies and researches [10–13], overuse/overload injuries are often the easiest injuries to prevent.

Fig. 19.1 Backhand with the elbow and forearm pronate

Fig. 19.2 Lower back hyperextension doing a smashes

From a biomechanical point of view, the elbow acts as a link in the upper limb kinetic chain transfer with racket. The most important lesion found in this region is lateral epicondylitis. This injury is a tendinopathy of the long radial extensor, the radial extensor of the little finger extensor, and/or common extensor digitorum by strong surges in the racquet sports. In paddle, the reverse is usually done due to the size of the racket so that the control and direction of the gesture depends on the dominant side of the player (Fig. 19.1). This, together with the lack of technique, can produce increased elbow injuries. Studies [9, 14, 15] conclude that lesions in the elbow are more related to improper technique. A change of racket, altering stroke technique, and exercising to build up muscle strength have been reported successful in alleviating the symptoms of paddle elbow and preventing recurrence [9, 14, 15].

Lower back pain [10–13] is probably the most common condition for a player in any sports. The lower back is constantly moving in all directions. This causes side bending, ligament extension, and rotation of the ball and socket joints (Fig. 19.2). Although pathology in the structure of the back is known in racket players, the most common cause of back pain is related to lumbar strain rather than to direct spinal pathology. Another common cause is muscular asymmetry. When a player has their right or left side stronger than the other, they might experience pain in the lower back area. That mostly happens if the area is underdeveloped [16, 17].

The knee can also be very exposed mostly in the case of abnormal bodyweight distribution and poor technique. The knee is stressed in flexion and valgus forces when the player hits the ball and in displacements, which can then lead to injuries (Fig. 19.3). The lateral collateral ligament and medial meniscus pathology were more frequent in racket players as compared to other sports [15, 18]. In addition, anterior cruciate ligament injury, patellofemoral pain (particularly in females), and patellar tendonitis ("jumper's knee") have been reported to be common in racket players [15, 18]. Beginners should stick to softer movements and practice rather than immediately jump to strength training.

Fig. 19.3 Forehand with knee valgus and flexion stress

Fig. 19.4 Recreational player

19.4 Prevention Strategies

Sport injuries are a common cause of disability for both professional and amateur players, leading to absence from competition or work. However, risk factors may differ depending of each population [19, 20]. For this reason, although previous studies [19] have examined other racquet sports, additional examinations on paddle, particularly the recreational players which comprise the larger paddle community, are needed (Fig. 19.4).

As previous experience in other racquet sports, examining injury and associated risk factors may be essential in the development of preventive measures in paddle. Such knowledge may assist in determining practical implications for equipment improvement and injury prevention [11, 13, 21]. After analyzing the main risk factors in this sport, it is necessary to create preventive programs including specific exercises [7, 10, 14, 20]. Physiotherapy can be useful not only as rehabilitation after injury but also to prevent from injuries, applying preventive programs. As it has been observed, programs for the scapulo-humeral, knee joints, and lumbar and/or pelvic area could be effective to avoid the risk of injuring [7]. Physiotherapy begins long before a lesion appears, and the ultimate goal is to prevent the damage. Risks of junior and adult recreational and elite paddle differs on volumes of exposure as well as the types of injury [10–13]. Changes in training and style of the modern game led to various overused injuries in the upper extremity and trunk. Return-to-play recommendations may vary depending on the age and the ability level of the player. Upper extremity injuries typically require a structured three-phase rehabilitation program, including core stabilization, kinetic chain integration, and functional strengthening (eccentric and isometric) [22, 23]. Lumbar spine injuries should involve rehabilitation in the pain-free direction and instruction on limiting extension in younger players (lumbar flexibilization, motor control of the multifidus and posterior fibers of psoas activation). Lower extremity injuries may be related generally to higher trauma and players at a competitive level (eccentric strengthening) [11, 13, 21–23]. Most conditions allow a successful return-to-play in paddle without surgical intervention. To reduce the incidence of acute change-direction knee injuries, incorporate traditional lower limb strengthening, particularly hip strength (e.g., squats, deadlifts) with single-leg plyometric exercises for dynamic control.

Paddle coaches must be prepared to enhance the quality and accuracy of training programs based on specific match activity and technical-tactical demands, according to stroke distribution and players' level. Finally, backhand strokes could be less invasive regarding injuries. In this sense, training exercises should propose a continuous diversification of movements, stokes (groundstrokes and volleys or smashes), and drills during training to prevent overuse musculoskeletal injuries [3–5, 8, 23, 24].

Further prevention include education of players, parents, and coaches about paddle injuries; interval musculoskeletal screening of players to identify problem areas before injuries occur; and adjustment of equipment including shoes, racquets, and balls as well as court surfaces [3, 21, 23, 25].

References

1. Sánchez-Alcaraz Martínez BJ. Historia del pádel. Materiales para la historia del deporte. 2013;11:57–60.
2. Courel-Ibáñez J, Alcaraz-Martínez BJS. The role of hand dominance in padel: performance profiles of professional players. Motricidade. 2018;14:33–41.
3. Courel-Ibáñez J, Sánchez-Alcaraz BJ, Muñoz Marín D. Exploring game dynamics in padel: implications for assessment and training. J Strength Cond Res. 2019;33:1971–7.
4. Courel-Ibáñez J, Sánchez-Alcaraz BJ, Cañas J. Game performance and length of rally in professional padel player. J Hum Kinet. 2017;55:161–9.
5. Shim J, Carlton LG, Kwon Y-H. Perception of kinematic characteristics of tennis strokes for anticipating stroke type and direction. Res Q Exerc Sport. 2006;77:326–33.
6. Courel-Ibáñez J, Cordero JC, Muñoz D, Sánchez-Alcaraz BJ, Grijota FJ, Robles MC. Fitness benefits of padel practice in middle-aged adult women. Sci Sport. 2018;33:291–8.
7. Chung KC, Lark ME. Upper extremity injuries in tennis players: diagnosis, treatment, and management. Hand Clin. 2017;33:175–86.
8. Sánchez-Alcaraz BJ, Courel-Ibáñez J, Cañas J. Groundstroke accuracy assessment in padel players according to their level of play. RICYDE Rev Int Ciencias del Deport. 2016;12:324–33.
9. Reid M, Elliott B, Crespo M. Mechanics and learning practices Associated with the Tennis forehand: a review. J Sport Sci Med. 2013;12:225–31.
10. Casuso C-LR, Holgado MJ. A comparison musculoskeletal injuries among junior and senior paddle-tennis players. Sci Sports Elsevier Masson SaS. 2015;30:268–74.
11. García-Fernández P, Guodemar-Pérez J, Ruiz-López M, Rodríguez-López ES, García- Heras A, Hervás-Pérez JP. Epidemiología lesional en jugadores españoles de pádel profesionales y amateur. Rev Int Med Cienc Act Fis Deporte. 2019;
12. Sánchez Alcaraz-Martínez BJ, Courel-Ibáñez J, Díaz García J, Muñoz Marín D. Descriptive study about injuries in padel: relationship with gender, age, players' level and injuries location. Rev Andaluza Med del Deport. 2019;12:29–34.
13. Castillo-Lozano R, Casuso-Holgado MJ. Incidence of musculoskeletal sport injuries in a sample of male and female recreational paddle-tennis players. J Sports Med Phys Fitness. 2017;57:816–21.
14. Gruchow H, Pelletier D. An epidemiologic study of tennis elbow. Incidence, recurrence and effectiveness of prevention strategies. Am J Sport Sci Med. 1979;7:234–8.
15. Perkins R, Davis D. Musculoskeletal injuries in tennis. Phys Med Rehabil Clin Am. 2006;17:609–31.
16. Courel-Ibáñez J, Herrera-Gálvez JJ. Fitness testing in padel: performance differences according to players' competitive level. Sci Sport. 2020;35:e11–9.
17. Sanchis-Moysi J, Idoate F, Izquierdo M, Calbet JA, Dorado C. The hypertrophy of the lateral abdominal wall and quadratus lumborum is sport-specific: an MRI segmental study in professional tennis and soccer players. Sport Biomech. 2013;12:54–67.
18. Majewski M, Susanne H, Klaus S. epidemiology of athletic knee in- juries: a 10-year study. Knee. 2006;13:184–8.
19. Courel-Ibáñez J, Sánchez-Alcaraz BJ, García S, Echegaray M. Evolution of padel in spain according to practitioners' gender and age. Cult Cienc y Deport. 2017;12:39–46.
20. Sánchez-Muñoz C, Muros JJ, Cañas J, Courel-Ibáñez J, Sánchez-Alcaraz BJBJ, Zabala M. Anthropometric and physical fitness profiles of world-class male padel players. Int J Environ Res Public Health. 2020;17:508.
21. Leone J. Title of subordinate document. Choosing suitable clothing is the prerogative to play paddle. https://www.italianpadel.com/clothing-and-equipment-for-playing-padel/. Accessed 28 Jun 2021.
22. Elliott B, Flesig G, Nicholls R, Escamilla R. Technique effects on upper limb loading in the tennis serve. J Sci Med Sport. 2003;6:76–87.
23. Elliott B. Biomechanics and tennis. Br J Sport Med. 2006;40:392–6.
24. Sánchez-Alcaraz Martínez BJ, Courel-Ibáñez J, Cañas J. Temporal structure, court movements and game actions in padel: a systematic review. Retos. 2018;33:308–12.
25. Priego JI, Melis JO, Llana-Belloch S, Pérezsoriano P, García JCG, Almenara MS, Priego Quesada JI, Olaso Melis J, Llana Belloch S, Pérez Soriano P, et al. Padel: a quantitative study of the shots and movements in the high-performance. J Hum Sport Exerc. 2013;8:925–31.

Running in Volleyball

20

Alberto Vascellari, Antonio Poser, Alex Rossi,
Terri Rosini, Rossano Bertocco,
and Giovanni Miale

20.1 Role and Characteristics of Running in Volleyball

Volleyball is a sport characterized by a small playing field: each of the two teams has available an area of 9×9 m. In addition to this space, there are other less used areas divided into two lateral areas of 9×5 m and one at the back of the court of 19×6.5 m.

During the volleyball game, there are very frequent breaks, and the action duration in high-level volleyball championships is on average less than 10 s for women and less than 7 s for men, with some rare actions lasting 45 s. Moreover, with the introduction of the video check, a system that allows referees to verify the correct interpretation of fouls, the pauses have become even longer, allowing long-term recoveries for players. At lower-level championships, longer actions can be observed, with average durations slightly higher.

Given these characteristics of the volleyball game, it is intuitive that straight running activity plays a marginal role. To evaluate the role of running in volleyball, ten matches of VakıfBank Spor Kulübü, a Turkish female volleyball team that has been a European champion for several times, were examined. VakıfBank Spor Kulübü is one of the few volleyball teams that utilizes a local positioning system (LPS) to track athletes' movements in their sports hall during competitions and training. This ultra-wideband technology (UWB)-based system consists in high-frequency cameras crossing information with miniaturized devices worn by the athletes (Kinexon Perform, Kinexon®) (Fig. 20.1). The Kinexon system has been developed for daily monitoring of players' physical load to optimize the periodization of training, prevent injuries, and organize the players' return to play [1].

In a randomized way, ten other selected matches from all the national and European championship matches played in the 2020/2021 season were examined to evaluate meters covered by the players grouped by position as well as the maximum speed reached and the average speed, expressed in km/h. The measurements begin with the first point until the end of the match; therefore, the warm-up phases are not taken into account.

The amount of space covered by volleyball players is the result of the following:

- Run up for the attack.
- Defense sprint.
- Lateral translocations for wall actions.

A. Vascellari (✉) · A. Poser · A. Rossi · T. Rosini
Kinè Physiotherapic and Orthopedic Center,
San Vendemiano, TV, Italy
e-mail: info@albertovascellari.it

R. Bertocco
Volley Team Club, San Donà di Piave, VE, Italy
e-mail: rossano@takeoff.pro

G. Miale
Vakifbank Sport Kulübü, Istanbul, Turkey

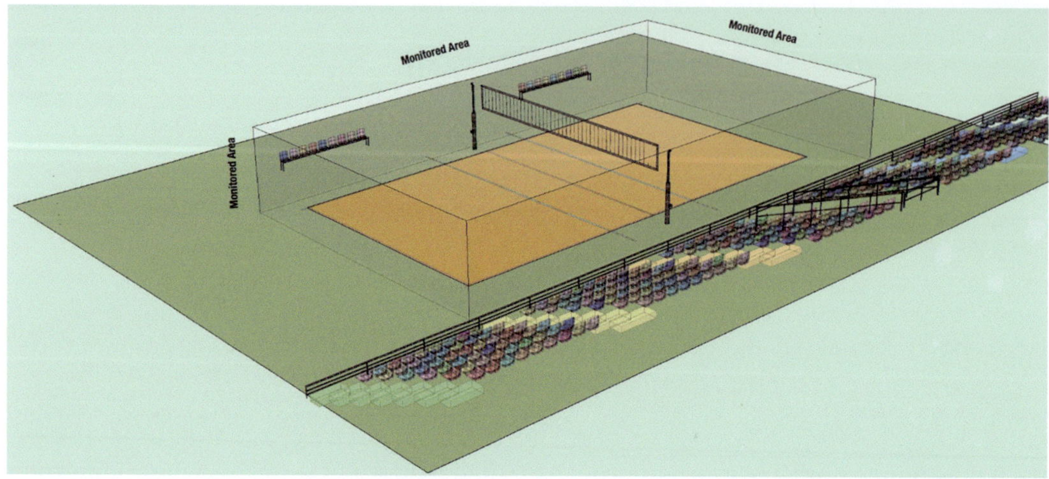

Fig. 20.1 Playing field area covered by high-frequency cameras crossing information with miniaturized devices worn by the athletes

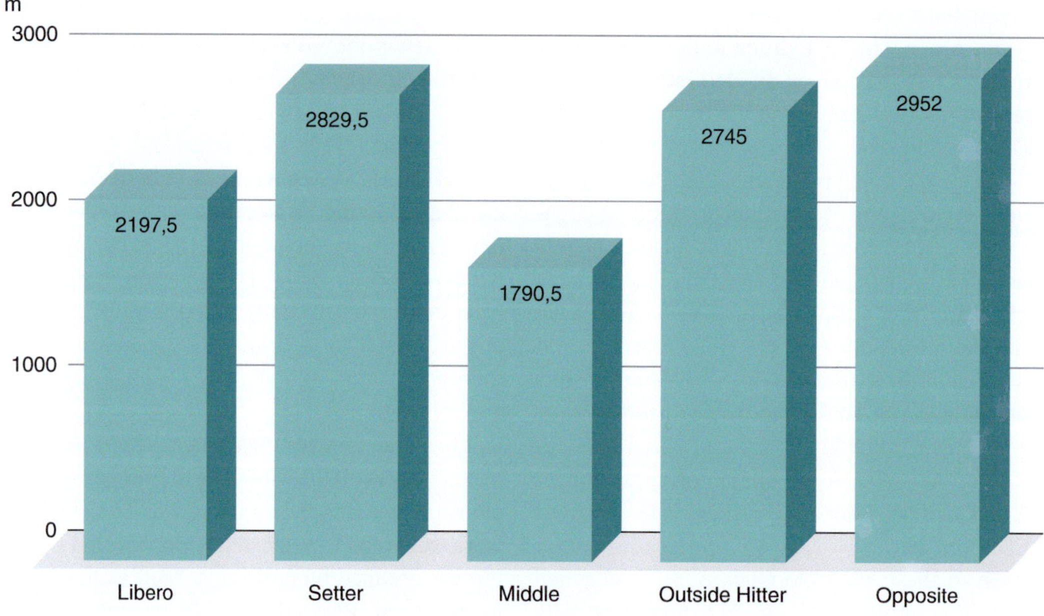

Fig. 20.2 Meters covered on average by players based on role (ten games analyzed)

- Preventive shifts, mostly performed by the setters.
- Other highly specific volleyball situations.

It has therefore been noticed that classic, frontal running rarely occurs.

The results of this study revealed that, depending on the role, very few kilometers are covered during a match: from a minimum of 1.2 km to a maximum of 4.6 km (Fig. 20.2) with average speeds ranging between 1.3 km/h and 3.7 km/h depending on the length of the game (Fig. 20.3). The maximum speed peaks are however high and describe the main explosive characteristic of this sport (Fig. 20.4). We recorded peaks of speeds close to 22 km/h.

The number of run up and accelerations were also counted for each role, founding an average

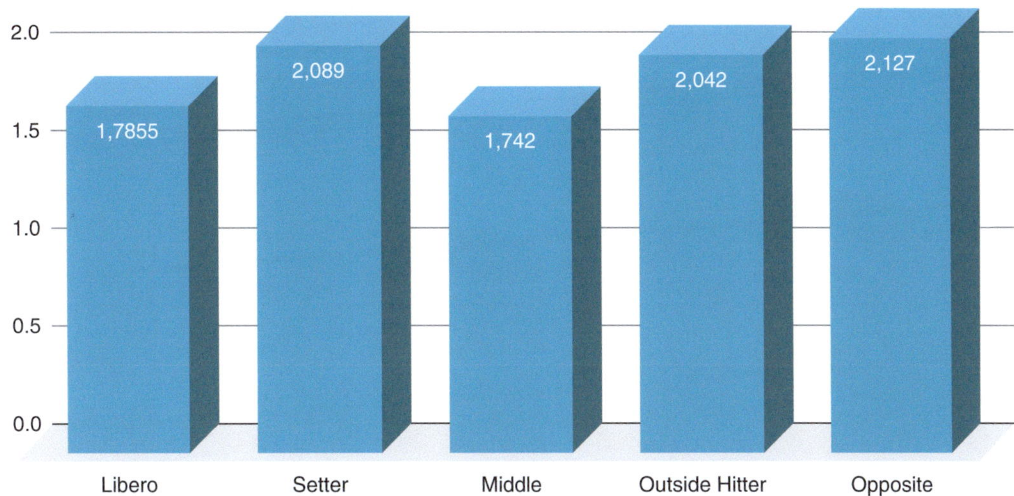

Fig. 20.3 Maximum speed reached by a player by type of role (ten games analyzed)

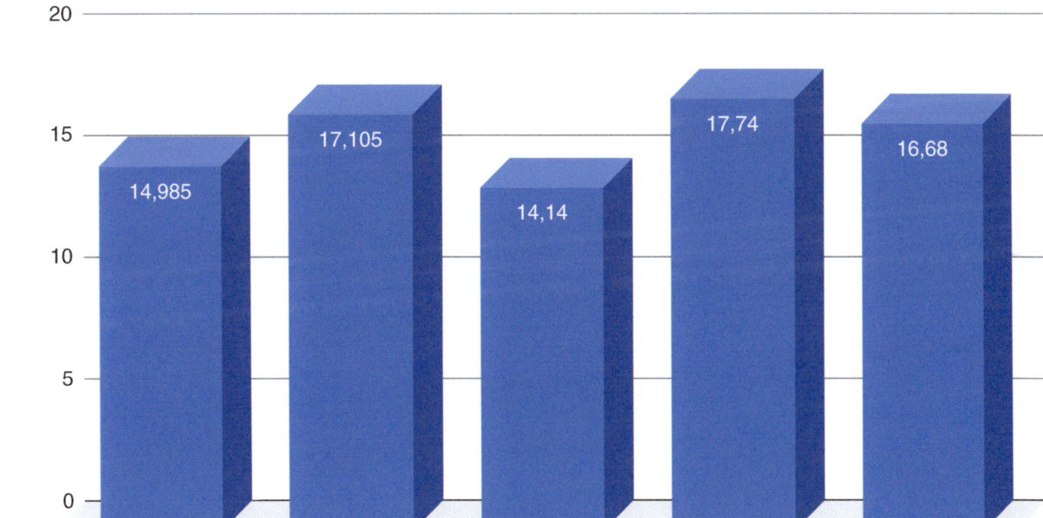

Fig. 20.4 Maximum speed reached by a player by type of role (ten games analyzed)

value of 50 for middles and 123 for outside hitters (Fig. 20.5).

Tables 20.1, 20.2, and 20.3 report average, minimum, and maximum values separated by different roles (Tables 20.1, 20.2, and 20.3).

Figures 20.6, 20.7, and 20.8 report the comparative graphs for each measure taken into account for each match based on the result and therefore on the overall duration. The matches took place with very different durations from a minimum of about 59 min to a maximum of 113 min.

In a similar study, Sánchez-Moreno et al. [2] analyzed the relationship between rally length

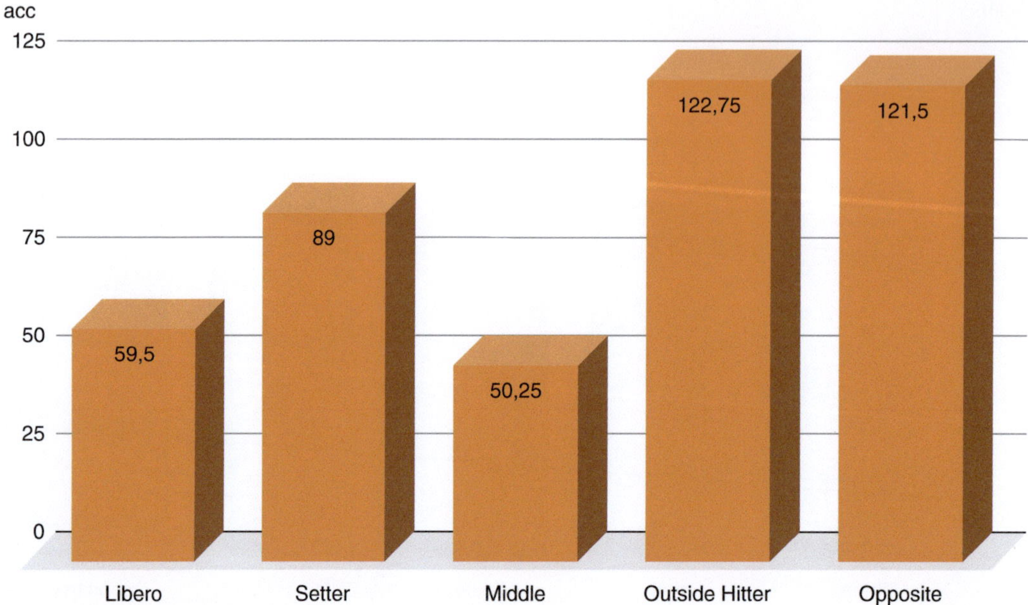

Fig. 20.5 Number of average "starts" accelerations performed with a standing start divided by role (ten games analyzed)

Table 20.1 Meters covered

	Libero	Setter	Middle	Outside hitter	Opposite
Average	2404	2903	2903	1766	3073
Minimum	1787	2008	2008	1217	2074
Maximum	3879	4450	4450	2863	4679

Table 20.2 Maximum speed

	Libero	Setter	Middle	Outside hitter	Opposite
Average	15.0	16.9	14.6	17.8	16.6
Minimum	13.2	14.7	12.5	14.4	14.7
Maximum	16.9	18.4	18.3	21.9	18.1

Table 20.3 Average speed

	Libero	Setter	Middle	Outside hitter	Opposite
Average	1.76	2.05	1.74	2.05	2.13
Minimum	1.38	1.75	1.38	1.93	1.97
Maximum	2.08	2.36	2.06	2.20	2.37

and rest times between rallies in high-level men's volleyball. Ninety-five matches from five of the premier worldwide competitions were analyzed (16,436 rallies). The average of rallies per set was 45.23 ± 6.15 (rally length: 4.99 ± 4.35 s; rest time: 29.02 ± 19.44 s; work-to-rest ratio 1:5.81 ± 0.17 s). The ratio of activity time to rest time was approximately 1/6; therefore volleyball has to be considered a high-intensity anaerobic sport.

Fig. 20.6 Average speed divided by role and match (ten games analyzed)

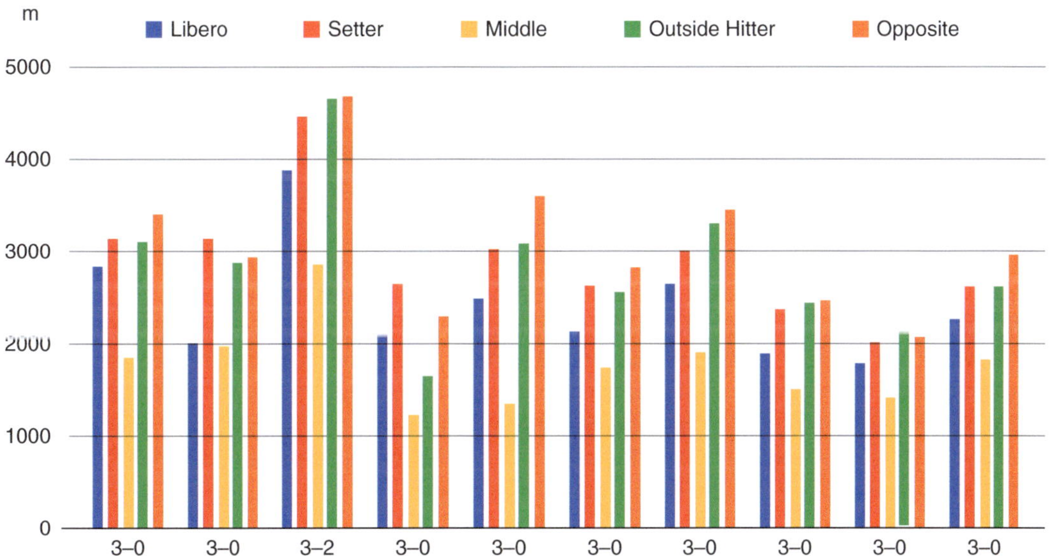

Fig. 20.7 Meters covered divided by role and match (ten games analyzed)

20.2 Specificities of Running in Beach Volleyball

Although the essential skills of indoor volleyball and beach volleyball are identical, important differences between the two disciplines do exist, including certain rules, the court dimensions, the composition of the playing surface, the environmental conditions in which the players must compete, and subtle differences in the size and weight of the indoor and outdoor balls.

Beach volleyball is characterized by a smaller playing field when compared to indoor volleyball, since each of the two teams plays on an area

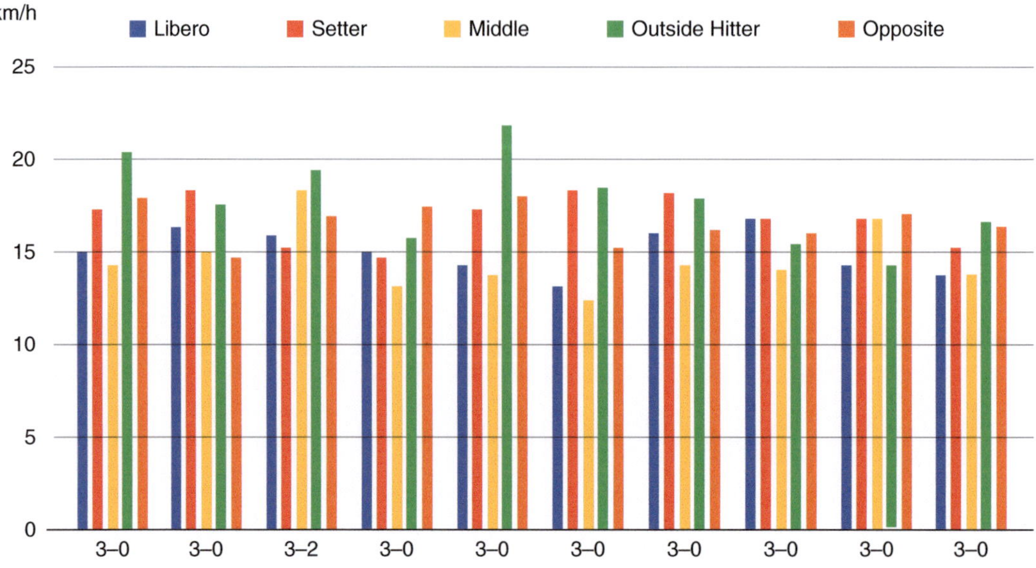

Fig. 20.8 Maximum speed divided by role and match (ten games analyzed)

of 8x8 meters. However, indoor volleyball has six players per team on the court, while beach volleyball teams consist of two players. This means that beach volleyball players are required to cover a larger area of the field.

As in indoor volleyball, beach volleyball is characterized by its intermittent nature, with brief periods of maximal or near maximal activity to longer periods of moderate- and low-intensity activity.

However the physiological demands of the two disciplines differ to some extent: the average indoor match during the 2005 FIVB World Grand Champions Cup lasted about 95 min, with nearly 165 rallies over the course of the contest, whereas the average match on the FIVB Beach Volleyball World Tour lasts about 50 min with 90 rallies contested [3].

During beach volleyball World Tour games, set duration is on average about 21–23 min, number of rallies per set is about 39–40, and the total rest time and rally duration are 17 min and 8.5 s, respectively [4, 5].

Physical actions and work-rest ratio in elite beach volleyball players have been recorded by Natali et al. during 12 volleyball matches of the beach volleyball 2016 World Tour [6]. The average rally length was 6.9 ± 4.0 s for men and 7.1 ± 3.9 for women, with rest time between rallies of 22.6 ± 16.0 s for men and 23.0 ± 17.9 for women. The ratio of rest time to activity time was 1:4.3 for men and 1:3.8 for women, indicating beach volleyball as a higher-intensity sport when compared to indoor volleyball [2].

20.3 Training Strategies

The role of running in the training protocols of volleyball players have been questioned since the specific tasks of the volleyball player do not include running, with the exception of short running sprint.

Volleyball is an activity that requires the development of strength, power, and speed [7]. Given these characteristics, most coaches and athletes agree that adopting an athletic conditioning program that implements these characteristics is effective and fundamental for both performance improvement and injury prevention. Coaches and athletes believe that performance and prevention programs should also be based on plyometric training, preferably managed by specialized personnel. In a recent research [8] on practices and perspectives of volleyball coaches and players, none of the respondents recommended running-based jobs although all considered strength- and conditioning-based work critical.

Moreover, the neuromuscular profile of volleyball players differs substantially from that of endurance and cross-country runners. In volleyball they have much higher strength and power values than runners. The differences seem to be related to the stimulation of specific muscle fibers and not to neuronal factors as there has been reported a greater size of the type II/I fiber area with no difference in the distribution of the fibers or in the speed of conduction [9]. Volleyball players were also taller, heavier, and with greater thigh volume than runners [9].

In volleyball, in addition to power and strength, it is also necessary to train agility, speed, and maximum aerobic power (VO2 max), since the volleyball player must develop an aerobic and anaerobic alactic energy metabolism (ATP-CP) to adequately manage short moments of high intensity associated with periods of low-intensity activity [10, 11].

One of the proposals for training and testing these mechanisms is training based on skill-based conditioning games with drills that simulate the movements required in the game, in particular rapid sprints, lateral movements associated with dips and jumps. The tests that are usually used to monitor these performances are the sprint on 5 m and on 10 m, the multistage fitness test on 20 meters to evaluate VO2 max and the T-test to evaluate agility [7, 10–14].

Training based on endurance and mid-distance running is therefore not considered effective because it does not reflect the metabolic and neuromuscular profile of volleyball players.

Given these premises, running activity has to be considered more important in the training sessions during the phases distant from the crucial moments of the season (**in-season**).

In the **off-season** period (the interval between one season and another), running is utilized to maintain a minimum aerobic condition. Generally athletes are required to perform 3–4 sessions per week of 30 min/1 h of low-speed running. In this phase it is possible to perform running training phases even on soft surfaces (sand, dirt).

At the beginning of the season (**pre-season**), running is used in some programs for the initial reconditioning, when the structures are not yet ready to support the specific workloads of the volleyball players (jumps) in high quantities. High-intensity work with short and repeated bursts is preferred, according to the principle of "high-intensity interval training" (HIIT) [15], used to optimize times and manage loads. A "typical" workout consists of 4–5 series of work lasting 4–5 min, with 2–3 min of complete recovery between the series. The goal is to work with a heart rate above the anaerobic threshold. The series are made up of short sprint lasting from 3 to 10 s, alternating with recoveries corresponding to the duration of the sprint in the case of a work/recovery ratio of 1/1, or with a recovery double compared to the duration of the sprint (ratio 1/2).

In the first standard training sessions, no changes of direction are used (to avoid overload). HIIT sessions are offered three times a week in the first 2 weeks of preparation and then move on to two times in the third week and only once in the fourth and fifth week.

The addition of changes of direction and specific paces is done progressively and goes hand in hand with the "technical" work. In general, the changes of direction are introduced from the second week, while the specific paces (various "drills" such as "X" drill, the "T" drill, the "W" drill, etc.) are added in the third week.

Skill-based training lasting from 6 to 8 weeks has been analyzed in several studies, in particular in women, and has shown to be a safe and effective training on the improvement of different performance parameters in particular in younger subjects. In general it showed to improve speed on 10 and 20 m and agility in the T-test as well as specific technical tasks [7, 10–14, 16].

To avoid overloads due to HIIT, it is generally placed as a separate training from other sessions, or as a completion at the end of technical training sessions.

When a sufficient general condition is reached, running gives way to sport-specific drills and is rarely revived throughout the season. It is generally used with injured athletes who are in the process of reactivating.

20.4 Injuries in Volleyball

Volleyball is a popular team sport that requires specific tasks, such as jumping, landing, blocking, and spiking the ball, that need to be combined with fast movements. These high demands from the musculoskeletal system lead to a significant overall injury burden associated with this sport, although it is considered a "noncontact" sport and players are separated by a net.

It is difficult to identify volleyball injuries associated with running. The main reasons lie in the fact that the most frequent acute injuries to which volleyball players are exposed occur as a consequence of traumatic events, such as knee or ankle sprains when landing from a jump or finger injuries during blocking. Furthermore, running is not a frequent gesture during the volleyball game phases, as previously reported.

Recent reports [17, 18] found a total incidence rate of musculoskeletal injuries in volleyball ranging from 1.7 to 10.7 injuries per 1000 playing hours reported by studies with a low risk of bias. The overall time loss (TL) injury rate in high school female volleyball have reported rates of 1.24/1000 AE [19] and 1.11/1000 AE [20], and the overall TL injury rate in female collegiate players have been reported to be 3.81/1000 AE [19].

Considering the incidence of both TL and non-time loss (NTL) injuries, Mcguine et al. [21] found in a girls' high school volleyball season an overall injury rate of 5.31 per 1000 AE, and TL injury rate of 3.8/1000 AE, with 71.6% of TL and 28.4% of NTL injuries. Baugh et al. in a study that included both TL and NTL injuries reported an injury rate of 7.07/1000 AE in female collegiate players [22].

A large proportion of acute TL injuries in volleyball are associated with player contact with the playing surface or with another person and ball contact during blocking [20, 22]. Overuse injuries, on the other hand, are the consequence of repeated gestures such as jumping or spiking the ball.

Baugh reported that injury rate was higher in the pre-season, when it is more frequent to use the run in the preparation phase, than in-season injury rates for women, while pre-season and in-season injury rates for men did not differ [22].

Most volleyball injuries are sustained to the lower extremity [18, 22–24]. The most common injury type is joint sprains, followed by muscle strains and contusions. Joint sprains affect mainly the ankle, finger/thumb, and knee, while most muscle strains are located in the lower back and thigh [18].

Whereas the ankle is the most commonly injured body part among TL injuries [18, 22–24], the knee comprised the largest proportion of NTL injuries [18, 22].

The majority of ankle and finger injuries are acute injuries. Conversely, knee, back, and shoulder injuries occur both as acute and as overuse injuries [24].

20.5 Injuries in Beach Volley

Evidence suggests that in beach volley there are less overuse lower limb injuries than in indoor volleyball. Bahr and Reeser investigated the injury pattern between athletes competing on the FIVB World Tour, documenting an acute time-loss injury rate of 3.1/1000 competition hours (2.9 for men and 3.3 for women) and 0.7/1000 training hours (0.8 for men and 0.7 for women) [25]. Knee injuries (30%), ankle injuries (17%), and finger injuries (17%) accounted for more than half of all acute time loss injuries. However, both male and female players reported a high prevalence of overuse injuries of the lower back, knee, and shoulder (25% of which resulted in missed training or competition) [25].

A recent study compared the epidemiology of injuries and time lost from participation between female National Collegiate Athletic Association (NCAA) Division I athletes who participate in indoor versus beach volleyball [26]. Athletes who participate in indoor volleyball were more likely to sustain an injury compared with beach volleyball athletes, with a total injury rate of 5.3 injuries per 1000 h played in indoor volleyball and 1.8 injuries per 1000 h played in beach volleyball. The knee was the most commonly observed site of injury in indoor volleyball, and the shoulder was the most common site of injury in beach volleyball. Although the rate of knee injury in indoor volleyball players was significantly higher compared with beach volleyball athletes, time lost from participation because of knee injury was significantly longer in beach volleyball. In addition, beach volleyball athletes had a significantly higher rate of abdominal muscle injury compared with indoor players. Shoulder injuries and low back injuries occurred at similar rates in beach and indoor volleyball, but time lost from participation because of these injuries was significantly shorter in indoor athletes [26].

The quality of landing could be one of the reason for fewer injuries and overuse conditions in beach volley when compared to indoor volleyball. Beach volleyball players land on softer surface and more often on both feet than indoor volleyball players, and this could result in decreases in peak forces during landings. Women landed more often on both legs than men probably because of the higher jumps and also for the tendencies of more second ball hits for men than women [27].

20.6 Prevention Strategies

As previously discussed, running training is not considered a priority in volleyball training protocols, and injuries to volleyball players usually are not associated with running activities. The choice to use running as training for volleyball players could instead have consequences on the injury rate and athletic performance.

First of all, it has to be defined which aspects of running are more functional to the specific activity of volleyball. Volleyball is an activity that requires the development of strength, power, and speed [7]. The ratio of rest time to activity time is approximately 1/6 and is therefore considered a high-intensity anaerobic sport [2].

It has previously been reported that most coaches and athletes agree that the most suitable training for increasing performance and prevention would be based on the development of strength and power, in particular plyometric training instead of running-based jobs [8].

Training volleyball players as mid-distance runners by introducing an endurance running program could alter their neuromuscular characteristics and make muscles and joints work according to a functional and metabolic demand to which they are not adapted—a choice that, if not adequately considered, could predispose the subject to an increased risk of injuries. Running is also an activity with an overuse injury rate higher than most other sports added together. Furthermore, one of the risk factors of running injuries is missing previous running experiences, and low running training volumes inevitably could happen in the case of volleyball players who cannot take too much time away from specific technical training [28–30].

Caution in the management of workloads that should be as individualized as possible must be placed in some subjects [28]:

- Female sex, especially if the subjects have amenorrhea, eating disorders, osteoporosis, or vitamin D deficiency and menopause, due to the risk of overuse-related injuries such as stress fractures.
- Athletes who have a history of previous traumatic and overuse injuries to the lower limbs, particularly if related to running.
- Older age.
- Athletes with biomechanical patterns that can alter the load on the lower limb by reducing the ability to absorb reaction forces in impact on the ground.

The tests based on short running sprint and drills that simulate the movements required in the game, in particular rapid sprints, lateral movements associated with dips and jumps, are used to evaluate the performance of athletes, as previously mentioned. These tests can be an excellent tool for monitoring the physiological abilities of athletes and are useful for preventing injuries [7, 10–14].

After a period of skill-based training, Noyes et al. [13] assessed the quality of the drop jump.

They hypothesized a reduction in the risk of injury to the lower limb thanks to the improvement in the alignment of the lower limb.

Skill-based training remains important, even with its limitations, in particular for those subjects who have functional test deficits because we can still train the physiological mechanisms associated with shooting and running without altering the neuromuscular profile and without incurring the high-risk injuries associated with running (incidence: 14.3–44.7% for short distance, 16.7–79.3% for long distance) [29].

To compensate and overcome the limitations found in skill-based training, plyometric training has proved effective, so much so that it is recommended by several authors as a more valid alternative [8, 16, 31].

Plyometric training has been shown to be a safe training to increase the different neuromuscular and physiological parameters as it also intervenes in the increase of joint proprioception by increasing the sensitivity of the neuromuscular spindles and in increasing the elongation capacity of the Achilles tendon by increasing its capacity of energy storage. These characteristics of plyometric training make it a valuable aid also in the prevention of injuries related to running and jumping with particular reference to overuse syndromes such as tendinopathies [32–34].

The prevention of injuries of the volleyball player should therefore be based on a correct strength development and prefer an adequate mix of plyometric training and high-intensity skill-based drills without requiring an endurance running training.

20.7 Specificities of Prevention Strategies in Beach Volleyball

Running on sand could be a preventive strategy and a safe way to train rather than a source of injury, but there are only few studies that investigate how to use effectively this surface for training. In beach volleyball, the choice to use running as a training must be carefully evaluated, and, in any case, it should preferably be performed on the sand to allow a safe and biomechanically similar activity to beach volleyball.

Training on softer surfaces like sand is considered a valuable alternative for sub-maximal aerobic conditioning in athletes returning from injury to congested competition schedule when we want continue training minimizing musculoskeletal strain. Brown et al. [35] found that in response to a matched-intensity exercise bout, markers of postexercise muscle damage may be reduced by running on softer ground surfaces. A significant time effect showed that myoglobin increased from pre- to post-exercise on grass ($p = 0.008$) but not on sand ($p = 0.611$).

Running on sand can also offer a higher energy cost (EC) and lower impact training stimulus compared with firmer and more traditional team sport training venues such as grass. The evidence shows greater net aerobic EC (~>1.5 times), anaerobic EC (~>2.5 times), and total EC (~>1.2–1.5 times) which are incurred when running at similar speeds on sand compared with grass.

Currently, the evidence suggests that the physiological and biomechanical adaptations unique to sand training can positively affect firm ground performance. Despite these suggestions, it is not known how the type of training performed, and the conditions of the sand surface used, can alter the resultant training stimulus experienced. For instance, softer and drier sand may be associated with higher EC values during exercise, since there is a greater degree of energy absorbed by the training surface. Also the granulation, moisture content, and/or the depth and consistency of the substratum on sand can all contribute to the stiffness recorded and affect the resulting energy cost. For running on sand, some researches show that the differences between sand and grass are lesser as running speed increase [36].

The training surface has also been shown to significantly influence the biomechanics of both running and jumping. The modification of the athletic gesture could therefore lead to an increased risk of injuries when you switch from an exclusive training on soft surfaces such as sand to more rigid surfaces. Even the performance of the athletes varies significantly to changing surfaces. The data show that both the height of the jump and the speed of the sprint in the sand are reduced. These elements should always be considered when planning training and recovering from injuries where sand work is heavily used [27, 37].

References

1. Fleureau A, Lacome M, Buchheit M, Couturier A, Rabita G. Validity of an ultra-wideband local positioning system to assess specific movements in handball. Biol Sport. 2020;37(4):351–7.
2. Sánchez-Moreno J, Afonso J, Mesquita I, Ureña A. Dynamics between playing activities and rest time in high-level men's volleyball. Int J Perform Anal Sport. 2016;16(1):317–31.
3. Reeser JC, Verhagen E, Briner WW, Askeland TI, Bahr R. Strategies for the prevention of volleyball related injuries. Br J Sports Med. 2006;40(7):594–600; discussion 599–600
4. Palao JM, Valades D, Ortega E. Match duration and number of rallies in men's and women's 2000–2010 FIVB world tour beach volleyball. J Hum Kinet. 2012;34:99–104.
5. Medeiros A, Marcelino R, Mesquita I, Palao JM. Physical and temporal characteristics of under 19, under 21 and senior male beach volleyball players. J Sports Sci Med. 2014;13(3):658–65.
6. Natali S, Ferioli D, La Torre A, Bonato M. Physical and technical demands of elite beach volleyball according to playing position and gender. J Sports Med Phys Fitness. 2019;59(1):6–9.
7. Gabbett T, Georgieff B. Physiological and anthropometric characteristics of Australian junior national, state, and novice volleyball players. J Strength Cond Res. 2007;21(3):902–8.
8. Weldon A, Mak JTS, Wong ST, Duncan MJ, Clarke ND, Bishop C. Strength and conditioning practices and perspectives of volleyball coaches and players. Sports Basel Switz. 2021;9(2):28.
9. Sleivert GG, Backus RD, Wenger HA. Neuromuscular differences between volleyball players, middle distance runners and untrained controls. Int J Sports Med. 1995;16(6):390–8.
10. Gabbett TJ. Do skill-based conditioning games offer a specific training stimulus for junior elite volleyball players? J Strength Cond Res. 2008;22(2):509–17.
11. Hedrick A. Training for high level performance in women's collegiate volleyball: part I training requirements. Strength Cond J. 2007;29(6):50–3.
12. Gabbett T, Georgieff B, Anderson S, Cotton B, Savovic D, Nicholson L. Changes in skill and physical fitness following training in talent-identified volleyball players. J Strength Cond Res. 2006;20(1):29–35.
13. Noyes FR, Barber-Westin SD, Smith ST, Campbell T. A training program to improve neuromuscular indices in female high school volleyball players. J Strength Cond Res. 2011;25(8):2151–60.
14. Trajković N, Milanović Z, Sporis G, Milić V, Stanković R. The effects of 6 weeks of preseason skill-based conditioning on physical performance in male volleyball players. J Strength Cond Res. 2012;26(6):1475–80.
15. Purkhús E, Krustrup P, Mohr M. High-intensity training improves exercise performance in elite women volleyball players during a competitive season. J Strength Cond Res. 2016;30(11):3066–72.
16. Gjinovci B, Idrizovic K, Uljevic O, Sekulic D. Plyometric training improves sprinting, jumping and throwing capacities of high level female volleyball players better than skill-based conditioning. J Sports Sci Med. 2017;16(4):527–35.
17. Bahr R, Bahr IA. Incidence of acute volleyball injuries: a prospective cohort study of injury mechanisms and risk factors. Scand J Med Sci Sports. 1997;7(3):166–71.
18. Bere T, Kruczynski J, Veintimilla N, Hamu Y, Bahr R. Injury risk is low among world-class volleyball players: 4-year data from the FIVB injury surveillance system. Br J Sports Med. 2015;49(17):1132–7.
19. Reeser JC, Gregory A, Berg RL, Comstock RD. A comparison of women's collegiate and girls' high school volleyball injury data collected prospectively over a 4-year period. Sports Health. 2015;7(6):504–10.
20. Kerr ZY, Gregory AJ, Wosmek J, Pierpoint LA, Currie DW, Knowles SB, et al. The first decade of web-based sports injury surveillance: descriptive epidemiology of injuries in US high school girls' volleyball (2005-2006 through 2013-2014) and national collegiate athletic association women's volleyball (2004-2005 through 2013-2014). J Athl Train. 2018;53(10):926–37.
21. McGuine TA, Post E, Biese K, Kliethermes S, Bell D, Watson A, et al. The incidence and risk factors for injuries in girls volleyball: a prospective study of 2072 players. J Athl Train. 2020.
22. Baugh CM, Weintraub GS, Gregory AJ, Djoko A, Dompier TP, Kerr ZY. Descriptive epidemiology of injuries sustained in national collegiate athletic association men's and women's volleyball, 2013-2014 to 2014-2015. Sports Health. 2018;10(1):60–9.
23. Agel J, Palmieri-Smith RM, Dick R, Wojtys EM, Marshall SW. Descriptive epidemiology of collegiate women's volleyball injuries: national collegiate athletic association injury surveillance system, 1988-1989 through 2003–2004. J Athl Train. 2007;42(2):295–302.
24. Verhagen EA, Van der Beek AJ, Bouter LM, Bahr RM, Van Mechelen W. A one season prospective cohort study of volleyball injuries. Br J Sports Med. 2004;38(4):477–81.
25. Bahr R, Reeser JC, Fédération Internationale de Volleyball. Injuries among world-class professional beach volleyball players. The Fédération

Internationale de volleyball beach volleyball injury study. Am J Sports Med. 2003;31(1):119–25.

26. Juhan T, Bolia IK, Kang HP, Homere A, Romano R, Tibone JE, et al. Injury epidemiology and time lost from participation in Women's NCAA division I indoor versus beach volleyball players. Orthop J Sports Med aprile. 2021;9(4):23259671211004544.

27. Tilp M, Rindler M. Landing techniques in beach volleyball. J Sports Sci Med. 2013;12(3):447–53.

28. Orejel Bustos A, Belluscio V, Camomilla V, Lucangeli L, Rizzo F, Sciarra T, et al. Overuse-related injuries of the musculoskeletal system: systematic review and quantitative synthesis of injuries, locations, risk factors and assessment techniques. Sensors. 2021;21(7).

29. van Poppel D, van der Worp M, Slabbekoorn A, van den Heuvel SSP, van Middelkoop M, Koes BW, et al. Risk factors for overuse injuries in short- and long-distance running: a systematic review. J Sport Health Sci. 2021;10(1):14–28.

30. Francis P, Whatman C, Sheerin K, Hume P, Johnson MI. The proportion of lower limb running injuries by gender, anatomical location and specific pathology: a systematic review. J Sports Sci Med. 2019;18(1):21–31.

31. Ramirez-Campillo R, García-de-Alcaraz A, Chaabene H, Moran J, Negra Y, Granacher U. Effects of plyometric jump training on physical fitness in amateur and professional volleyball: a meta-analysis. Front Physiol. 2021;12:636140.

32. Bonacci J, Chapman A, Blanch P, Vicenzino B. Neuromuscular adaptations to training, injury and passive interventions: implications for running economy. Sports Med. 2009;39(11):903–21.

33. Barnes KR, Kilding AE. Strategies to improve running economy. Sports Med. 2015;45(1):37–56.

34. Pardos-Mainer E, Lozano D, Torrontegui-Duarte M, Cartón-Llorente A, Roso-Moliner A. Effects of strength vs. plyometric training programs on vertical jumping, linear sprint and change of direction speed performance in female soccer players: a systematic review and meta-analysis. Int J Environ Res Public Health. 2021;18(2).

35. Brown H, Dawson B, Binnie MJ, Pinnington H, Sim M, Clemons TD, et al. Sand training: exercise-induced muscle damage and inflammatory responses to matched-intensity exercise. Eur J Sport Sci. 2017;17(6):741–7.

36. Binnie MJ, Dawson B, Pinnington H, Landers G, Peeling P. Sand training: a review of current research and practical applications. J Sports Sci. 2014;32(1):8–15.

37. Sanchez-Sanchez J, Martinez-Rodriguez A, Felipe JL, Hernandez-Martin A, Ubago-Guisado E, Bangsbo J, et al. Effect of natural turf, artificial turf, and sand surfaces on sprint performance. a systematic review and meta-analysis. Int J Environ Res Public Health. 2020;17(24):E9478.

Running in Handball

21

Leonard Achenbach, Lior Laver, Romain Seil,
and Kai Fehske

Handball derived from the year 1897 which was introduced as "Torball." Other than in football, the player was supposed to throw the ball into the goal. In 1917, field handball developed out of "Torball" with the first German Championship in the year 1921. After founding the International Handball Federation in 1927, field handball became part of the Olympic Games in 1936. The International Handball Federation was founded in 1946, and since the 1972 Olympic Games in Munich, handball became an Olympic sport.

Handball is one of the most popular types of sports in Europe. There are currently approximately 170 members of the IHF including 795,000 teams and an estimated 25 million players worldwide. The game is played by two teams, each consisting of six field players, one goalkeeper, and seven substitutes. Team handball is a dynamic and physically demanding sport, particularly because of the intensive body contact between the players.

Catching the trend of event sports, beach handball has been established over the last years. Using the beach as a playground and adapted match rules, beach handball is made fast and spectacular to be more attractive to the spectators.

21.1 Role and Characteristics of Running in Handball and Beach Handball

Modern professional handball is a very fast ball game, which is characterized by a high number of quick high-intensity movements interspersed with regular periods of recovery [1–5]. Handball is characterized by fast changes between offensive and defensive play, while all players can participate in defense and offense except the goalkeeper and unlimited substitutions are allowed at all times, with no interruptions. Apart from set offense against set defense, fast break situations are quite common in the game. Pace changes such as acceleration and deceleration as well as one-on-one situations are frequent and accompanied by passing and throwing moves with or without body contact. Throwing is often

L. Achenbach (✉)
Department of Orthopedics, König-Ludwig-Haus, University Wuerzburg, Wuerzburg, Germany

L. Laver
Department of Orthopedics, Sports Medicine Unit, Hillel Yaffe Medical Center (HYMC), Hadera, Israel

Rappaport Faculty of Medicine, Technion (Israel Institute of Technology), Halifax, Israel

R. Seil
Clinic for Orthopedics, Centre Hospitalier Luxembourg-Clinique d'Eich, Luxemburg, Luxemburg

K. Fehske
Department of Trauma-, Hand-, Plastic- and Reconstructive Surgery, University of Wuerzburg, Wuerzburg, Germany
e-mail: fehske_k@ukw.de

contested with contact, which is allowed and is an integral part of the game. The free substitutions contribute to the high speed over the full duration of the matches. As contact is allowed and is an integral part of the game with less restrictive rules regarding contact, punishment is subjective to referee decisions, while severe fouls can be punished with either a yellow card, a 2-min suspension, or a direct red card. Protective gear is not commonly used in handball, apart from selected braces and mouth guards. There is no limit for foul play, yet after sanctioned fouls and three "2-min" suspensions by the same athlete, the athlete is automatically suspended with a red card for the rest of the game. Handball comprises several types of movements that place moderate-to-high demands on the intermittent endurance running capacity of players, and matches are marked by many brief periods of high-intensity running. During high-intensity running, players carry out a high number of change-of-direction movements [6].

Beach handball is a relatively new type of sports which has been derived from the indoor handball and is played on a beach or some form of sand. The number of players is small which demands more from the beach sports players compared to the indoor variant. While this sport has been played informally on beaches for many years, the introduction of their official beach variants codified rules for the games. This was a major foundation and has helped the sports to grow rapidly in popularity. The sand enables players to perform acrobatic moves, which are awarded two points when resulting in a goal. Contact is not allowed, and fouls are punished more strictly by suspension or disqualification than in indoor handball.

21.2 Training Strategies

Strength, excellent physical conditioning, agility, optimal muscle activation and control, and good throwing technique are the main components for today's elite handball player. In addition, well-trained cognitive and tactical skills are consequential requirements in order to compete at a high level. Handball is characterized by rapid changes of game speed, and therefore athletes have to be well trained in both aerobic and anaerobic areas to be able to compete over 60 min. Male athletes show moderate to high blood lactate values post-match which show the demand on the anaerobic energy system. Furthermore, workload during matches are up to 70–80% VO_2 max.

Team managers need to be aware of the growing frequency of these high-intensity phases during matches to train their players accordingly. Performance is based on the complex interaction of the cardiovascular and respiratory systems. Training workloads are applied to athletes with the goal of inducing positive physiological changes and maximizing performance. The various biological adaptations induced by (appropriate) training increase the athletes' capacity to accept and withstand load. The aim of load management is thus to optimally configure training, competition, and other load to maximize adaptation and performance with a minimal risk of injury. Load management therefore comprises the appropriate prescription, monitoring, and adjustment of external and internal loads. The two most frequent training methods are high-intensity intermittent training and small-sided games [7].

21.3 Prevention Strategies for Handball and Beach Handball

Substitution is unlimited in handball. For the medical staff, this makes continuous surveillance of all players during the match very difficult. Medical staff members are therefore advised to seek a place close to the substitution bench with a good possibility of close monitoring and good communication with all players.

The body sites most frequently, severely, and acutely injured in indoor handball are the knee, hand, shoulder, and ankle [8]. For each body site, a distinct injury mechanism could be identified [9]. Ligamentous injuries of the knee, such as anterior cruciate ligament tears, occur most frequently during noncontact, side-cutting maneuvers, or landings. Metacarpal and finger fractures are sustained mainly through collision with opponents or direct trauma from the ball. Acute shoulder injuries, such as dislocation or AC joint injuries, occur due to pulls from opponents or direct falls onto the shoulder after contact. Ankle sprains are common due to frequent landings and direction changes as well as high friction between the shoe and the playing surface, while foot-to-foot contacts are also common.

Knowledge regarding prevention of especially acute lower extremity injuries has improved substantially in the last 15 years. By means of reducing previously identified risk factors, injury risk and thus injury incidence could be shown to be reducible by implementing regular, neuromuscular injury prevention exercises into practices and before matches. By means of strengthening exercises, such as the Nordic hamstring exercise and plyometric-oriented neuromuscular exercises, the injury incidence of noncontact ACL tears could be reduced to 50% [10, 11]. Narrow cutting techniques, landings on the toes, and knee-over-toe position are found to be techniques which decrease the likelihood for ACL tears [12]. For ankle sprains, bracing and taping and decreasing lateral shoe friction combined with neuromuscular proprioceptive exercises have been shown to be effective to reduce the risk of secondary ankle sprains [13–15].

In beach handball, prevention strategies should be especially directed at ankle sprains and skin injuries of the feet and toes [16]. The sand as playing surface needs to be considered in beach sports. While official tournaments have official guidelines, amateur tournaments may have a wide range of the sand offered for practice and competition. The coarseness of the sand differs between each court, and practice on the specific match court is recommended to let players adapt to the specificities of the match sand. Thicker, heavier sand may allow a faster match play with faster cutting and higher jumps, while lighter sand may be more forgiving to hard landings.

The skin of the players, especially of the foot and ankle, should be adapted to the coarseness of the specific match sand, while players with susceptible skin may wear foot and ankle protections, such as socks, if allowed. Shoes are not allowed in beach handball while elastic binding or ankle guards around the ankles or feet may be used as injury prevention and foot protection. Taping is allowed in beach handball, and the tape's color can be chosen regardless of the player's shirt color. Before playing, the sand should be checked for dangerous parts, such as hidden broken glass or shells.

The risk of skin damage due to excessive sun exposure can be reduced by seeking shade under an umbrella, tree, or any other shelter. Sunscreen, sunglasses, and protective clothing is recommended for players that are not used to sun exposure. Players should be taught guidelines for UV and sun exposure, and signs and symptoms of heat illness should be checked regularly. Player's hygiene should be encouraged to prevent infections from sand worms, such as nematodes.

In some beach sports, athletes may switch from the indoor or grass team sports to the beach variants during summer. This may happen due to leisure and fun as well as selection for a team competing in national and international championships, such as the European Beach Handball Championships in which more than 95% of all athletes are recruited only 6–2 weeks before the tournament from indoor team handball. This places a unique substantial short-term change in the musculotendinous structures and thus to the training and match loads.

To reduce the high number of overuse injuries for both sports, injury prevention programs from other ball sports may be used as they have been developed very sports-specific and can be applied easily to both indoor handball and beach handball. However, the effectiveness of these exercise programs on the reduction of overuse injury, such as shoulder overuse injury, is debated, to date. Further prevention strategies include the implementation of monitoring training and match load and adapting the individual training exercises and match actions for each player.

References

1. Michalsik LB, Aagaard P. Physical demands in elite team handball: comparisons between male and female players. J Sports Med Phys Fitness. 2015;55:878–91.
2. Michalsik LB, Aagaard P, Madsen K. Locomotion characteristics and match-induced impairments in physical performance in male elite team handball players. Int J Sports Med. 2013;34:590–9.
3. Michalsik LB, Madsen K, Aagaard P. Match performance and physiological capacity of female elite team handball players. Int J Sports Med. 2014;35:595–607.
4. Michalsik LB, Madsen K, Aagaard P. Physiological capacity and physical testing in male elite team handball. J Sports Med Phys Fitness. 2015;55:415–29.
5. Michalsik LB, Madsen K, Aagaard P. Technical match characteristics and influence of body anthropometry on playing performance in male elite team handball. J Strength Cond Res. 2015;29:416–28.
6. Póvoas SC, Seabra AF, Ascensão AA, Magalhães J, Soares JM, Rebelo AN. Physical and physiological demands of elite team handball. J Strength Cond Res. 2012;26:3365–75.
7. Iacono AD, Eliakim A, Meckel Y. Improving fitness of elite handball players: small-sided games vs. high-intensity intermittent training. J Strength Cond Res. 2015;29:835–43.
8. Luig P, Krutsch W, Nerlich M, Henke T, Klein C, Bloch H, Platen P, Achenbach L. Increased injury rates after the restructure of Germany's national second league of team handball. Knee Surg Sports Traumatol Arthrosc. 2018;26:1884–91.
9. Luig P, Krutsch W, Henke T, Klein C, Bloch H, Platen P, Achenbach L. Contact - but not foul play - dominates injury mechanisms in men's professional handball: a video match analysis of 580 injuries. Br J Sports Med. 2020;54:984–90.
10. Achenbach L, Krutsch V, Weber J, Nerlich M, Luig P, Loose O, Angele P, Krutsch W. Neuromuscular exercises prevent severe knee injury in adolescent team handball players. Knee Surg Sports Traumatol Arthrosc. 2017;26(7):1901–8.
11. Olsen OE, Myklebust G, Engebretsen L, Holme I, Bahr R. Exercises to prevent lower limb injuries in youth sports: cluster randomised controlled trial. BMJ. 2005;330:449.
12. Kristianslund E, Faul O, Bahr R, Myklebust G, Krosshaug T. Sidestep cutting technique and knee abduction loading: implications for ACL prevention exercises. Br J Sports Med. 2014;48:779–83.
13. Fehske K, Lukas C. Ligamentous ankle injury: an underestimated trauma? Sportverletz Sportschaden. 2020;34:147–52.
14. Lysdal FG, Bandholm T, Tolstrup JS, Clausen MB, Mann S, Petersen PB, Grønlykke TB, Kersting UG, Delahunt E, Thorborg K. Does the Spraino low-friction shoe patch prevent lateral ankle sprain injury in indoor sports? A pilot randomised controlled trial with 510 participants with previous ankle injuries. Br J Sports Med. 2021;55:92–8.
15. Møller M, Zebis MK, Myklebust G, Lind M, Wedderkopp N, Bekker S. "Is it fun and does it enhance my performance?" Key implementation considerations for injury prevention programs in youth handball. J Sci Med Sport. 2021;24(11):1136–42.
16. Achenbach L, Loose O, Laver L, Zeman F, Nerlich M, Angele P, Krutsch W. Beach handball is safer than indoor team handball: injury rates during the 2017 European Beach handball championships. Knee Surg Sports Traumatol Arthrosc. 2018;26:1909–15.

Trail Running

22

Gian Luigi Canata, Valentina Casale,
and Nico Valsesia

22.1 Role and Characteristics of Trail Running

Trail running is a sport activity which combines running and hiking, whenever a steep gradient is present. It shows several common characteristics with both mountain and fell running, even though it may include paved surfaces (but no more than the 25% of the entire route) and does not necessarily need a significant amount of ascent or navigating skills. For example, a race on a sandy beach trail can be considered trail running.

It consists of different natural off-road terrain (forest paths, dirt roads, but also sand, snow trails, track foot paths) and in different environments (mountains, forests, plains, deserts), in warm climates on trails, even if winter racing has been gaining popularity lately (Figs. 22.1a and b).

Paths or tracks must be relatively easy to follow. The route may or may not be marked but usually allows route choice. Aid stations are often located at checkpoints, and kit may or may not be required to be carried [1].

Trail running races do not have limits on distance or elevation gain or loss, and the properly marked course must represent a logical discovery of a region [2].

Most of the trail races are one-stage events, where runners are timed over the entire competition, including stops at aid stations. However, multiday stage races also exist; they can be either segmented into daily events of a specified distance or time or staged so that competitors can run as far as they want, over a set course or a set number of days. This usually happens in ultra-trail or endurance trail races.

In 2013, the International Trail Running Association (ITRA) was founded to promote trail running worldwide and regulate this discipline.

In 2015, trail running was recognized by World Athletics as a discipline of athletics. Nowadays, the ITRA is World Athletics' partner for the management of trail running.

According to the distance and the elevation gain to be run, some distinctions can be done:

- Trail races: less than 42 km long and a mean of 3000 m elevation gain route.
- Ultra-trail races: more than 42 km long and/or more than 4000 m elevation gain route.
- Endurance trail races: more than 320 km long and more than 10,000 m elevation gain route, such as the Tor des Géants race.

G. L. Canata (✉) · V. Casale
Centre of Sports Traumatology, Koelliker Hospital,
Torino, TO, Italy
e-mail: studio@ortosport.it

N. Valsesia
International Endurance Runner and Mountain
Cyclist, A.S.D. Bikeadventures,
Borgomanero, NO, Italy
e-mail: info@bikeadventures.it

© The Author(s), under exclusive license to Springer-Verlag GmbH, DE, part of Springer Nature 2022
G. L. Canata et al. (eds.), *The Running Athlete*, https://doi.org/10.1007/978-3-662-65064-6_22

Fig. 22.1 (**a**, **b**) Trail running training. ©Dino Bonelli

For standard races, ITRA classifies them by "km-effort." The km-effort measure is the sum of the distance expressed in kilometers and a hundredth of the vertical gain expressed in meters, rounded to the nearest whole number (e.g., the km-effort of a race of 65 km and 3500 m ascent is 65 + 3500/100 = 100). Races are then classified as follows [3]:

Category	Km-effort
XXS	0–24
XS	25–44
S	45–74
M	75–114
L	115–154
XL	155–209
XXL	≥210

ITRA allocates a corresponding number of points (i.e., ITRA points), from 0 to 6 to these seven different categories, in consideration of distance and elevation gain and loss, and then evaluating the race's effort/difficulty. In this way, routes can be assessed and classified, also providing an accurate idea of the effort involved.

It is not possible in fact to compare directly two race times, since each race takes place on diverse surfaces, conditions, and technical skills. For this reason, ITRA has developed a Performance Index (PI) to compare the speed of all trail runners, at all levels. It is calculated by obtaining the weighted mean of up to the five best scores achieved by a runner over the previous 36 months (for the general index or the index by category). It is built on a scale up to a maximum of 1000 points.

A further classification system is the ITRA score. It is given to each runner for every race they complete. Each person's runner file profile consists of many different scores, corresponding to each race completed. It is based only on the finish time, not on the finish position, and also considers the specific characteristics of the race runners who have participated in (distance, elevation gain and loss, average altitude).

Other trail running disciplines like the skyrunning have been lately gaining popularity, especially for the elevation gains rather than the distance to be run. An example are the vertical races consisting of routes less than 5 km long but with a minimum of 1 km of elevation gain.

Therefore, if timing is essential in road running, other features become important in trail running, such as being outdoor on mountain trails, and above all the positive and negative gradients that represent the real challenge for runners (Fig. 22.2).

Furthermore, trail running can be considered less traumatic than road running. While the pace of a road running race is continuous due to

Fig. 22.2 Endurance champion Nico Valsesia during uphill running. ©Dino Bonelli

the uniform path—without huge variations in elevation gain or route surface—in trail running every step must be pondered instead, to avoid sprains or sudden falls and to measure energies according to the elevation gain faced at that moment. The result is a more careful running mode.

In fact, during uphill running, it is not unusual for trail runners to cover some stretches slowing down their pace until even walking, as well as using aids such as trekking or hiking poles.

Conversely, when running downhill, speed increases and consequently the risk of sprain injuries (Fig. 22.3).

22.1.1 Trail Running Equipment: What Athletes Need to Run?

Runners often carry some amount of food, water, clothing, and extra gear in longer outings, particularly when trail running. How to carry this stuff depends on the weather, location, terrain, length of the run, and the level of the runner's experience.

When running long distances (about 3 h of running or more), especially in remote areas, race belts, handled bottles, or a waistpack can be inadequate to bring the required amount of food, water, and gear. In those cases, a running vest or backpack can fit the extra essentials [4].

The main consideration on trail-running wear pertains to shoes, which belong to the basic equipment for a running athlete.

Fig. 22.3 Endurance champion Nico Valsesia during downhill running. ©Dino Bonelli

Although this topic has been extensively addressed in Chap. 10, it can be useful to deepen the main specific features of trail running shoes.

They differ from road running footwear in several aspects [5]:

- Grip on rugged terrain: Soles are lugged to enhance traction, helping runners move more sure-footedly over mud, roots, gravel, dirt, and rock slabs.
- Foot protection: Specific internal and external features help shield feet from impact with roots and stones. Long-lasting upper materials withstand abrasion and tears.
- Stiff construction: Trail running footwear should prevent excessive foot rotation. Furthermore, pronation control is not essential in trail running, because it requires a shorter, more variable stride. Several runners often make use of insoles for overpronation; however they should keep in mind that an overcorrection may predispose to ankle sprains, hip pain, or back pain in this type of running.

- Cushioning goes from barefoot to minimal, moderate, and maximum levels, according to the padding thickness of the midsole. Maximum cushioning should ensure, in theory, a reduction in fatigue on the high-mileage run, but high-level trail runners report an increased risk of joint instability in the front plane when landing on uneven or inclined surfaces. About the effects of a thick midsole on the passive and active stabilizer of the ankle, please refer to the chapter "Shoes for running."

Compression garments such as knee-high compression socks and/or compression short tights have recently gained popularity among long-distance running, because of their reported capacity to improve runner's performance and facilitate recovery. In particular, they are supposed to reduce tissue vibrations and muscle fatigue during uphill and downhill sections [6–8]. It has been reported in fact that during trail running, extensive muscle oscillations due to repeated and prolonged downhill sections may affect the muscle function and the contractile apparatus [8].

Several publications on the influence of such garments have been published, with contradictory results [9]. For this reason, Engel et al. calculated the effect sizes associated with several markers related to performance and recovery in a systematic review [9]. From the final outcomes that emerged, using compression garments neither has any beneficial effect on racing performance (from 400 m sprint to marathon) nor exerts any influence on the execution of strength-related tasks during recovery from running. However, considering postexercise pain, damage, and muscle inflammation, effect size values in the existing literature report large positive effects of compression.

The use of poles in trail running is becoming increasingly popular [10]. They help runners to maintain a more upright posture while climbing, especially when they are fatigued and also to shift partially the workload from the legs to the arms, thus reducing localized fatigue in the legs and enabling a sustainable pace.

Vertical races benefit from their use, because they help greatly on the vertical changes. Training with poles is fundamental for arms to acquire skill, strength, and stamina. Their use may be optional, moving uphill or in both uphill and downhill sections, and this is the reason why many trail runners avoid them, considered as an extra gear to be carried downhill and flat sections. Furthermore, the pole's extension must be adjusted according to the slope of the route, so that the arms make a 90-degree bend at the elbows when holding the pole with the tip on the ground near the foot. As a result, the steeper the slope, the more runners shorten poles to avoid distress to the arms, shoulders, back, and neck.

The characteristics of the hand grips are also important [11]. Grip materials can include cork, which resists moisture from sweaty hands, decreases vibration, and conforms to the hands' shape; foam, which absorbs moisture and is the softest to the touch; or rubber, that insulates hands from cold, shock, and vibration, but is more likely to chafe or blister sweaty hands so is less suitable for warm-weather races. Instead of the straight grip handles commonly used, the ergonomic ones with a 15- to 45-degree corrective angle have recently been gaining popularity for allowing the wrists to keep a neutral and comfortable position.

22.2 Training Strategies

A wide range of plans and approaches to training for trail running exist, according to the level of the runner, the distance to be run, and the terrain the race will take place in.

However, some characteristics and steps should be presented in every training scheme:

- An initial physical assessment is particularly recommended, especially from a cardiovascular point of view. In fact, a close relationship between the duration and speed of run, the amount of time spent training, and the likelihood of heart disorders has been demonstrated [12, 13]. Training triggers physiological adaptations, including the heart and circulatory system [14]. The start of a competition is usually associated with extreme cardiovascular efforts, and the several transient changes commonly observed in blood test results may suggest heart damage or a cytoprotective reaction to long-term stress (e.g., increases in troponin and N-pro brain natriuretic protein(BNP)) [15]. For these reasons, a heart rate (HR) monitoring is strongly advisable.

The heart rate varies among runners due to the heart size, individual background, and genetics. Every runner should know and monitor its heart rate during training and competition, especially during the uphill sections. The goal in fact is to keep it constant, even walking in some sections. On the contrary, this aspect is less relevant on the descent.

- A variety of running tempos and terrains. During training periods, it is suggested to vary the type of run and the terrain during the week. You should shift from "easy run" (60–65% of maximal heart rate, building cardiovascular base miles) to "speed run" ($\geq$ 90% of max HR during speed-paced runs in intervals of 1–5 min, always preceded by a warm-up run and followed by a cooldown run), to "tempo run" (85–88% of max HR during tempo-paced runs in intervals longer than 15 min, always preceded by a warm-up run and followed by a cooldown run). Hill running workout is also essential to improve core stability by modifying the center of mass. Trail running in fact requires much more core activation than road running does, because of different terrains, unstable footing, and extended elevation gain and loss.
- Dynamic warm-up, along with post-workout cooldown and stretching.
- Gradually building up mileage until the runner is at or near his event distance. It is important to consider that a rapid increase in the number of miles run per week may enhance the risk of injury. It has been well reported in fact that the onset and the development of running-related injuries is strongly associated to insufficient management of training loads

in respect to load capacity [16–19]. Also training at different paces is an effective way to work on endurance [20].

- Nutrition and hydration planning.

Cross-training is particularly important for improving strength and flexibility, reducing muscle imbalances, and contributing to muscle recovery [21]. Sports disciplines like cycling, swimming, and cross-country skiing can help improve cardiovascular health and develop strength and endurance. Furthermore, cross-country skiing is a great alternative to wintertime running, especially during the "off-season" period.

Even spending some time in the gym can increase strength and physical abilities. In particular, resistance training with body weight exercises like push-ups, pull-ups, squats, lunges, sit-ups, and planks can improve overall strength. Trail runners can benefit most from single-leg exercises (e.g., single-leg deadlift, lateral lunge, pistol squat, and back-elevated single-leg bridge) [22]. Anti-rotation exercises strengthen the core stability, especially against the counterrotation of the thoracic and the lumbar spine that develops during the act of running. Some examples are dumbbell twists and ball bridge twists. Finally, plyometric exercises are highly effective in training the fast-twitch muscle fibers, improving strength, and building explosive power for increasing running speed and endurance. Great examples are pogo jumps, single-leg pogo jumps, box jumps, and lateral jumps.

22.3 Prevention Strategies

Trail running-related injuries mainly consist in dermatological (such as blisters, chafing, superficial wounds, and frostbite or hypothermia signs) and musculoskeletal disorders, especially ankle sprains in flat or downhill sections [23]. Yamato et al. defined musculoskeletal injuries in running as "musculoskeletal pain in the lower limbs that causes a restriction or stoppage of running (distance, speed, duration, or training) for at least 7 days or three consecutive scheduled training sessions or that requires the runner to consult a physician or other health professional" [24].

If considering ultra-endurance running, which includes different running events by distance (>42 km) or time (>6 h) and multiday/multistage events, an excessive stress on the musculoskeletal system develops during both training and competitions. It has been reported in fact that the average training loads in ultra-endurance running are between 66 and 83 km/week in adults and around 57 km/week in youth athletes [25]. These training demands can cause overuse injuries once they exceed the adaptive mechanisms, especially affecting lower limbs. Acute injuries are more frequent during competitions instead.

The ankle is a frequently injured anatomical site. Tendinopathy of the dorsiflexors of the foot and the paratendinopathy/tenosynovitis of the tendons passing nearby the extensor retinaculum of the ankle are some of the most reported long-distance running-related injuries [25].

The lower leg is often involved (with pathologies such as medial tibial stress syndrome and chronic exertional compartment syndrome), as well as the foot (with plantar fasciitis) and the knee (with patellofemoral pain syndrome and meniscal injuries) [25].

Prevention strategies are then essential to guarantee the right protection against the development of running-related injuries. The adaptive mechanisms must be trained and strengthened, and several programs have been proposed, especially by online format [26].

They consider the following:

- Personal Factors
  - Age: The higher age, the higher risk of injuries and the higher importance not to run with pain.
  - Weight: In both overweight and underweight conditions, a preconditioning phase is suggested, with partial weight-bearing sports.
  - Previous injuries: Starting at a very low level after an injury is highly recommended.
  - No previous experience with running or other sports: A preconditioning phase of walking is suggested.
- Training Aspects
  - Distance-to-run per week to be gradually increased.
  - Frequency of training sessions per week.
  - Risk of overtraining: Rest periods are recommended.
  - Running surface, ideally choosing soft surfaces rather than others.
  - Stretching at every training session.
  - Cross-training sessions are recommended.
- Biomechanics Factors
  - Running cadence to be gradually increased. Furthermore, taking short, quick steps can reduce the impact of each stride.
  - Foot strike: Runners usually adopt different foot strike patterns as slope changes from level running to uphill running to downhill running. In trail running in fact, runners do not adopt a single-foot strike pattern, because they must adapt to uneven surfaces [27]. To effectively switch from the midfoot to rearfoot to forefoot, specific strengthening exercises for the foot and ankle muscles should be performed [26].
- Equipment
  - Shoes: Neutral, stabilizing, and motion control shoes are suggested. Foot pronation is considered as a strong risk factor for running-related injury, and the right choice of footwear has been suggested as a strong intervention to prevent such injuries [28]. However, recent evidence has contradicted the widespread belief that moderate foot pronation is associated with high risk of injuries among novice runners. Another traditional position recently contradicted is that supinated feet are best served in neutral shoes, highly pronated feet in motion control shoes, and neutral feet in neutral or stability shoes [29]. Another important consideration regards minimalistic shoes. They have been shown to decrease the risk of running-related injuries, as well as to increase foot muscle size, strengthening

the feet [30]. Nevertheless, the reduced support caused by such footwear can enhance the demand on the foot, promoting a more anterior strike pattern and a consequent increased demand on the calf [31]. Therefore, the advice is a gradual switch to minimalistic shoes, not to use inlays unless some injuries have already occurred [26].

22.4 Conclusions

Trail running is perfectly adapting to the current times, seeking care for the environment where we live. With a proper training, every athlete may breathe freely in wonderful landscapes. A simple equipment and endless paths are all that is needed to improve our well-being by selecting the distance according to the current physical form.

References

1. Rowell S, Dodds W. Trail and mountain running. Crowood; 2013.
2. Trail running. 2021. https://www.worldathletics.org/disciplines/trail-running/trail-running. Accessed 20 may 2021.
3. World Athletics Series Regulations. Book – C2.1. Part VIII – Cross Country, Mountain and Trail Races (in force from 1 November 2019). 2020. p. 129–133. https://www.worldathletics.org/about-iaaf/documents/book-of-rules. Accessed 20 May 2021.
4. How to pack your running gear. 2021. https://www.rei.com/learn/expert-advice/trail-running-gear.html?series=intro-to-trail-running. Accessed 25 May 2021.
5. How to choose Trail-Running shoes. 2021. https://www.rei.com/learn/expert-advice/trail-running-shoes.html. Accessed 18 June 2021.
6. Bieuzen F, Brisswalter J, Easthope C, et al. Effect of wearing compression stockings on recovery after mild exercise-induced muscle damage. Int J Sports Physiol Perform. 2014;9:256–64.
7. Friesenbichler B, Stirling LM, Federolf P, et al. Tissue vibration in prolonged running. J Biomech. 2011;44:116–20.
8. Vercruyssen F, Gruet M, Colson SS, et al. Compression garments, muscle contractile function, and economy in trail running. Int J Sports Physiol Perform. 2017;12:62–8.
9. Engel FA, Holmberg HC, Sperlich B. Is there evidence that runners can benefit from wearing compression clothing? Sports Med. 2016;46:1939–52.
10. Halpin E, Wells MG. How to choose and use trekking poles and hiking staffs. 2021. https://www.rei.com/learn/expert-advice/trekking-poles-hiking-staffs.html. Accessed 24 May 2021.
11. Koop J. The pros and cons of poles for trail running and Ultrarunning. 2021. https://trainright.com/poles-trail-ultrarunning-running/. Accessed 24 May 2021.
12. Mont L, Sambola A, Brugada J, et al. Long-lasting sport practice and lone atrial fibrillation. Eur Heart J. 2002;23:477–82.
13. Zaidi A, Sharma S. Arrhythmogenic right ventricular remodelling in endurance athletes: Pandora's box or Achilles' heel? Eur Heart J. 2015;36:1955–7.
14. Pluim BM, Zwinderman AH, van der Laarse A, et al. The athlete's heart. A meta-analysis of cardiac structure and function. Circulation. 2000;101:336–44.
15. Kim TJ, Shin YO, Lee YH, et al. The effects of running a 308 km ultra-marathon on cardiac markers. Eur J Sport Sci. 2014;14:S92–7.
16. Soligard T, Schwellnus M, Alonso JM, et al. How much is too much? (part 1) international committee consensus statement on load in sport and risk of injury. Br J Sports Med. 2016;50:1030–41.
17. Meeusen R, Duclos M, Foster C, et al. Prevention, diagnosis, and treatment of the overtraining syndrome: joint consensus statement of the European College of Sport Science and the American College of Sports Medicine. Med Sci Sports Exerc. 2013;45:186–205.
18. Bertelsen ML, Hulme A, Petersen J, et al. A framework for the etiology of running-related injuries. Scand J Med Sci Sports. 2017;27:1170–80.
19. Nielsen RO, Bertelsen ML, Möller M, et al. Training load and structure-specific load: applications for sport injury causality and data analyses. Br J Sports Med. 2018;52:1016–7.
20. Running: increasing distance and endurance. 2021. https://www.rei.com/learn/expert-advice/trail-running-distance-endurance.html. Accessed 25 May 2021.
21. How to cross-train for trail running. 2021. https://www.rei.com/learn/expert-advice/cross-training-for-trail-running.html. Accessed 26 May 2021.
22. Hobbs N. How to build strength to improve your trail running. 2020. https://trailrunner.com/trail-news/how-to-build-strength-to-improve-your-trail-running/. Accessed 23 May 2021.
23. Matos S, Silva B, Clemente FM, et al. Running-related injuries in Portuguese trail runners: a retrospective cohort study. J Sports Med Phys Fitness. 2021;61:420–7.
24. Yamato TP, Saragiotto BT, Lopes AD. Consensus definition of running-related injury in recreational runners: a modified Delphi approach. J Orthop Sports Phys Ther. 2015;45:375–80.
25. Scheer V, Krabak B. Musculoskeletal injuries in ultra-endurance running: a scoping review. Front Physiol. 2021;12:664071.
26. Fokkema T, de Vos RJ, van Ochten J, et al. Preventing running-related injuries using evidence-based online

advice: the design of a randomised-controlled trial. BMJ Open Sports Exerc Med. 2017;3:e000265.

27. Giandolini M, Pavailler S, Samozino P, et al. Foot strike pattern and impact continuous measurement during a trail running race: proof of concept in a world-class athlete. Footwear Sci. 2015;7:127–37.

28. Richards CE, Magin PJ, Callister R. Is your prescription of distance running shoes evidence-based? Br J Sports Med. 2009;43:159–62.

29. Nielsen RO, Buist I, Partner ET, et al. Foot pronation is not associated with increased injury risk in novice runners wearing a neutral shoe: a 1-year prospective cohort study. Br J Sports Med. 2014;48:440–7.

30. Chen TL, Sze LK, Davis IS, et al. Effects of training in minimalist shoes on the intrinsic and extrinsic foot muscle volume. Clin Biomech. 2016;36:8–13.

31. Ridge ST, Johnson AW, Mitchell UH, et al. Foot bone marrow edema after a 10-wk transition to minimalist running shoes. Med Sci Sports Exerc. 2013;45:1363–8.

Beat Knechtle and Pantelis T. Nikolaidis

Triathlon is a multisport discipline consisting of swimming, cycling, and running. Considering the Ironman distance, the swim split is 3.8 km, the cycling split 180 km, and the running split 42.195 km. The sequence of the disciplines has an influence on the performance in the marathon. In the beginning of the Ironman triathlon with the "Ironman Hawaii," overall performance time decreased rapidly from 1981 but has remained stable since the late 1980s [1]. Between 1983 and 2012, the overall top ten men and women improved their swimming (only men), cycling, running, and overall race times [2]. From 1988 to 2007, linear regression analyses showed that the change in swimming, cycling, running, and overall race time for both males and females was less than 1.4% per decade, except for females' running time, which decreased by 3.8% per decade [1]. Interestingly, the age of annual top ten female and male triathletes in the "Ironman Hawaii"

increased over the last three decades, while their performances improved [3].

Apart from elite athletes, also master athletes improved their performance in Ironman triathlon [4–6]. Men older than 44 years and women older than 40 years significantly improved their performances in the "Ironman Hawaii" in the three disciplines and in overall race time [6]. This is most probably due to the age-related performance decline in Ironman triathlon. A study investigating more than 300,000 Ironman triathletes recommended that women should start competing in Ironman triathlon before the age of 30 years and men before the age of 35 years to achieve their personal best Ironman race time [7].

It has been shown that the cycling split seems to be the Ironman triathlon discipline that most improved overall race times and is also the discipline with the greatest influence on the overall race time of elite men and women in the Ironman World Championship, the "Ironman Hawaii" [8]. An analysis of more than 300,000 Ironman triathletes competing in more than 250 Ironman races showed that the fastest Ironman triathletes were the relatively the fastest in running and transition times [9]. However, it has also been shown that cycling and running presented similar contributions (~40%) in the Ironman triathlon, when top athletes were analyzed [10]. These disparate findings are explained by the different samples that were investigated.

B. Knechtle (✉)
Medbase St. Gallen Am Vadianplatz,
St. Gallen, Switzerland

Institute of Primary Care, University of Zurich,
Zurich, Switzerland
e-mail: beat.knechtle@hispeed.ch

P. T. Nikolaidis
Exercise Physiology Laboratory, Nikaia, Greece

School of Health and Caring Sciences, University of
West Attica, Athens, Greece
e-mail: pnikolaidis@uniwa.gr

Regarding race tactics in an Ironman triathlon, athletes should focus on saving energy during swimming and cycling for the running split [9]. It has been shown that Ironman triathletes slowed down during cycling and running (i.e., positive pacing) [11]. By analyzing split times to assess the pacing strategies of the top 100 finishers of the cycling part of 13 Ironman races and of the running part of 11 Ironman races taking place in 2014, a continuous decrease in speed was observed during cycling in 10 of 13 races and during running in all 11 races [12].

23.1 Preparation for an Ironman Triathlon

In order to achieve the optimum performance, and partly the importance of previous experience in an Ironman triathlon—especially in the run split—the athlete should specifically prepare for the race regarding training, mental preparation, and nutrition. Distribution of training intensity within and between different sessions is an important aspect of training [13]. For successful Ironman triathletes, training distances appear to be a more important factor for competitive success than training paces [14]. Apart from physical preparation, also mental preparation is important. The athlete needs to break down the monumental task into smaller pieces. In other terms, the athlete must sit down and break the required training into a logical progression of monthly goals. Mental toughness seems to contribute positively to Ironman competitors' exertion of the cognitive and emotional control necessary to experience flow and perform better [15].

Also, nutrition is important in the preparation of an Ironman triathlon [16]. Depending upon the performance level of an Ironman triathlete, specific metabolic strategies should be considered in order to minimize the endogenous carbohydrate cost associated with exercise at competitive intensities [16]. Since the stores for endogenous carbohydrates are limited (i.e., muscle and liver glycogen), the goal of the Ironman preparation should be to minimize the oxidation of endogenous carbohydrate while performing during training at competitive intensities. A further option is a low-carbohydrate diet. Furthermore, periodized training with low-carbohydrate availability combined with exogenous carbohydrate supplementation during the race might be most appropriate for elite and top-amateur Ironman triathletes who elicit very high rates of energy expenditure [16]. It has been shown that Ironman race time was inversely related to carbohydrate intake during the marathon run for males but not for females, suggesting that increasing carbohydrate ingestion during the run may have been a useful strategy for improving Ironman performance in male triathletes [17].

23.2 The Aspect of Anthropometry

It has also been shown that anthropometric characteristics are important for overall race time and the marathon split, especially in male Ironman triathletes [18, 19]. Percent body fat was related to overall race time in male Ironman triathletes [18]. The somatotype of a male Ironman athlete defines for 28.6% variance in the Ironman race performance [19]. Athletes not having an ideal somatotype of 1.7-4.9-2.8 could improve their performance by altering their somatotype [19]. Lower rates in endomorphy (i.e., athletes with short stature, high levels of muscle mass and body fat), as well as higher rates in ectomorphy (i.e., athletes who are small, thin, and have a flat chest, small shoulders, and a faint definition), resulted in a significant better Ironman race performance [19]. The impact of somatotype was the most distinguished on the run discipline for male Ironman triathletes and had a much greater impact on overall race time than the quantitative training effort [19]. The somatotype component of endomorphy and the fat mass percentage were highly correlated with the percentage change in maximum oxygen uptake in the run segment [20].

23.3 Personal Best Marathon Time

Furthermore, individual marathon time is important for a fast Ironman race time. In male Ironman triathletes, the personal best marathon time was significantly and positively related to the run split time in an Ironman triathlon [21, 22]. It has been shown that faster running during training and a fast personal best time in a marathon and in an Olympic distance triathlon were both associated with a fast Ironman race time [21]. Although it has been shown that the personal best marathon time was an independent predictor for overall Ironman race time in male triathletes, male marathoners and male Ironman triathletes differ in their anthropometric and training characteristics [21]. Also in female Ironman triathletes, there are differences in both anthropometry and training between recreational female Ironman triathletes and recreational female marathoners and different predictor variables for race performance in these two groups of athletes [23].

23.4 Injuries and Preventive Measures Before and During an Ironman Triathlon

Injuries are very common among triathletes [24–27] and restrict the athletes from training and competition. The distance of the triathlon races seems to be of importance. Injuries occur more often in Ironman triathletes than in Half Ironman triathletes [28]. Overuse injuries occur in triathletes during preseason and during race season [28, 29] mostly in running than in cycling [25, 30]. Overuse was the reported cause in 41% of the injuries [31]. The injury rate is higher during race season than during training [25, 29]. The exposure rate is at 2.5 per 1000 training hours. During race season, the injury exposure rate is at 4.6 per 1000 training hours [29].

The most frequently affected sites were the ankle/foot, thigh, knee, lower leg, and back [31]. In some cases, elite triathletes can also suffer stress fractures (e.g., lower limbs, sacrum) [13]. Different risk factors such as sex, age, and morphological characteristics, such as body height, body weight, and body mass index, have been investigated whether they correlate with an injury [24]. Increased years of triathlon experience, high running training volume, history of previous injury, and inadequate warm-up and cooldown regimes appeared to have individual associations with injury incidence [29]. Performance level and weekly training hours seemed also to be related to injuries in triathlon [32]. Whether age or sex seems to influence the prevalence of injuries seems to depend on the distance of the triathlon race [30, 32].

The association between training volume and injury has shown, however, inconsistent results [24]. The large number of training sessions and overall volume undertaken by triathletes to improve fitness and performance can also increase the risk of injury, illness, or excessive fatigue [13]. Short- and medium-term individualized training plans, periodization strategies, and work/rest balance are necessary to minimize interruptions to training due to injury, illness, or maladaptation [13]. Incorporating strength training to complement the large amount of endurance training in triathlon can help to avoid overuse injuries [13].

Experience seems important in the prevention of injuries. The likelihood of an injury was positively associated with experience in a triathlon (i.e., number of finished races) [31]. It has been shown that elite triathletes were also more efficient runners [33] [31]. Injuries seem to occur more often during competing than during training [30, 31] with 5.4 injuries per 1000 h of training and 17.4 per 1000 h of competition [31].

Exercise-associated muscle cramps are frequent. Triathletes with exercise-associated muscle cramps were significantly taller and heavier, had faster Ironman race times despite being of similar performance level, and predicted and achieved a faster overall race time during an Ironman triathlon. There was an association among a positive family history for exercise-associated muscle cramps, a history of tendon and/or ligament injuries, and a self-reported history of exercise-associated muscle cramps [34].

The most frequent are muscle cramping, heat illness, postural hypotension, excessive exposure to ultraviolet radiation, musculoskeletal injuries and trauma, gastrointestinal problems as well as postrace bacterial infection, immunosuppression, sympathetic nervous system and psychological exhaustion, and hemolysis [35]. Musculoskeletal injures are reported to be the second most frequent cause behind gastrointestinal issues [36].

It seems that injuries are more often during the run than during the cycling split [37]. In 10,197 individual starters in triathlon races over different distances, injuries were predominantly sustained during the run (38.4%) and cycle (14.3%) splits. Lower limb injuries (59.5%) and abrasions (28.6%) were the most common site and nature of injury, respectively [30].

Death can occur during triathlon practice. In a competition, the sudden death rate is 1.5 occurrences for every 100,000 participations and mostly swim-related [38].

23.5 Metabolic Changes

An Ironman triathlon leads to disturbances in fluid (i.e., dehydration) and electrolyte metabolism (i.e., exercise-associated hyponatremia) mainly due to heat problems [39–42]. In the Ironman triathlon, dehydration and heat problems are complicated by hyponatremia [39].

In hot climates, proper attention should be paid to these risks, and the medical staff should be prepared [37, 43]. In Hawaii where the World Championship "Ironman Hawaii" is held, heat illness can definitely compromise athletic performance [44]. During an Ironman triathlon, body core temperature increases significantly [45, 46]. A recent study showed that body core temperature increased from 36.62 ± 0.17 °C pre-race to 38.55 ± 0.64 °C postrace after the marathon and remained elevated 60 min after the event at 38.65 ± 0.41 °C [46].

Generally, athletes finishing an Ironman triathlon are dehydrate or hypo-hydrated [39, 40]. The incidence of exercise-associated hyponatremia depends upon the distance of the race and the state of hydration [41]. It has been shown in Ironman triathletes that percentage change in body weight was linearly related to postrace serum sodium concentrations but unrelated to

postrace rectal temperature, or performance, in the marathon [47].

The loss in body mass during an Ironman triathlon is primarily due to fluid loss [45]. A body mass loss of up to 3% is tolerated by well-trained triathletes during an Ironman competition in warm conditions without any evidence of thermoregulatory failure [45]. However, a loss in solid mass is due to a decrease in both fat and lean mass [48]. The loss in lean mass was explained by a decrease in muscle density, as an indicator of glycogen loss, and increases in several indicators for dehydration (i.e., hematocrit, hemoglobin, electrolytes, urine-specific gravity, and urine osmolality) [48].

23.6 Musculoskeletal Changes

Several studies reported an increase in biomarkers of skeletal muscle damage such as creatine kinase [49–53], myoglobin [51–54], creatinine [52, 54, 55], and lactate dehydrogenase [53] after an Ironman triathlon. These increases were generally higher in male compared to female triathletes [54]. Women seem to tolerate better the effects of prolonged strenuous physical activity than men [54]. This partially massive increases are mainly due to the eccentric muscle damage in the run split [55]. It has been shown that muscular strength and jump height decreased significantly after an Ironman triathlon [56].

Apart from markers of skeletal muscle damage, also markers of myocardial damage such as cardiac troponin T (cTnT) [57, 58] and troponin I (cTnI) [50, 58] increased after an Ironman triathlon. Ultra-endurance exercise may cause myocardial damage as indicated by biochemical cardiac-specific markers and echocardiography. However, the increase in these markers does not reflect myocardial damage. The troponin rise was associated only with echocardiographic evidence of abnormal left ventricular function, but not with muscle damage [57]. The cellular nature of this damage and whether it is transient or permanent is unclear [58]. Ultra-endurance exercise does not result in sustained myocardial injury in Ironman athletes [50, 57].

23.6.1 Fact Box

The order of the triathlon disciplines has an effect on the marathon split in the Ironman triathlon. Since the cycling split is the most important split in an Ironman triathlon, any Ironman triathlete should focus on saving energy during swimming and cycling for the running split.

Fast running during training, previous experience with a fast personal best time in marathon and a fast personal best time in an Olympic distance triathlon and low body fat are the most important predictors for a fast Ironman race time in male triathletes.

References

1. Lepers R. Analysis of Hawaii ironman performances in elite triathletes from 1981 to 2007. Med Sci Sports Exerc. 2008;40(10):1828–34.
2. Rüst CA, Knechtle B, Rosemann T, Lepers R. Sex difference in race performance and age of peak performance in the ironman triathlon world championship from 1983 to 2012. Extreme Physiol Med. 2012;1(1).
3. Gallmann D, Knechtle B, Rüst CA, Rosemann T, Lepers R. Elite triathletes in 'Ironman Hawaii' get older but faster. Age. 2014;36(1):407–16.
4. Lepers R, Knechtle B, Stapley PJ. Trends in triathlon performance: effects of sex and age. Sports Med. 2013;43(9):851–63.
5. Stiefel M, Knechtle B, Lepers R. Master triathletes have not reached limits in their ironman triathlon performance. Scand J Med Sci Sports. 2014;24(1):89–97.
6. Lepers R, Rüst CA, Stapley PJ, Knechtle B. Relative improvements in endurance performance with age: evidence from 25 years of Hawaii ironman racing. Age. 2013;35(3):953–62.
7. Käch IW, Rüst CA, Nikolaidis PT, Rosemann T, Knechtle B. The age-related performance decline in ironman triathlon starts earlier in swimming than in cycling and running. J Strength Cond Res. 2018;32(2):379–95.
8. Barbosa LP, Sousa CV, Sales MM, Olher RR, Aguiar SS, Santos PA, et al. Celebrating 40 years of ironman: how the champions perform. Int J Environ Res Public Health. 2019;16(6).
9. Knechtle B, Käch I, Rosemann T, Nikolaidis PT. The effect of sex, age and performance level on pacing of ironman triathletes. Res Sports Med. 2019;27(1):99–111.
10. Figueiredo P, Marques EA, Lepers R. Changes in contributions of swimming, cycling, and running performances on overall triathlon performance over a 26-year period. J Strength Cond Res. 2016;30(9):2406–15.
11. Wu SSX, Peiffer JJ, Brisswalter J, Nosaka K, Lau WY, Abbiss CR. Pacing strategies during the swim, cycle and run disciplines of sprint, olympic and half-ironman triathlons. Eur J Appl Physiol. 2015;115(5):1147–54.
12. Angehrn N, Rüst CA, Nikolaidis PT, Rosemann T, Knechtle B. Positive pacing in elite ironman triathletes. Chin J Physiol. 2016;59(6):305–14.
13. Etxebarria N, Mujika I, Pyne DB. Training and competition readiness in triathlon. Sports (Basel). 2019;7(5):101.
14. Gulbin JP, Gaffney PT. Ultraendurance triathlon participation: typical race preparation of lower level triathletes. J Sports Med Phys Fitness. 1999;39(1):12–5.
15. Meggs J, Chen MA, Koehn S. Relationships between flow, mental toughness, and subjective performance perception in various triathletes. Percept Mot Skills. 2019;126(2):241–52.
16. Maunder E, Kilding AE, Plews DJ. Substrate metabolism during ironman triathlon: different horses on the same courses. Sports Med. 2018;48(10):2219–26.
17. Kimber NE, Ross JJ, Mason SL, Speedy DB. Energy balance during an ironman triathlon in male and female triathletes. Int J Sport Nutr. 2002;12(1):47–62.
18. Knechtle B, Wirth A, Baumann B, Knechtle P, Rosemann T. Personal best time, percent body fat, and training are differently associated with race time for male and female ironman triathletes. Res Q Exerc Sport. 2010;81(1):62–8.
19. Kandel M, Baeyens JP, Clarys P. Somatotype, training and performance in ironman athletes. Eur J Sport Sci. 2014;14(4):301–8.
20. Sellés-Pérez S, Fernández-Sáez J, Férriz-Valero A, Esteve-Lanao J, Cejuela R. Changes in Triathletes' performance and body composition during a specific training period for a half-ironman race. J Hum Kinet. 2019;67(1):185–98.
21. Knechtle B, Wirth A, Rosemann T. Predictors of race time in male ironman triathletes: physical characteristics, training, or prerace experience? Percept Mot Skills. 2010;111(2):437–46.
22. Rüst CA, Knechtle B, Knechtle P, Rosemann T, Lepers R. Personal best times in an olympic distance triathlon and in a marathon predict ironman race time in recreational male triathletes. Open Access J Sports Med. 2011;2:121–9.
23. Rüst CA, Knechtle B, Knechtle P, Rosemann T. A comparison of anthropometric and training characteristics between recreational female marathoners and recreational female ironman triathletes. Chin J Physiol. 2013;56(1).
24. Kienstra CM, Asken TR, Garcia JD, Lara V, Best TM. Triathlon injuries: transitioning from prevalence to prediction and prevention. Curr Sports Med Rep. 2017;16(6):397–403.
25. Zwingenberger S, Valladares RD, Walther A, Beck H, Stiehler M, Kirschner S, et al. An epidemiological investigation of training and injury patterns in triathletes. J Sports Sci. 2014;32(6):583–90.

26. Cipriani DJ, Swartz JD, Hodgson CM. Triathlon and the multisport athlete. J Orthop Sports Phys Ther. 1998;27(1):42–50.
27. Strock GA, Cottrell ER, Lohman JM. Triathlon. Phys Med Rehabil Clin N Am. 2006;17(3):553–64.
28. Rimmer T, Coniglione T. A temporal model for nonelite triathlon race injuries. Clin J Sport Med. 2012;22(3):249–53.
29. Burns J, Keenan AM, Redmond AC. Factors associated with triathlon-related overuse injuries. J Orthop Sports Phys Ther. 2003;33(4):177–84.
30. Gosling CM, Forbes AB, McGivern J, Gabbe BJ. A profile of injuries in athletes seeking treatment during a triathlon race series. Am J Sports Med. 2010;38(5):1007–14.
31. Korkia PK, Tunstall-Pedoe DS, Maffulli N. An epidemiological investigation of training and injury patterns in British triathletes. Br J Sports Med. 1994;28(3):191–6.
32. Egermann M, Brocai D, Lill CA, Schmitt H. Analysis of injuries in long-distance triathletes. Int J Sports Med. 2003;24(4):271–6.
33. Laurenson NM, Fulcher KY, Korkia P. Physiological characteristics of elite and club level female triathletes during running. Int J Sports Med. 1993;14(8):455–9.
34. Shang G, Collins M, Schwellnus MP. Factors associated with a self-reported history of exercise-associated muscle cramps in ironman triathletes: a case-control study. Clin J Sport Med. 2011;21(3):204–10.
35. Dallam GM, Jonas S, Miller TK. Medical considerations in triathlon competition: recommendations for triathlon organisers, competitors and coaches. Sports Med. 2005;35(2):143–61.
36. Turris SA, Lund A, Bowles RR, Camporese M, Green T. Patient presentations and medical logistics at full and half ironman distance triathlons. Curr Sports Med Rep. 2017;16(3):137–43.
37. Jeong J, Yang HR, Kim I, Kim JE. Medical support during an Ironman 70.3 triathlon race. F1000Res. 2017;6:1516.
38. Vleck V, Millet GP, Alves FB. The impact of triathlon training and racing on athletes' general health. Sports Med. 2014;44(12):1659–92.
39. Hiller WDB, O'Toole ML, Fortess EE, Laird RH, Imbert PC, Sisk TD. Medical and physiological considerations in triathlons. Am J Sports Med. 1987;15(2):164–7.
40. Sousa CV, Da Aguiar SS, Dos Olher RR, Sales MM, De Moraes MR, Nikolaidis PT, et al. Hydration status after an ironman triathlon: a meta-analysis. J Hum Kinet. 2019;70(1):93–102.
41. Hiller WDB. Dehydration and hyponatremia during triathlons. Med Sci Sports Exerc. 1989;21(5): S219–S21.
42. Lockett LJ. Hydration-dehydration, heat, humidity, and "cool, clear, water". Sports Med Arthrosc Rev. 2012;20(4):240–3.
43. Laird RH, Johnson D. The medical perspective of the Kona ironman triathlon. Sports Med Arthrosc Rev. 2012;20(4):239.
44. Gordon S. Heat illness in Hawai'i. Hawaii J Med Public Health. 2014;73(11 Suppl 2):33–6.
45. Laursen PB, Suriano R, Quod MJ, Lee H, Abbiss CR, Nosaka K, et al. Core temperature and hydration status during an ironman triathlon. Br J Sports Med. 2006;40(4):320–5.
46. Olcina G, Crespo C, Timón R, Mjaanes JM, Calleja-González J. Core temperature response during the marathon portion of the ironman world championship (Kona-Hawaii). Front Physiol. 2019;10.
47. Sharwood K, Collins M, Goedecke J, Wilson G, Noakes T. Weight changes, sodium levels, and performance in the south African ironman triathlon. Clin J Sport Med. 2002;12(6):391–9.
48. Mueller SM, Anliker E, Knechtle P, Knechtle B, Toigo M. Changes in body composition in triathletes during an ironman race. Eur J Appl Physiol. 2013;113(9):2343–52.
49. Nyborg C, Melau J, Bonnevie-Svendsen M, Mathiasen M, Melsom HS, Storsve AB, et al. Biochemical markers after the Norseman Extreme Triathlon. PLoS One. 2020;15(9):e0239158.
50. La Gerche A, Boyle A, Wilson AM, Prior DL. No evidence of sustained myocardial injury following an ironman distance triathlon. Int J Sports Med. 2004;25(1):45–9.
51. Neubauer O, König D, Wagner KH. Recovery after an ironman triathlon: sustained inflammatory responses and muscular stress. Eur J Appl Physiol. 2008;104(3):417–26.
52. Carlsson J, Ragnarsson T, Danielsson T, Johansson T, Schreyer H, Breyne A, et al. Biochemical changes after strenuous exercise – data from the Kalmar Ironman. Lakartidningen. 2016;113.
53. Chia-Ching WU, Huang TH. The effects of a 226-km ironman triathlon race on bone turnover in amateur male triathletes. J Sports Med Phys Fitness. 2019;59(10):1709–15.
54. Danielsson T, Carlsson J, Schreyer H, Ahnesjo J, Siethoff LT, Ragnarsson T, et al. Blood biomarkers in male and female participants after an ironman-distance triathlon. PLoS One. 2017;12(6).
55. Suzuki K, Peake J, Nosaka K, Okutsu M, Abbiss CR, Surriano R, et al. Changes in markers of muscle damage, inflammation and HSP70 after an ironman triathlon race. Eur J Appl Physiol. 2006;98(6):525–34.
56. Mueller SM, Knechtle P, Knechtle B, Toigo M. An ironman triathlon reduces neuromuscular performance due to impaired force transmission and reduced leg stiffness. Eur J Appl Physiol. 2015;115(4):795–802.
57. Tulloh L, Robinson D, Patel A, Ware A, Prendergast C, Sullivan D, et al. Raised troponin T and echocardiographic abnormalities after prolonged strenuous exercise - the Australian ironman triathlon. Br J Sports Med. 2006;40(7):605–9.
58. Rifai N, Douglas PS, O'Toole M, Rimm E, Ginsburg GS. Cardiac troponin T and I, electrocardiographic wall motion analyses, and ejection fractions in athletes participating in the Hawaii ironman triathlon. Am J Cardiol. 1999;83(7):1085–9.

Running in Parkour

24

Dan Edwardes and Francesco Feletti

24.1 Introduction

The history of the terminology in parkour is often obscure and dependent on anecdotal evidence. However, the progression of names of the discipline is relatively well established as beginning with 'le parcours,' a pre-existing French term meaning 'the route' or 'the course,' which was then replaced by 'l'Art du Deplacement,' the first name assigned by the founders to identify their discipline as something particular. The name of parkour was adopted in the late 1990s when one of the group, David Belle, broke away and chose to use this new term to differentiate himself, and then the terms freerunning or le freerun appeared around 2003 during the filming of the seminal Channel 4 documentary for British television, *Jump London.*

The terms *parkour* and *freerunning* have been utilized in different ways at different stages of its development to refer to or emphasize different aspects of the practice.

D. Edwardes
Dipartimento di Diagnostica per Immagini - Ausl della Romagna, U.O. Radiologia - Ospedale S. Maria delle Croci, Ravenna, Italy
e-mail: dan@parkourgenerations.com

F. Feletti (✉)
Dipartimento di Medicina Traslazionale e per la, Romagna, Università degli Studi di Ferrara, Ferrara, Italy
e-mail: feletti@extremesportmed.org

In particular, some scientific literature has attempted to differentiate parkour from freerunning by asserting that the former is focused on the search for efficiency and usefulness of movements, with a particular emphasis on safety, while in freerunning the goal would be purely the search for aesthetic or expressive movement.

Specifically, it is often held that parkour would not incorporate flips or other acrobatic movements that pursue aesthetic purposes that are not functional to completing paths [1, 2].

However, from a historical perspective, it is evident that these names originated as simply different ways to refer to essentially the same activity, with minor variations on the theme.

For ease of communication, we will utilize the term *parkour* to refer to the activity as a whole, simply because this has become the most widely recognized name for it around the world and has been the term used to formalize the practice on a sports governance level in various territories.

Parkour originated in the suburbs of Paris, France, in the 1990s. It began as the pastime of a group of youth, concentrated around the Belle and Hnautra families, who evolved the practice organically through a combination of play, exploration, challenge and self-discovery.

Parkour is a movement practice based around overcoming obstacles in one's environment with only the human body's abilities. The activity utilises foundational human locomotion patterns, including running, jumping, brachiating, crawling, rolling, swinging and climbing, the regular practice of which results in the development of highly functional physical capabilities founded on strength, endurance, power, agility and fitness. It primarily relies on a constraints-led approach to motor skill improvement, enabling and prizing adaptation and variability to solve movement problems, navigate the terrain and accomplish complex and demanding physical tasks.

Despite parkour's founders often describing it as merely running, jumping and climbing, they drew a set of guiding principles and concepts around which parkour took shape on many influences.

These influences ranged from the physical ideals of athlete-firefighter Raymond Belle and Georges Hebert's method naturelle, the discipline and philosophy of the martial arts, the training of athletics and acrobatics, to elements of popular culture such as Japanese anime and manga, superhero comics and action movies.

The early practitioners valued resilience, discipline, durability and the practical application of movement and espoused a philosophy of altruism, community and friendship. To 'be strong to be useful' was a central creed adopted by the founders and early generations, and the practice of parkour for show or to impress others was typically deemed base and tasteless.

In the following two decades, parkour has seen rapid and varied evolution due principally to the rise of more immediate communication technology and platforms such as YouTube and social media software, enabling the instant worldwide dissemination and adaptation of movements, techniques, methodologies and concepts.

The spectacular nature of parkour, primarily when performed at height in urban environments, saw it quickly co-opted by almost all forms of commercial media, including film, television and advertising, and regularly used as a device in brand marketing and product launches across multiple industries.

Seminal moments that raised widespread awareness included the theatrical release *Yamakasi* (2001), directed by Luc Besson, which starred many of the founding members; Carbon Media's *Jump London* (2003) documentary for the UK television's Channel 4, and the French film *District 13* (2004) which starred one of the founding members, David Belle. Finally, it is worth mentioning parkour's inclusion in the opening scene of the James Bond film *Casino Royale* (2006).

By the early 2010s, parkour was often referred to as one of the fastest-growing 'lifestyle sports' in the world, with official numbers [3] in England reaching 100,000 practitioners—roughly twice the number of skateboarders, three times the number of Judoka and four times the number of triathletes. Global estimates have the number of practitioners (also called traceurs) at perhaps one- to two-million, though, with very little actual data on numbers outside of England, this is extremely hard to quantify.

Large practising communities, given loose organization by the rise of hundreds of parkour clubs, organizations, associations and influencer channels, began to take shape in countries as diverse as the UK, China, Russia, France, the USA, Denmark, Taiwan, Sweden, Australia, Brasil, Finland and more. The UK saw the earliest attempt to formalize the activity, which became a recognized sport in 2016.

Attempts to create a competitive model of parkour, though fraught with controversy due to any form of competition being seen by the majority of the community as antithetical to the core philosophies of the activity, began as early as 2007 with Redbull's Art of Motion event, followed in 2008 by the Barclaycard Freerunning World Championship in the UK, before national parkour communities, principally in the USA and Europe, began formulating their competitive formats.

Under the umbrella of the International Gymnastics Federation, most recently, the gymnastics associations have striven to appropriate parkour as one of their disciplines and even attempted to have it included under the gymnastics banner within the 2021 Olympics (declined by the IOC). This misappropriation of their discipline has been vehemently opposed by the global parkour communities because parkour is neither historically derived from gymnastics nor is it similar conceptually, practically or philosophically beyond some movement pattern overlap at a superficial level.

As a result of its meteoric growth in popularity, parkour has inevitably seen many variations arise in its relatively short lifetime, with different interpretations of its core concepts, goals, function and even history proliferating around the world.

24.2 Role and Characteristics of Running in Parkour

24.2.1 Characteristics of the Sport in Terms of Performance

At its core, and indeed, at its origin, parkour is a concept of functional, practical movement. The discipline's declared goal is to overcome any obstacle in one's path using the human body's capabilities solely, leading to the necessary improvement of foundational human locomotion patterns such as running, jumping, climbing, vaulting, brachiating, rolling and crawling.

The word 'functional' recalls the evolutionary function of the human body to move across variable terrain in order to reach desired destinations, escape threats or explore new environments—as opposed to individual functions of the musculo-skeletal system, which are so often the focus of sports science, fitness training and strength and conditioning programmes.

As a result, there is no ideal set of anthropometric characteristics for a parkour practitioner—instead, the ideal parkour practitioner is someone who finds and practices the optimal methods for his or her build. However, it can be observed that experienced parkour practitioners tend to develop a highly favourable strength-weight ratio which enables them to more easily propel and carry their body weight through space and over obstacles and reduces the relative impact sustained by regular landings from height.

Common athletic gestures within parkour, learned and regularly utilized by the majority of practitioners worldwide, would include the following:

- Passement: vaulting and diving patterns to cross low obstacles at speed.

- Saut de détente: running jumps, typically from one raised object to another.
- Saut de precision: any jump that requires a precise landing, such as to a round handrail or a thin wall.
- Passe muraille: sudden redirection of momentum to propel the individual up walls and vertical surfaces.
- Balancer: swinging motion to release while hanging from overhead bars or branches to cross the distance to another obstacle.
- Roulade: rolling patterns, most typically over one shoulder and diagonally across the back to enable an individual to disperse impact after a landing from a height.
- Grimper: quick, dynamic climbing movements to ascend or descend tall obstacles, often interspersed with jumps, swings and landings.

See also https://www.youtube.com/watch?v= SMppD-bUNWo&ab_channel=11consolable [4].

Two decades of growth within the parkour community has led to innumerable variations on the core functional movement patterns, involving a heavy focus on acrobatic skills and a tendency towards aesthetic goals over the original utilitarian emphasis.

It is also fundamental to understand the psychological processes common to the practice of parkour, which, when carried out in its native outdoor environment, offers no protective clothing or gloves and no soft surfaces or safety matting. The lack of any protection emphasizes the nature of risk and the importance of the individual's ability to assess and manage risk and judge precisely one's capabilities to not attempt something far beyond one's current physical capacity.

The emotion of fear plays a decisive role in determining how much of one's own potential an individual can reach in parkour, and regular practice of parkour will bring about familiarity with fear which then enables a practitioner to learn to understand and manage their fear responses.

Parkour practitioners seem to show quite peculiar neuromuscular characteristics that undoubtedly deserve to be considered and further studied to implement effective sport-specific training programs. In particular, parkour mainly involves plyometric and eccentric exercises, often performed with significant mechanical stress due to high-drop jumps [5].

Traceurs could exhibit neuromuscular adaptation closed to what can be defined as an 'ultra-power profile' [5]. In addition, the practice of parkour can induce significant changes in the neuromuscular profiles not only on muscle properties but also up to the central command. The excellent maximal force production capability of parkour athletes might be underlain, in part, by greater central activation and descending command, lower antagonist contribution and a more remarkable ability to modulate spinal network efficiency.

It should also be interesting to assess how these athletes deal with the high neuromuscular fatigue rate that such power-specific adaptations can induce. Indeed, while they can develop higher maximal force, ultra-power athletes may, on the contrary, exhibit more significant fatigue during sustained contractions than untrained individuals [5].

24.2.2 The Role of Running in Parkour

From a biomechanical point of view, running is crucial in parkour. The kinetic chain involved in the running gait, generating the phases of the stride and the stretch-reflex elasticity required to manage and utilize the ground reaction force, is activated in the great majority of parkour movement training, which is often focused on completing specifically formulated 'lines' or 'routes' as efficiently and effectively as possible. The completion of parkour lines necessitates running, typically between applying unique parkour athletic gestures or motor skills. For example, to complete a typical parkour routine, an individual may begin by running towards a wall, vaulting it and then transitioning immediately into a running gait again to accelerate a complete running jump before leaping to a bar to swing and release, again landing into a running gait to finish with another vault.

While these momentary applications of the running gait may be brief, they are critical in enabling the efficient completion of each section of the 'line,' providing the acceleration, velocity and momentum required to jump or vault far enough. Thus, the ability to transition instantly into a running gait, accelerating or decelerating as required by a particular 'line,' is a skill in which most practitioners will become highly competent, whether intentionally trained or spontaneously talented, due to simple adaptation constraints.

Moreover, running in parkour has specific features; in parkour, running is widely used on a horizontal surface and oblique or vertical surfaces [6]. In the complex urban landscape environment, stairs, tracks, walls and building surfaces are common obstacles, and it is common to climb facades that are too high to jump simply using the running with different inclinations and vertically [7]. For this purpose, a distinctive running transition is often used where the legs act in a coordinated sequence, one as the last contact on the horizontal surface and the other as the initial contact on the vertical surface. This is notable for the aggressive nature of the abrupt transition at speed and for the elegance with which accomplished parkour athletes (traceurs) negotiate this challenge [8].

24.2.3 Running Training in Parkour

Running has originally been a core aspect of parkour training. The utilitarian philosophy that drove the nascent stages of the discipline valued simple, pure tests of functional capacities and fitness such as how far one could run, how quickly one could ascend an obstacle and how easily one could reach a particular location (e.g. a rooftop or the opposite bank of a river).

Aerobic fitness and muscular endurance were therefore seen as crucial to a practitioner's development, and distance running was a traditional element within the training, although not approached in a particularly sport-scientific manner. Running remains a fundamental element of training among those communities that have maintained a strong emphasis on the original concepts of parkour. However, recently, younger communities tend to focus more on the aesthetic, explosive movements within parkour, do not typically practice distance running of any sort, and many will actively avoid it.

At the time of writing, running as a skill in and of itself is practised by only a minority of the parkour community worldwide.

24.2.4 Physiological Characteristics of the Sport

Parkour is a physiologically complex activity, and a complete professional exploits the different energy systems in a combined and synergistic way.

Training should regularly include prompt, explosive actions that draw on the phosphagen anaerobic system and more extended sequences of movements that exploit the anaerobic glycolytic metabolic system. Many parkour communities also maintain the practice of endurance challenges that require a robust aerobic system.

A predominance of the use of the glycolytic system has been demonstrated in a 2017 study of adolescents [9]. However, it must be acknowledged that the type of parkour training currently prevalent in many groups of adolescent males heavily focuses on shorter routes combining explosive movements to the exclusion of any of the original endurance-based training carried out by the early generations of practitioners. However, more experienced practitioners might argue that these groups only practice one specific aspect of parkour and neglect other significant elements that make the discipline what it is.

Considering the muscular efforts required in commonly practised parkour movements, it is clear that explosive power is highly valued and is the aspect that develops fastest in most practitioners, simply due to the volume of jump-based practice that occurs. High levels of power and endurance then arise due to the repetition of this form of practice over months and years, meaning most practitioners can perform explosive movements such as maximum or near-maximum distance jumps hundreds of times across a single training session.

Due to the high levels of power resistance, it follows that commensurate strength training, which supports the practitioner in safely absorbing multiple impacts, should be standard. However, this is not the case in most communities, particularly younger practising communities. It is observable that young male parkour athletes primarily develop power levels disproportionate to their strength levels, leading to a greater likelihood of overuse injuries, tendonitis and joint damage, the vast majority of which result from the impact on landing [10].

The conceptual ideal in parkour training is to achieve overall physical preparation, meaning no weak link in any of the muscle groups required for practical movement skills. However, there are biases and dominances common to many parkour athletes due to the intense focus on specific movement skills above others, such as jumping and climbing. As a result, many parkour practitioners develop a clear dominance in the anterior muscle chain of the legs, primarily due to the eccentric loading placed upon those muscles during landings to slow/absorb impact, as well as a significantly developed pulling chain in the upper body resulting from climbing up and over obstacles regularly.

24.2.5 Training Strategies

Parkour training varies considerably across practising communities due to environmental, architectural differences, length of practice, influence/lineage of practice and cultural norms. As a result, it is difficult, to some degree, to generalize parkour training strategies. However, some general training principles can be identified.

Firstly, being that parkour is a concept to develop the optimal movement capabilities for each specific individual, a 'one-size-fits-all' training approach should be avoided. It follows that the general training principle of 'individualization' has a crucial role in parkour. Training must be prescribed according to individual performance capacity and predispositions such as age, sex, anthropometric factors, force/velocity profiles and injury/recovery status [11–13].

Secondly, an essential difference between parkour and many traditional sports is the increased variety of athletic gestures and the near-endless variations of their combination. At the same time, as with any athletic pursuit, the primary requirement to move safely and well in parkour is to possess sufficient strength, power and mobility to perform the parkour movements and perform them multiple times, much the same. Thus the postures and positions of support that a parkour practitioner's body will be subjected to across a training period are typically more than a sport or activity with far more specific constraints, such as those played on a flat pitch with minimal multiplanar movement required. This increase in variety and variation leads to a requisite increase in variability—the body's ability to adapt existing movement patterns to new conditions, such as different distances, surfaces, heights, angles and light levels. Accordingly, a significant imbalance in muscular strength levels within the body can negatively impact performance as a whole and significantly increase the risk of acute and overuse injuries. A healthy balance across the musculoskeletal system is essential to maintain sufficient capacity for the demands of parkour.

24.2.6 Programming in Parkour

Very few communities follow prescribed programmes of training in parkour, beyond the organic progression encouraged by more experienced practitioners. Simple heuristics such as 'First, do it; then do it well; then do it fast and well' and 'Once is never; twice is luck; three times is skill' have been the training mantras of many generations of practitioners.

On the one hand, these verbal cues may be examples of appropriate coaching communication which in the specific context of this sport population may be effective to obtain the required results in terms of motor pattern, force and power production [11]. On the other hand, this apparently overly simplistic approach however involves the application of a methodological

strategy basic in the clear focusing on 'whole-skill' development followed by the refinement of the quality of the movement before it is applied at great speed.

Such an approach benefits from being easy to comprehend and to transmit to other practitioners.

It is important to understand that as the vast majority of parkour practitioners are not following cycles of periodization in order to peak for performance at specific events of times of the year, the desired programming might be and often is very different from competition-based sports. Programming for longevity, health and practical skill acquisition over a lifetime is clearly distinct from the programming created for elite performance.

The majority of parkour practitioners use programming types that can be described as self-regulating programming and project-based programming.

Self-regulating programming is where a practitioner follows their body/mind feelings when choosing what to focus on during a given training session.

This means that in the case of discomfort to the lower limbs, an athlete might choose to focus on upper body strength or endurance training, simply following the natural cues of the body. While it of course can be argued that this is a form of non-programming, it is evident that this is an extremely common approach in parkour as it reflects the underlying philosophy of this sport. At the same time, it can be noted that it underlies the basic concepts of competency-based progression typical in the training of more traditional sports, according to which athletes should not progress to more challenging tasks of training until they master the underpinning principles [14].

Project-based programming is also widespread within the parkour community and revolves around the strategy of selecting a specific physical goal or feat the individual wishes to be able to perform—for example, a free-standing handstand—and then devoting a significant portion of one's practice time to the achievement of that goal, until it is reached.

This approach is similar to the training approach adopted in other extreme sports freestyle disciplines, where tricks are often learned through trial and error. Completion of the goal then leads to the practitioner selecting another project and beginning the process again.

Only a minority of practitioners—usually either beginners with access to a highly experienced coaching structure or experienced athletes who seek ever-increasing levels of performance—devise more structured programmes of periodization focusing on specific attribute development, such as maximal strength or explosive jump programmes.

Many benefits from introducing running as a regular practice for parkour participants would support and improve any practitioner's general physical preparation. For example, the increase in speed and power resulting from regular sprint training has apparent benefits for the explosive movements of parkour.

Considering that the use of the glycolytic metabolic system is predominant in many practising communities worldwide, explosive running activities such as sprint training would be highly beneficial in building the explosive power required for many parkour movements such as jumps and vaults and in supporting the development of the anaerobic system. An excess of endurance-based running might reduce explosivity and muscle mass, leading to a decrease in performance of those aspects of parkour movements that require both power and strength to perform safely and efficiently, such as vaults and jumps.

The ADAPT Qualifications global parkour coaching certification programme, for example, requires a candidate to be able to complete a 5-kilometre run in under 25 min as part of the Level 2 Coach Certification. This achievement is considered sufficient to demonstrate a healthy standard of aerobic fitness and running capacity and the ability to self-programme to achieve a goal for those unable to complete the task, but by no means requires excessive running training [15].

In conclusion, the practice of running would almost always be beneficial to parkour practitioners, and primarily if they pursue the 'origi-nal' goals of being competent in general human movement tasks.

Specific training for parkour must be directed on the motor patterns generally used to overcome the obstacles which the athlete may encounter, such as running, climbing, swinging, vaulting, rolling, crawling and jumping [6, 16–18]. Specifically, traceurs can train themselves to run on challenging surfaces such as ground lines, curbs or rails to improve movement control and increase the safety margin [6].

This approach improves the level of automatism connected to the more excellent reliability of the postural system and a reduced need for attentional resources [6, 19, 20]. In parkour, the central nervous system directs the body's centre of gravity relative to the support base and coordinates postural movement patterns in response to available data, receptive and biomechanical forces [6, 21–23].

During running, the plantar skin receptors considered a dynamometric map for guiding bipedal stability [24] are in intimate connection with various surfaces. In running training, recreating postural condition characteristic of parkour, like unstable surfaces such as foam and dynamic conditions, may help to improve postural control [6].

24.2.7 Prevention Strategies

Prevention strategies in parkour have broadly two goals:

1. Developing the physical attributes required to perform the movements safely and to help prevent both acute and overuse injuries during practice—often referred to as 'body armour' by the original communities.
2. Improving technique and form to reduce the chance of failing a movement and help to absorb better the impacts and strains placed upon the body during movement.

Training methodologies vary among communities as to the emphasis placed upon these two goals, with some groups focussing more on building a solid and resilient body while others prefer a focus on improving techniques to avoid the need for resilience. Progressive, gradual parkour practice should lead to a natural improvement in both areas.

A typical primary cause of injury is attempting to progress too far, too fast. Young male practitioners are prone to perform the biggest jumps possible before their bodies can safely absorb the resulting impacts, leading very commonly to overuse injuries of the knee, ankle and lower back. Tendons and ligaments take much more time, roughly four times as long, to strengthen than muscles, and so it is commonly the connective tissue that will give way first in such practitioners, often leading to tendinitis problems.

Training methods that include a focus on progressive increases in maximal strength and joint mobility and flexibility, combined with an improvement in landing technique, posture and the anticipation of impact (enabling the timely pre-firing of the relevant impact absorption chains within the body), will reduce the chances of both acute and overuse injuries. Specific adaptations such as increased ankle dorsiflexion and improved hip mobility can help the body adapt to unusual landing positions and thus improve the capacity to endure the novel strains and demands that might arise during parkour practice.

The importance of endurance, both general muscular and aerobic, is often understated regarding injury prevention in parkour. Parkour training by nature involves extended periods of repetition of dynamic movements which require power, balance, stability and strength. Fatigue quickly undermines these capacities, leading to an increased likelihood of a significant postural failure over time, resulting in injury. Improving muscular endurance, both local and global, enables an individual to perform more repetitions before fatigue sets in, thus reducing the overall chance of an injury, and so is recommended as an injury prevention strategy.

The early practitioners included this form of injury prevention naturally by insisting upon significant muscular endurance and strength practice within every training session. Not only did this build the desired 'body armour' coveted by all early exponents, but it also developed the individual's capacity to perform explosive movements more safely over long periods. It is notable that many of the early practitioners of parkour, some with over 25 years of regular, extreme training, are still athletically high-functioning individuals relatively free of injury, even well into their 40s. Many communities of younger practitioners who have not followed a similar protocol have not fared as well.

Running practice could bring many benefits to injury prevention in parkour. It increases dynamic stability and balance during general movement; it increases contralateral coordination, which facilities power production and thus the ability to manage impact through the elastic slings of the body; it maintains healthy hip function and range of motion, which supports almost every explosive jump or vault pattern in parkour. Trail running improves ankle, hip and knee coordination and mobility and postural responsiveness, which directly benefits the variability required in parkour.

However, while running has clear benefits to a parkour practitioner, it would be wrong to imply that it is more impactful in injury prevention or skill competence than other training strategies and methods. For example, jump precision/accuracy and landing form might be the most critical elements to improve when it comes to avoiding common parkour injuries, and various resistance training modalities—such as weighted squat training—are likely more important in terms of helping the body develop strength and thus the capacity to absorb impact safely.

Ultimately it is the excellent practice of parkour that progressively reduces the chance of injury in parkour.

24.3 Conclusions

Parkour is a young sport in tumultuous evolution, challenging to study due to how it is interpreted in various ways among the different groups of participants.

Running is one of the basic athletic gestures of parkour. However, there are quite peculiar characteristics in this sport because it is linked to other motor patterns in lines of extremely complex acrobatic movements; moreover, it is sometimes practised on unstable, oblique and even vertical planes. The role of running in parkour is universally recognized; some groups of participants reject a systematic approach to training; and running as a form of preparation for the practice of sport is not codified.

Future research should address the physiology of parkour and the physical and performance qualities required by the complex movements of this sport. Research should also estimate the role of sports running training and the optimal ways of running in this sport-specific context.

Contribution Statement FF conceived the paper, gathered data from literature and created the first outline.

DE formulated the main conceptual ideas discussed in the paper based on his unique experience as an athlete and athletic trainer.

Both FF and DE discussed the results and contributed to the final manuscript.

References

1. Puddle DL, P.S. Maulder ground reaction forces and loading rates associated with parkour and traditional drop landing techniques. J Sports Sci Med. 2013;12:122–9.
2. Derakhshan N, Machejefski T. Distinction between parkour and freerunning. Chin J Traumatol. 2015;18(2):124. https://doi.org/10.1016/j.cjtee.2015.07.001.
3. The Sport England 'Active Lives' Survey collates information on the participation of all sports across the country every year. https://www.sportengland.org/know-your-audience/data/active-lives
4. Parkour, literally. https://www.youtube.com/watch?v=SMppD-bUNWo&ab_channel=l1consolable.
5. Grosprêtre S, Gimenez P, Martin A. Neuromuscular and electromechanical properties of ultra-power athletes: the traceurs. Eur J Appl Physiol. 2018;118(7):1361–71. https://doi.org/10.1007/s00421-018-3868-1.
6. Jabnoun S, Borji R, Sahli S. Postural control of parkour athletes compared to recreationally active subjects under different sensory manipulations: a pilot study. Eur J Sport Sci. 2019 May;19(4):461–70. https://doi.org/10.1080/17461391.2018.1527948.
7. Taylor JET, Witt JK, Sugovic M. When walls are no longer barriers: perception of wall height in parkour. Perception. 2011;40:757–60. https://doi.org/10.1068/p6855.
8. Croft JL, Schroeder RT, Bertram JEA. Determinants of optimal leg use strategy: horizontal to vertical transition in the parkour wall climb. J Exp Biol. 2019;222(Pt 1):jeb190983. https://doi.org/10.1242/jeb.190983.
9. Dvorak M, Eves N, Bunc V, Balas J. Effects of parkour training on health-related physical fitness in male adolescents. Open Sports Sci J. 2017;10:132–40.
10. Wanke EM, Thiel N, Groneberg DA, Fischer A. Parkour—"Kunst der Fortbewegung" und ihr Verletzungsrisiko [Parkour—"art of movement" and its injury risk]. Sportverletz Sportschaden. 2013;27(3):169–76. German. https://doi.org/10.1055/s-0033-1350183.
11. Haugen T, Seiler S, Sandbakk Ø, Tønnessen E. The training and development of elite Sprint performance: an integration of scientific and best practice literature. Sports Med Open. 2019;5(1):44. https://doi.org/10.1186/s40798-019-0221-0.
12. Kraemer WJ, Adams K, Cafarelli E, Dudley GA, Dooly C, Feigenbaum MS, et al. American College of Sports Medicine position stand. Progression models in resistance training for healthy adults. Med Sci Sports Exerc. 2002;34(2):364–80. https://doi.org/10.1097/00005768-200202000-00027.
13. Morin JB, Samozino P. Interpreting power-force-velocity profiles for individualised and specific training. Int J Sports Physiol Perform. 2016;11(2):267–72. https://doi.org/10.1123/ijspp.2015-0638.
14. Schmidt RA, Wrisberg CA. Motor learning and performance: a situation based learning approach. 4th ed. Champaign: Human Kinetics; 2008.
15. A.D.A.P.T. (Art du Déplacement And Parkour Teaching) Qualifications. https://www.adaptqualifications.com. Accessed July 2021.
16. Atkinson M. Parkour, anarcho-environmentalism, and poiesis. J Sport Soc Issues. 2009;33:169–94. https://doi.org/10.1177/0193723509332582.
17. Geyh P. Urban free flow: a poetics of parkour. M/C J. 2006;9:3. http://journal.media-culture.org.au/0607/06-geyh.php
18. Mould O. Parkour, the city, the event. Environ Plan. 2009;27:738–50. https://doi.org/10.1068/d11108.
19. Logan G. Automaticity, resources, and memory: theoretical controversies and practical implications. Hum Factors. 1988;30(5):583–98. https://doi.org/10.1177/001872088803000504.
20. Milton JG, Small SS, Solodkin A. On the road to automatic: dynamic aspects in the development of expertise. J Clin Neurophysiol. 2004;21(3):134–43. https://doi.org/10.1097/00004691-200405000-00002.
21. Allum J, Bloem B, Carpenter M, Hulliger M, Hadders-Algra M. Proprioceptive control of posture: a review of new concepts. Gait Posture. 1998;8:214–42. https://doi.org/10.1016/S0966-6362(98)00027-7.
22. Buchanan J, Horak F. Emergence of postural patterns as a function of vision and translation frequency. J Neurophysiol. 1999;81(5):2325–39. https://doi.org/10.1152/jn.1999.81.5.2325.
23. Dietz V, Trippel M, Ibrahim IK, Berger W. Human stance on a sinusoidally translating platform: balance control by feedforward and feedback mechanisms. Exp Brain Res. 1993;93(2):352–62. https://doi.org/10.1007/BF00228405.
24. Kavounoudias A, Roll R, Roll JP. Foot sole and ankle muscle inputs contribute jointly to human erect posture regulation. J Physiol. 2001;532(Pt 3):869–78. https://doi.org/10.1111/j.1469-7793.2001.0869e.x.

Running in Field hockey

25

Martin Häner, Wolf Petersen, Joel Mason,
Stefan Schneider, and Karsten Hollander

25.1 Characteristics of the Sport

Field hockey, an Olympic sport since 1908, is played on a 100 × 60 yard (91.4 × 55 m) pitch. Players use a stick in different length made of wood or a combination of carbon fiber and fiberglass. A sprinkled artificial turf is typically used as surface. The aim of the game is to shoot the ball (made of hard plastic, 22.4–23.5 cm in circumference, approximately 160 grams) into the opponent's goal (2.14 m high, 3.66 m wide). Eleven players play against each other (ten field players and one goalkeeper per team), and the players can be substitute in and out at any time (called "interchanging") with a playing time of 4 × 15 min per game. A team consists of 16–18 players, depending on the competition. According to the international hockey federation (Fédération Internationale de Hockey, FIH), around 30 million people from 137 national federations play field hockey on all five continents. Due to many rule changes over the past few years, the game has increased significantly in running and ball speed. Among other things, a so-called self-pass was recently introduced, in which a player can immediately start running with the ball after a foul without playing a pass. Similar to football, field hockey is played with a goalkeeper, defenders, midfielders, and strikers. The number varies depending on the tactics chosen by the teams. Field hockey is played by recreational athletes on the one hand and as a professional sport on the other. In professional sport, world championships and Olympic Games for women and men are held every 4 years. Similar to other team sports, the number of participating countries will be increased in order to give more countries the chance to qualify for international competitions such as world championships. Sixteen countries are currently qualifying for world championships and 12 countries for the Olympic Games, both for women and men. A specialty of hockey is that especially in Europe, indoor hockey is also very popular and played by most of the field hockey players during the season break in the European winter. It is played on a playing surface as large as a handball field, six vs six. However, this form of play is not an Olympic sport and is rarely

M. Häner (✉) · W. Petersen
Department of Orthopedics and Trauma Surgery,
Martin Luther Krankenhaus, Berlin, Germany
e-mail: martin.haener@jsd.de; Wolf.Petersen@jsd.de

J. Mason
Institute of Sports Science, University of Jena,
Jena, Germany
e-mail: joel.mason@uni-jena.de

S. Schneider
Orthocentrum Hamburg, Hamburg, Germany
e-mail: dr.schneider@oc-h.de

K. Hollander
Institute of Interdisciplinary Exercise Science and
Sports Medicine, MSH Medical School Hamburg,
Hamburg, Germany
e-mail: karsten.hollander@medicalschool-hamburg.de

© The Author(s), under exclusive license to Springer-Verlag GmbH, DE, part of Springer Nature 2022
G. L. Canata et al. (eds.), *The Running Athlete*, https://doi.org/10.1007/978-3-662-65064-6_25

played in warmer countries. Therefore, the following chapter will focus on field hockey.

25.2 Physiological and Biomechanical Demands

Field hockey requires many different physical abilities. Good sprinting skills, as well as endurance, strength and agility, are required for running. Since the game is partly played in a forward flexed and semi-rotated body posture, good stability, especially of the core, is also required. Furthermore, many small, quick changes of direction on a relatively hard surface have a great stress on the joints, tendons, and ligaments, particularly of the lower extremities and the lower back.

25.3 Running Demands for Different Playing Positions

Unfortunately, there is little scientific literature on the different running demands in field hockey on the different playing positions. McGuinness et al. [1] observed that players run about 5500 m per game, with just under 600 m run at high speed. The defenders ran the greatest distance (6170–6643 m), while in comparison midfielders ran 5626–6931 m and strikers 4700–6154 m. It is important to mention, however, that the strikers showed the greatest relative distance of 70–124 m per minute in contrast to midfielders 79–113 m per minute and defenders 79–110 m per minute [1]. These differences can be explained by the fact that the playing time varies between the positions due to the possibility of interchanging, and this must therefore also be taken into account [2]. As a further analysis of running on the field, it is remarkable that strikers play at high intensities for longer periods of the match than midfielders and defenders (8% vs. 6% and 6%) [3]. It should be noted, however, that strikers have more time to recover because their playing intervals are usually shorter [4]. These findings can be important

for the training control and indicates that the training should be adapted to the match requirements according with position on the field.

25.4 Epidemiology and Etiology of Injuries

25.4.1 Epidemiology

As seen above, hockey is a physically demanding sport with increased risks for contact (ball, opponent, and stick) and noncontact injuries (quick changes in direction and frequent acceleration and deceleration phases) [5]. When looking at injury rates from large tournaments, such as the Olympic Games, field hockey is typically seen in the upper tercile of injury rates of all Olympic sports. For example, in the 2008 Summer Olympic Games in Beijing, field hockey was the sport with the third highest injury rate. Over the course of the 2008 Olympic Games, 20.4% of all player were injured [6], with more than 90% of these injuries occurring in competition as opposed training. Similar findings were reported from the 2012 Summer Olympic Games in London, in which field hockey placed fourth among all sports with 17% of all players injured but less in-competition injuries (71.0%) [7]. In the following Olympic Games 2016 in Rio, hockey placed on the tenth position (12% of all players injured), again with a fivefold increased risk for in-competition injuries than training injuries [8]. However, for the 2016 Olympic Games, the highest number of fractures were registered for hockey and rugby (three each) when compared to all sports. When looking at other major international tournaments, 0.7 injuries per match (95%CI 0.5–1.0) were reported for female players and 1.2 injuries per match (95%CI 0.8–1.7) for males [9]. This corresponds to 29.1 (18.6–39.7) injuries per 1000 player match hours (female) and 48.3 (30.9–65.8) injuries per 1000 player match hours (male). Most tournament injuries occurred at the head/face and thigh/knee.

As tournaments are only a small portion of the life of a field hockey player, prospective

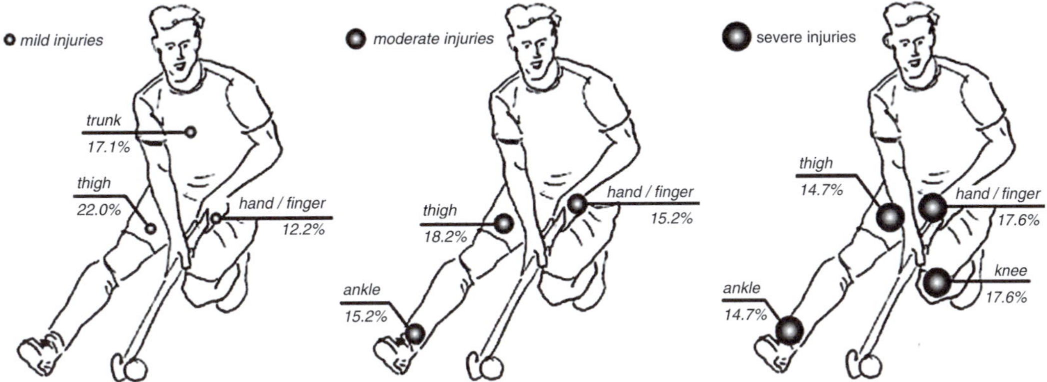

Fig. 25.1 Most common mild, moderate, and severe injuries in field hockey over one season. Reproduced from Hollander et al. (2018) [10]

studies from regular season are relevant to estimate the complete injury epidemiology. Over a complete season, injury incidence rates range between 4.0 and 11.8 injuries per 1000 player hours, with typically higher incidence rates for matches than for practice [10, 11]. Most injuries occurred at the lower limb (thigh, knee, ankle, groin) and were overuse injuries or contusions/hematomas [5, 10, 11]. Figure 25.1 demonstrates the most common mild, moderate, and severe injuries in field hockey according to Hollander et al. [10]. Furthermore, secondary analysis of prospective data revealed that hamstring injuries and muscle strain injuries had the highest burden of injuries [11].

25.4.2 Etiology

There is conflicting evidence regarding the leading mechanisms of injury in field hockey, which likely stems from differences in injury reporting protocols between different levels of play and between different formats of competition.

During international hockey tournaments, most injuries are the result of contact, with 52% of injuries in women and 37% of injuries in men, as result of being hit by the ball which is the first leading mechanism [9]. Contact with the stick is the second leading mechanism of injury in major tournaments, accounting for between 14 and 25%

of injuries, followed by collisions with other players [9]. About half of tournament injuries occur in the circle [9], and 13.9% of all match injuries occur in relation to a penalty corner [12] in which the leading location of injury is the head and face.

Throughout a regular hockey season, noncontact injuries appear to be more frequent [10] and cause a higher injury burden than contact injuries [11]. The relatively high rate of muscle strain injuries, particularly of the hamstring, is purported to arise from the exposure to a high frequency of high-velocity running as well as accelerations and decelerations, which also explains why goalkeepers experience drastically lower rates of hamstring injury than outfield players [11]. It should also be noted that contact injuries remain a consistent threat to players during a regular season, accounting for the highest injury severity [11]. A number of additional factors may contribute to the risk of injury, including playing a high number of games, having an older age, and demonstrating relatively poor or asymmetrical dynamic balance, which have been shown to jointly influence the risk of sustaining an injury throughout a regular season [13].

Standardized injury reports should be implemented across all competition formats so that all types of injury and all severities of injury are accurately registered, and so the risks of injury through both contact and noncontact mechanisms can be mitigated appropriately.

25.5 Training Strategies

The structure of the sport-specific athletic training and field hockey training is similar to that of other ball sports such as football, basketball, or handball. Regardless of age and performance class in the athletic area, it should consist of strength, endurance, speed, interval, and jump training. This training contents can either be carried out separately from training with the stick or integrated into field hockey training. The website of the German Hockey Association (Deutscher Hockey Bund DHB) provides a helpful overview of the training content with an extra portal for trainers and also refers to the different age groups [14]. Particular attention should be paid to core training, as the forward flexed and semi-rotated body posture puts a lot of strain on the lumbar spine compared to other sports (Fig. 25.2 and b).

Depending on age and performance class, the scope and frequency of the training vary greatly. On recreational level, field hockey is usually trained twice a week (1.5–2 h per unit) plus once athletic training. The more professional and performance-oriented the sport becomes, the training can be increased to up to two training units a day for players who, for example, are active in the respective club team in the league as well as in the national team.

Many studies have shown that overall field hockey has a moderate intensity during a game and game types during training [2, 3, 15]. This can be explained by the fact that the game intervals are relatively short (due to the interchanging), and therefore there are always breaks between high intensities during the game. Therefore, the focus in training should be on aerobic conditioning in order to train that the body recovers quickly and can break down any lactate that occurs as quickly as possible [16, 17]. Unfortunately, there are very few studies on different athletic training and its effects in field hockey. A study by Chapman et al. [18] showed that a 3000-m interval-based conditioning program in amateur field hockey players leads to significantly better endurance after 20 weeks of training. In order to be able to make valid statements about this, however, more studies on this topic are certainly necessary.

25.6 Medical Team Support

Due to the bent body posture while playing, the trunk and the lower extremity are exposed to local stress during field hockey compared to other field sports. Athletes mainly suffer injuries affecting the lower limb and the hand. Therefore, special attention should be focused on this regions of the body during sports medicine examinations and physiotherapeutic treatments. In cooperation with the coaches, appropriate exercises should be integrated into the training and conditioning sessions as well as the recommendation to use protective wear.

Fig. 25.2 (a and b) Typical posture when playing the hockey. Forward flexed and semi-rotated posture increases the stress on the lower back

Further, the medical team should be aware of bleeding wounds. During training and game sessions, the players frequently get hurt by the ball or the stick. Contusions, hematomas, abrasions, and lacerations are common [10]. The equipment of the doctor's case should therefore contain suture sets, staplers, wound closure strips, and cool packs. Immunization of all players against tetanus should be up to date. To reduce the rate of these injuries, goalkeepers wear full body protection, and field players wear shin guards and mouthguards, and many players wear gloves and a few using a suspensory.

25.7 Specific Rehab and Return to Play

Most injuries in field hockey are minor with a return to play of less than 7 days [19]. However, more severe injuries can occur and may require a sport-specific return to play schedule. Most severe injuries affect the bones, followed by ruptures and strains.

For bony injuries, rehabilitation differs between acute traumatological fractures and overuse bone stress injuries. Acute traumatological fractures need urgent attention and may be treated conservatively or may require surgical treatment depending on the type, location, displacement, or accompanying injuries [20]. Bone stress injuries can typically be treated conservatively, and a high number of athletes are able to return to sports [20]. However, the return-to-sport time depends on the MRI-based grading and typically ranges from 42 to 99 days. Furthermore, injuries to trabecular-rich bones, such as the calcaneus, femoral neck, or pelvis, take longer time to heal than cortical-rich bones, such as the tibia and metatarsals [20].

For ligamentous injuries, the anterior cruciate ligament (ACL) rupture is an injury with a high burden of disease (high incidence coupled high severity) and typically needs specific and individualized rehabilitation strategies. While the early operative treatment is not cost-effective for all patients, many professional athletes will undergo a surgical repair [21]. There is good evidence that postoperative rehabilitation should incorporate immediate knee mobilization, closed kinetic chain exercises, cryotherapy, and neuromuscular training [22]. In contrast, passive motion and functional bracing is to be avoided. Early weight bearing (within the first week)) is also encouraged, depending on a patient's toleration [22]. A return to play decision should be taken on an interdisciplinary approach taking in consideration graft healing and clinical/functional progression (including scores such as KOOS, IKDC, or Lysholm scores), psychological evaluation (ACL-RSI), and a functional return-to-sport assessment. The return-to-sport criteria may include isokinetic strength and functional performance (single hop, triple hop, crossover hop, 6-m timed hop). While there has been progress in the scientific literature regarding return-to-sport decisions after ACL repair, there is no one-size-fits-all recommendation for all sports [22]. Especially for field hockey, there are no specific recommendations in return to sports, and, due to the high demands, the decision should be taken according with the clinical and physical parameters in a gradual, step-by-step, way. Apart from the ACL rupture, there are no evidence-based recommendation for an association of functional performance and return-to-play decision [23]. However, in recent years, there have been recommendations published by the German professional sport statutory accident insurance (Verwaltungs-Berufsgenossenschaft (VBG)) for ACL rupture as well as ankle sprains and concussion [24, 25]. While these are not specific for field hockey, they should be taken into consideration.

25.8 Prevention Strategies

As already mentioned, field hockey has changed significantly in the last few decades due to further development of the stick and artificial turf as well as the changing of increasing physical demands on the players with the need of injury prevention significantly increasing. Noncontact injuries occur most frequently on the lower extremities (e.g., ankle sprains and thigh muscle strains) [10,

26]. Therefore, special attention should be paid to the correct warm-up before every training session and matches. In addition to running and active stretching exercises, the warm-up program should also include proprioceptive and neuromuscular exercises. A study by Verhagen et al. [27] demonstrated that training on a balance board, for example, could reduce the likelihood of ankle sprain. Another study showed that poor peak dorsiflexor torque at the ankle was associated with an increased incidence of ankle injuries in elite female field hockey players [28]. Back pain can also be reduced through targeted training in field hockey [29, 30].

There is very little literature on specific warm-up programs in field hockey to reduce the risk of injury. Barboza et al. [31] examined a warm-up program (developed by the Dutch hockey federation (KNHB) and VeiligheidNL) in field hockey with a view to reducing injuries, their severity, and burden. No significant difference in the reduction of injuries or their severity could be shown. However, the burden of injuries on players' field hockey participation in the intervention group could be reduced (difference of 8.42 (95%CI = 4.37, 12.47) days lost per 1000 player hours of field hockey). Probably the best studied warm-up program is FIFA11 + (football). In a systematic review and a meta-analysis, Thorborg et al. [32] have shown that a standardized warm-up program could reduce the incidence of football injuries by 39%. The German Knee Society (Deutsche Knie Gesellschaft, DKG) has also developed a prevention program ("Stop-X") to reduce the occurrence of severe knee injuries [33]. This program is divided into four main prevention areas: (1) functional diagnostics, (2) running exercises, (3) mobilization and activation, and (4) neuromuscular training. With various videos and related explanations (unfortunately only in German so far), even amateur athletes can independently carry out a prevention program. Since knee injuries occur relatively rarely in field hockey, ACL or PLC injuries can represent very serious injuries, and prevention should be stimulated.

In field hockey, further studies are necessary in order to substantiate existing initial findings and to find out further patterns of development.

Existing prevention programs from other sports could then be adapted to the specific requirements of field hockey or even developed on their own.

References

1. McGuinness A, Malone S, Petrakos G, Collins K. Physical and physiological demands of elite international female field hockey players during competitive match play. J Strength Cond Res. 2019;33:3105–13.
2. Vescovi JD, Frayne DH. Motion characteristics of division I college field hockey: female athletes in motion (FAiM) study. Int J Sports Physiol Perform. 2015;10:476–81.
3. Macutkiewicz D, Sunderland C. The use of GPS to evaluate activity profiles of elite women hockey players during match-play. J Sports Sci. 2011;29:967–73.
4. Lythe J, Kilding AE. Physical demands and physiological responses during elite field hockey. Int J Sports Med. 2011;32:523–8.
5. Barboza SD, Joseph C, Nauta J, van Mechelen W, Verhagen E. Injuries in field hockey players: a systematic review. Sport Med. 2018;48:849–66.
6. Junge A, Engebretsen L, Mountjoy ML, Alonso JM, Renström PAFH, Aubry MJ, Dvorak J. Sports injuries during the summer Olympic games 2008. Am J Sports Med. 2009;37:2165–72.
7. Engebretsen L, Soligard T, Steffen K, u. a. Sports injuries and illnesses during the London summer Olympic games 2012. Br J Sports Med. 2013;47:407–14.
8. Soligard T, Steffen K, Palmer D, u. a. Sports injury and illness incidence in the Rio de Janeiro 2016 Olympic summer games: a prospective study of 11274 athletes from 207 countries. Br J Sports Med. 2017;51:1265–71.
9. Theilen TM, Mueller-Eising W, Bettink PW, Rolle U. Injury data of major international field hockey tournaments. Br J Sports Med. 2016;50:657–60.
10. Hollander K, Wellmann K, Eulenburg CZ, Braumann KM, Junge A, Zech A. Epidemiology of injuries in outdoor and indoor hockey players over one season: a prospective cohort study. Br J Sports Med. 2018;52:1091–6.
11. Rees H, McCarthy Persson U, Delahunt E, Boreham C, Blake C. The burden of injury in field hockey: a secondary analysis of prospective cohort data. Scand J Med Sci Sport. 2021;31:884–93.
12. Theilen TM, Green M, Mueller-Eising W, Rolle U. Incidence of penalty corner injuries in international field hockey. Res Sport Med. 2022;30(2):193–202. https://doi.org/10.1080/15438627.2020.1862842.
13. Mason J, Wellmann K, Groll A, Braumann KM, Junge A, Hollander K, Zech A. Game exposure, player characteristics, and neuromuscular performance influence injury risk in professional and youth field hockey play-

ers. Orthop J Sport Med. 2021;9(4):2325967121995167. https://doi.org/10.1177/2325967121995167.

14. Athletik | DHB. https://trainer.hockey.de/athletik. Accessed 24 Mai 2021.

15. Gabbett TJ. GPS analysis of elite women's field hockey training and competition. J Strength Cond Res. 2010;24:1321–4.

16. Jennings DH, Cormack SJ, Coutts AJ, Aughey RJ. International field hockey players perform more high-speed running than nationallevel counterparts. J Strength Cond Res. 2012;26:947–52.

17. Sell KM, Ledesma AB. Heart rate and energy expenditure in division i field hockey players during competitive play. J Strength Cond Res. 2016;30:2122–8.

18. Chapman DW, Newton MJ, McGuigan MR. Efficacy of interval-based training on conditioning of amateur field hockey players. J Strength Cond Res. 2009;23:712–7.

19. Rees H, Shrier I, McCarthy Persson U, Delahunt E, Boreham C, Blake C. Transient injuries are a problem in field hockey: a prospective one-season cohort study. Transl Sport Med. 2020;3:119–26.

20. Hoenig T, Tenforde AS, Strahl A, Rolvien T, Hollander K. Does magnetic resonance imaging grading correlate with return to sports after bone stress injuries? A systematic review and meta-analysis. Am J Sports Med. 2021;363546521993807.

21. Eggerding V, Reijman M, Meuffels DE, Van Es E, Van Arkel E, Van Den Brand I, Van Linge J, Zijl J, Ma Bierma-Zeinstra S, Koopmanschap M. ACL reconstruction for all is not cost-effective after acute ACL rupture. Br J Sport Med. 2022;56:24–8.

22. Andrade R, Pereira R, Van Cingel R, Staal JB, Espregueira-Mendes J. How should clinicians rehabilitate patients after ACL reconstruction? A systematic review of clinical practice guidelines (CPGs) with a focus on quality appraisal (AGREE II). Br J Sports Med. 2020;54:512–9.

23. Vereijken A, Aerts I, Jetten J, Tassignon B, Verschueren J, Meeusen R, van Trijffel E. Association between functional performance and return to performance in high-impact sports after lower extremity injury: a systematic review. J Sport Sci Med. 2020;19:564–76.

24. Bloch H, Klein C, Kühn N, Luig P. (PDF) Return to Competition - Test manual for assessment of the ability to play after an acute lateral ankle sprain injury. https://www.researchgate.net/publication/336835897_Return_to_Competition_-_Test_manual_for_assessment_of_the_ability_to_play_after_an_acute_lateral_ankle_sprain_injury. Accessed 5 Juni 2021.

25. Bloch H, Klein C, Luig P, Riepenhof H. Return to competition: safe return to sport. Trauma Berufskrankheit. 2017;19:26–34.

26. Murtaugh K. Field hockey injuries. Curr Sports Med Rep. 2009;8:267–72.

27. Verhagen E, Van Der Beek A, Twisk J, Bouter L, Bahr R, Van Mechelen W. The effect of a proprioceptive balance board training program for the prevention of ankle sprains: a prospective controlled trial. Am J Sports Med. 2004;32:1385–93.

28. Naicker M, McLean M, Esterhuizen TM, Peters-Futre EM. Poor peak dorsiflexor torque associated with incidence of ankle injury in elite field female hockey players. J Sci Med Sport. 2007;10:363–71.

29. Twomey L. Spinal mobility and trunk muscle strength in elite hockey players. Aust J Physiother. 1988;34:123–30.

30. Fenety A, Kumar S. Isokinetic trunk strength and lumbosacral range of motion in elite female field hockey players reporting low-back pain. J Orthop Sports Phys Ther. 1992;16:129–35.

31. Barboza SD, Nauta J, Emery C, Van Mechelen W, Gouttebarge V, Verhagen E. A warm-up program to reduce injuries in youth field hockey players: a quasi-experiment. J Athl Train. 2019;54:374–83.

32. Thorborg K, Krommes KK, Esteve E, Clausen MB, Bartels EM, Rathleff MS. Effect of specific exercise-based football injury prevention programmes on the overall injury rate in football: a systematic review and meta-analysis of the FIFA 11 and 11+ programmes. Br J Sports Med. 2017;51:562–71.

33. Trainingsübungen für Prävention und Rehabilitation von Knieverletzungen. https://www.stop-x.de/uebungen/?art=praevention. Accessed 24 Mai 2021.

Running in Cycling

26

George A. Komnos and Jacques Menetrey

26.1 Introduction

Cycling and running are both aerobic activities but are also considered to be two different means of exercising. There is a lot of amateur or recreational athletes who like to combine or alternate between these two sports. However, things are most complicated for professional cyclists, who have to be very familiar with the advantages of incorporating running sessions in their training and with the possible risks as well.

In general, running can be a beneficial cross-training activity for cyclists. Nevertheless, several factors, such as the type of cycling someone does, the training phase, and demographic data define the frequency and intensity of running in a cyclist. Misuse can lead to an injury or performance impairment. Notably, running can contribute a lot to a cyclist's training and performance. Proper application of cross-training can definitely improve performance. Among others, it can conduce to ameliorate cardiovascular and exercise tolerance. Moreover, used as a cross-training exercise, it can provide a completely different aerobic experience leading to increased muscular strength and endurance.

Cycling may be considered a workout for the lower body, but it can improve overall muscle strength and enhance stamina. Specific muscles, such as the quadriceps, gluteus, and core muscles, are used and strengthened more during cycling. Cycling is a nonimpact sports; therefore cyclists need to train their muscles and joints to smoothly switch from one type of training to the other (running). Generally, we can assume that both cycling and running are good for overall health, cardiovascular system, and weight loss, with few particular differences between them (Table 26.1).

G. A. Komnos
Centre de Medecine du Sport et de l'Exercice, Swiss Olympic Medical Center, Hirslanden Clinique la Colline, Geneva, Switzerland
e-mail: Jacques.Menetrey@hirslanden.ch

J. Menetrey (✉)
Centre de Medecine du Sport et de l'Exercice, Swiss Olympic Medical Center, Hirslanden Clinique la Colline, Geneva, Switzerland

Orthopaedic Surgery Service, University Hospital of Geneva, Geneva, Switzerland
e-mail: jacques.menetrey@hcuge.ch

Table 26.1 Overview of general similarities and differences between cycling and running [1, 2]

- Both improve cardiovascular fitness
  - Increase maximum oxygen volume (VO_2 max)
  - Decrease heart rate
- Running is more calorie-consuming than cycling
- Weight bearing applied is less in cycling, adding less pressure on the knees, hips, and spine
- Cycling is less prone to injuries
  - Better for recovery
  - Better for starters
- Running enhances immune system more, while cycling enhances muscle tone more

26.2 Pathophysiology and Biomechanics

Running and cycling seem to result in different physiological and metabolic responses having different physiological effects [3]. It has been demonstrated that fat oxidation is higher in running than in cycling at the same percentage of maximum oxygen uptake (VO_2 max) [4]. Furthermore, plasma lactate concentrations are reported to be higher in cycling than running at the same sub-maximal workload [4, 5]. Subsequently, if lactate concentration defined the intensity of running and cycling, then the metabolic demand would be higher in running. Capostagno et al. demonstrated that that fat oxidation is higher during running than cycling at certain relative intensities expressed as either maximum workload (%WL max) or maximum oxygen uptake (%VO_2 max) [6]. Several theories may justify this process. Catecholamine responses, which activate lipolysis, may differ between running and cycling at the same relative intensity and result in higher rates of fat oxidation in running. Nevertheless, this is not always the case, since Nieman et al. reported almost equal levels of norepinephrine and epinephrine after 2 and a half hours of cycling and running training at the same intensity [7]. Another proposed theory suggests that the workload during cycling may be distributed in fewer muscle fibers than in running, making muscle strain more localized and producing a higher relative workload on the muscles during cycling [3]. In this way, the energy requirement of each fiber is higher during cycling. This leads to higher oxidation of carbohydrates and greater production of lactate with a decrease in fat oxidation [5]. Although recreational cyclists have higher maximal oxygen uptake (VO_2 max) values during running, this does not always apply to professional cyclists [8].

Exercise economy is defined as the whole rate of oxygen or metabolic energy consumption for a specified exercise task. The economy of the mode of exercise used has been investigated in an attempt to define if particular modes of exercise enhance cycling economy (CE). Although this has been investigated for running, the mode-specific trainability of the cycling economy is not so clear yet. Therefore, there is still the question of whether specific training for running improves CE. Millet et al. [9] evaluated the physiological differences between running and cycling and proposed that regarding VO_2 max and lactate threshold, there is a greater training transfer from running to cycling than vice versa. In other words, cyclists are more benefited from running exercise than runners from cycling ones. This was also reported by Swinnen et al. [8]. They found that the running test was a more intense effort for the cyclists compared to the runners or triathletes. Moreover, they concluded that cyclists could replace some of the time-consuming cycling training with shorter sessions of running training without impairment of their CE.

Aerobic exercise is performed through both cycling and running, but with different muscles stimulated in each case. Muscle biomechanics are distinctly different between running and cycling. There is a different posture and different muscles engaged and forces applied [10]. During running, muscles act almost isometrically, and eccentric muscle actions play an important role, while in cycling they undergo mainly concentric contractions (muscle shortening) [11]. Subsequently, the mechanical cost of the muscle work varies considerably. Muscle fibers are more severely damaged in running than cycling, applying a greater strain to the muscles and the whole body for the same exercise period. Another issue is that in running, no neuromuscular connections specific for cycling are developed.

One of the mechanisms that may contribute to better performance in cycling with running training sessions is through the enhancement of transformation of muscle fibers type II to type I. It is reported that general endurance training, like running and cycling, promotes muscle fiber transformation from fast (type II) to slow (type I) fibers [12]. As running promotes this fiber transformation, this can lead to better performance in cycling, as a positive correlation between the percentage of type I slow muscle fibers and gross cycling efficiency has been reported [13].

26.3 Benefits of Running for Cyclists

As aforementioned, from a physiological point of view, running can potentially improve VO_2 max and optimize other cardiovascular demands better than cycling [9]. Furthermore, running provides a quicker workout than cycling, burning more fat in less time [6]. Of note, there are certain benefits of running for cyclists. These mainly consist of an increase in bone density, increased cardiovascular and muscular strength and exercise tolerance, and psychological benefits.

26.3.1 Bone Density

Cycling is a nonimpact sport without weight-bearing stress. As a result, it cannot increase bone density. There is data showing that endurance athletes, such as cyclists, exhibit significantly lower bone mineral density since the biomechanics of cycling may not adequately stimulate bone formation. On the contrary, running activates a series of reactions that include the muscles and strengthen the bones, decreasing the possibility of subsequent osteoporosis or osteopenia [14]. Furthermore, this lowers the possibility of injuries and bone fractures while aging. Mathis et al. [15] suggested that cyclists should participate in weight-bearing training, such as running, in order to increase or maintain the bone mineral density of the lumbar spine and hip.

26.3.2 Increased Cardiovascular, Muscular Strength, and Exercise Tolerance

Furthermore, running can provide a completely different aerobic experience for a cyclist, adding muscular endurance. The period of training is of paramount importance in this case. It is preferable to add running to a cyclist's training program at the beginning of the season, in an attempt to enhance cardiovascular fitness. Besides, running uses more muscle mass than cycling. This mainly applies to mountain bikers and cyclo-cross racers whose races have high demands of muscular strength and endurance.

26.3.3 Psychological Benefits

Running can also be used as a break of everyday practice. It can help the athlete to take a break from the main sport and exercise but also relax by doing a different sport. Besides, it can be utilized as an alternative means of exercising in traveling or in other circumstances where no access to bike equipment exists. Furthermore, it can be used as a mind-clearing means, which is not always easy during cycling, which demands a high and continuous amount of focus and caution.

Other advantages of running for cyclists include the development of balance skills and proprioception, the fact that it can act as a good postural activity and that it is much more easily performed anywhere.

26.4 Negative Impact of Running on Cyclists

Nevertheless, running in cyclists has also been correlated to an increased risk for injury. In a study including 739 cyclists, overuse or fatigue injury within the last year was related to engaging in running or swimming training [16]. Nonetheless, combining running and cycling without a proper periodization of the training can lead to overload, fatigue, and subsequent injury [17].

Past injuries should always be considered. In cases that a cyclist has a past injury related to running, then he/she can use another form of aerobic training activity such as swimming. Weight and age are two essential parameters to take into consideration if someone wants to enhance his/her cycling performance. Younger athletes can undergo more workouts, which is not always evitable in older athletes who need more time to recover. Older age and increased weight can negatively impact health and cycling performance.

26.5 Running in Different Modes of Cycling

26.5.1 Running for a Road Cyclist

In this group, running can contribute to the phase of total body conditioning during off-seasons. It assists in maintaining a good condition of the cardiovascular system and tolerance. Nevertheless, it should be better to quit during advanced training sessions, due to the high risk of injury.

26.5.2 Running for a Mountain Bike Rider

In mountain bike cyclists, running can also play an essential role as an alternative form of aerobic endurance training in the phase of total body conditioning, but it belongs to a more specific training for this particular group in preparation for the portage zone. More generally, it can improve body balance and neuromuscular coordination, which is required in this type of cycling for better performance and a lower risk of injuries.

26.5.3 Running for a Cyclo-Cross Rider

In cyclo-cross, running when carrying the bike is one important part of the sports. Therefore, running session should be part of the general and specific training. These running sessions should be considered as a part of the regular off- and in-season training cycle in cyclo-cross. Running on

an uneven ground is recommended. The rider should also add technical sessions where he practices the jump-off and jump-on the bike and run with the bike on his shoulder.

Cyclo-cross racing represents the perfect merge of the two activities, running and cycling, and requires significant upper body strength and cardiovascular conditioning. Nevertheless, injuries are usually minor [18]. The most common injuries are similar to cycling injuries including acromioclavicular ligament injuries, head injuries, and concussion [18].

References

1. Nieman DC, Luo B, Dréau D, et al. Immune and inflammation responses to a 3-day period of intensified running versus cycling. Brain Behav Immun. 2014;39:180–5.
2. Hanon C, Dorel S, Delfour-Peyrethon R, et al. Prevalence of cardio-respiratory factors in the occurrence of the decrease in oxygen uptake during supramaximal, constant-power exercise. Springerplus. 2013;2:651. Epub ahead of print 2013. https://doi.org/10.1186/2193-1801-2-651.
3. Achten J, Venables MC, Jeukendrup AE. Fat oxidation rates are higher during running compared with cycling over a wide range of intensities. Metab Clin Exp. 2003;52:747–52.
4. Knechtle B, Müller C, Willmann F, et al. Fat oxidation in men and women endurance athletes in running and cycling. Int J Sports Med. 2004;25:38–44.
5. Carter H, Jones AM, Barstow TJ, et al. Oxygen uptake kinetics in treadmill running and cycle ergometry: a comparison. J Appl Physiol. 2000;89:899–907.
6. Capostagno B, Bosch A. Higher fat oxidation in running than cycling at the same exercise intensities. Int J Sport Nutr Exerc Metab. 2010;20:44–55.
7. Nieman DC, Nehlsen-Cannarella SL, Fagoaga OR, et al. Effects of mode and carbohydrate on the granulocyte and monocyte response to intensive, prolonged exercises. J Appl Physiol. 1998;84:1252–9.
8. Swinnen W, Kipp S, Kram R. Comparison of running and cycling economy in runners, cyclists, and triathletes. Eur J Appl Physiol. 2018;118:1331–8.
9. Millet GP, Vleck VE, Bentley DJ. Physiological differences between cycling and running: lessons from triathletes. Sports Med. 2009;39:179–206.
10. Novacheck TF. The biomechanics of running. Gait Posture. 1998;7:77–95.

11. Bijker KE, de Groot G, Hollander AP. Differences in leg muscle activity during running and cycling in humans. Eur J Appl Physiol. 2002;87:556–61.

12. Wang YX, Zhang CL, Yu RT, et al. Regulation of muscle fiber type and running endurance by PPARδ. PLoS Biol. 2004;2(10):e294. Epub ahead of print October 2004. https://doi.org/10.1371/journal.pbio.0020294.

13. Horowitz JF, Sidossis LS, Coyle EF. High efficiency of type I muscle fibers improves performance. Int J Sports Med. 1994;15(3):152–7.

14. Shanb A, Youssef E. The impact of adding weight-bearing exercise versus nonweight bearing programs to the medical treatment of elderly patients with osteoporosis. J Fam Community Med. 2014;21:176.

15. Mathis SL, Farley RS, Fuller DK, et al. The relationship between cortisol and bone mineral density in competitive male cyclists. J Sports Med. 2013;2013:1–7.

16. Priego Quesada JI, Kerr ZY, Bertucci WM, et al. A retrospective international study on factors associated with injury, discomfort and pain perception among cyclists. PLoS One. 2019;14(1):e0211197. Epub ahead of print January 1, 2019. https://doi.org/10.1371/journal.pone.0211197.

17. Stanley J, D'Auria S, Buchheit M. Cardiac parasympathetic activity and race performance: an elite triathlete case study. Int J Sports Physiol Perform. 2015;10:528–34.

18. Evenson L. Cyclo-cross racing: a two-wheeled steeplechase. Phys Sportsmed. 1982;10:204–6.

Running in Alpine Skiing

27

Felix Mayr, Lukas Willinger, and Philipp W. Winkler

27.1 Characteristics of Running and Its Role in Alpine Skiing

There are different types of running with varying demands on the musculoskeletal system, neuromuscular coordination, cardiovascular system, and metabolism. Combining different types of running in a structured and diversified training program may improve running performance and prevent overuse running-related injuries. Depending on the running distance, runs are categorized as short-distance ($\leq$15 km), long-distance (21.1–42.195 km), and ultra-long-distance (>42.195 km) [1, 2]. Depending on the running pace, sprint, tempo, and interval runs can be distinguished from base and recovery runs. While the former focus on the anaerobic zone, the latter are characterized by an aerobic training. In addition, there are differences in the running terrain, so that a distinction can be made between on-road and off-road running. Off-road running, also called trail running, sky running, or mountain running, has become increasingly popular in recent years and is characterized by running on unpaved trails in wooded or alpine terrain [3]. Running on unpaved trails offers many benefits, such as avoiding repetitive impacts on the lower limb from hard concrete surfaces and improving neuromuscular coordination. Given the high speed in alpine skiing, neuromuscular coordination is of utmost importance as the load on the lower limb muscles can change within milliseconds during downhill skiing. Therefore, off-road running should be an inherent part of a diversified training program prior to an upcoming ski season. However, downhill sections are commonly part of off-road running trails and are characterized by eccentric quadriceps contractions which have been shown to cause exercise-induced muscle damage. Microstructural damage to myofibrils may ultimately compromise muscle function and stamina and cause neuromuscular fatigue. Consequently, it is recommended to avoid downhill trails prior to ski races/days [4].

Consideration of the timing of the most common injuries in alpine skiing underscores the importance of a well-structured endurance training in preparation for an upcoming ski season. In World Cup alpine skiers, it has been shown that the knee is the most frequently injured body part (39.9% of all injuries), regardless of sex [5]. In ski-related knee injuries that are hospitalized, the anterior cruciate ligament (ACL) is the most frequently injured part of the knee in alpine skiing

F. Mayr (✉) · L. Willinger
Department for Trauma Surgery, Klinikum rechts der Isar, Technical University of Munich,
Munich, Germany
e-mail: Lukas.Willinger@mri.tum.de

P. W. Winkler
Department for Orthopaedic Sports Medicine, Klinikum rechts der Isar, Technical University of Munich, Munich, Germany

[6]. A systematic video analysis of 69 injuries in World Cup alpine skiing found that 46% of injuries occurred in the last quarter of the race course [7]. In addition, a study that longitudinally assessed professional alpine skiers based on quantitative magnetic resonance imaging (i.e., T2* mapping) showed an increase in T2* relaxation time of the ACL over an entire ski season, indicating a higher risk of fatigue injury to the ACL at the end of the competitive season [8]. Consequently, it is assumed that muscular and ligament fatigues are major risk factors for injuries in alpine skiers. Accordingly, a structured aerobic running exercise program to increase muscular endurance may reduce the risk of fatigue injuries.

27.2 Training Strategies

Physical demands on modern alpine skiing athletes are high and require a broad range of capabilities within all physiological systems such as maximal strength, anaerobic and aerobic endurance, and motor skills (balance, coordination, and mobility). Each of the four disciplines of alpine ski racing (downhill, super-G, giant slalom, slalom) demonstrates different loading patterns with regard to duration (between 45 s and 150 s) and muscle activation patterns. Athletes are commonly specialized in either speed or technical disciplines in order to being able to meet physiological requirements for either of them. Alpine ski racing is classified as a mainly anaerobic activity with maximal demands on the circulatory system and isometric leg muscle contraction for almost the entire race duration [9].

Professional alpine skiers typically train around 150 days each year. The training is divided into a preparation period between April and October and the competitive season during the winter months [10]. The competition period is mainly composed of participating in official trainings and contests as well as active recovery session in between. The goal is to maintain physical fitness and engage in recovery and rehabilitation. In contrary, the preparation season is typically used for improving the physical condi-

tion to meet the broad physical demands of the sports during the competitive season. Elite athletes perform between 10 and 14 training sessions a week including endurance, strength, agility, and stability sessions. Within this framework and depending on the season, the training is divided into off-snow and on-snow training sessions. In particular, during summer months, the training is focused on gaining maximal strength and power as well as anaerobic endurance with a lasting effect.

The training of athletes is complex and involves various dimensions such as strength and endurance during the same period. Furthermore, it has to be considered that when off- and on-snow training are combined, fatigue from off-snow training must not compromise technical skills training on skis. The combination of technical and physical training and the attempt to improve the athletes' performance overall is one of the most difficult aspect of training schedules.

Ski athletes have a high endurance capacity compared to the normal population which reflects the training program and the demands of the sport. Elite athletes reach high VO_2 max values between 60 and 70 ml/kg/min [9].

In order to increase the aerobic and anaerobic capacity of alpine skiers, running has been shown to be an effective training modality [11]. Concerning sport-specific endurance training, high-intensity interval training (HIIT: short- and high-intensity intervals of exercise alternating with recovery phases) has been shown to best match the physical requirements of alpine ski racing with regard to loading duration and energy consumption. It is a time efficient alternative to traditional continuous endurance training [12] and results in superior adaptions concerning performance markers. It increases both aerobic and anaerobic capacities which are important for alpine ski racers to improve competitive performance. Studies have shown that separated blocks of strength and endurance training lead to a superior overall improvement of performance compared to combined training [11–14]. Short-duration HIIT blocks (e.g., 2-week shock cycle) were sufficient to increase VO_2 max by 4.3–6% [13, 14]. In both studies, HIIT was carried out with non-skiing-specific modalities like ergometer cycling or running [11, 13]. Running induced similar or even superior response in VO_2 compared to alpine ski racing-specific HIIT modes using a ski ergometer [11].

27.3 Prevention Strategies

Alpine skiing inspires a large number of people regardless of their age and sex. All people are moving comparable loads. However, the constitution, state of training, and technical requirements are different for each individual alpine skier. An essential task of the skiing athlete is to combine physical and mental requirements. A high level of endurance is important to prevent overload and ski-related injuries. The key is to improve muscle strength and coordination. Therefore, there is a need for preparatory measures such as warm-up exercises. Individual goals are key for success and safety in recreational and competitive skiing [15].

27.3.1 Injury and Injury Patterns in Alpine Skiing

Ten percent of German skiers sustain injuries that require medical attention annually [16]. The knee joint is the most commonly injured part of the body, followed by the hands and shoulders (Fig. 27.1). In nearly half of all severe injuries, the anterior cruciate ligament is involved [17].

To develop efficient prevention strategies, one must understand the different pathomechanisms of ski injuries [17].

A typical pathomechanism in recreational skiing is the effect of rapidly turning direction on the lower extremity in relation to the torso. When now the flexed knee makes an external rotation of the lower leg, the medial side of the knee opens ("valgus stress"), so that the native ligament can no longer resist the acting forces. Also, internal rotation with an outward tilt ("varus stress") is possible. This mechanism often leads to ACL injuries also at minor speed.

In recreational skiing the most common pathomechanisms for ACL injury is the so-called phantom foot. In this situation, the skier is out of balance. The body position is in layback with the hips below the knees. The upper body generally faces the downhill ski and the uphill arm is back. The knee is forced into internal rotation in a deeply flexed position, when the inside edge of the downhill ski tail engages the snow surface. The ski acts as a lever to twist or bend the knee. Therefore, this mechanisms for injury is called the "phantom foot" [18].

In regard of ski racing, we focus on three major groups of pathomechanisms for injuries. The most common one is the slip catch where the outer ski catches the inside edge. So the outer knee is forced into internal rotation and valgus. There is the same pathomechanism in the dynamic snowplow [19]. The third mechanism is typical for jumps. After the jump the athlete lands backward, trying to bring the body's center of gravity frontward again to regain balance. The combination of tibiofemoral compression, the "boot-induced anterior drawer" and the quadriceps anterior drawer, strains the ACL [17].

27.3.2 Running as a Prevention Tool in Alpine Skiing

27.3.2.1 Interval Training and Long-Distance Run

As initially mentioned, an essential prerequisite for the practice of alpine skiing at the lowest possible risk is general endurance and strength training. Not only the metabolism has to be able to mobilize enough energy to meet the specific demands but must also provide for a muscular milieu, concerning pH, that can enable required fine motor skills [17]. Therefore, interval training and long-distance runs play an important role for prevention. Interval training, depending on its characteristics, aims to improve strength endurance, speed endurance, lactate tolerance, lactate breakdown, maximum oxygen uptake, or even the feeling of speed (e.g., competition pace). Furthermore, interval training is helping

Fig. 27.1 Distribution of ski injuries on body parts (Photo: Cornelia and Paul Schmidt)

to improve and to economize movement sequences (inter- and intramuscular coordination) [20]. As mentioned before, running plays a subordinate role in alpine skiing; therefore interval training is not only having a preventive effect. It is involving risks for inexperienced athletes. This high burden on the organism may lead to damage especially to the musculoskeletal system in inadequately trained athletes. For example, the athlete can suffer injuries to the Achilles tendon, as the adaptation to training stimuli due to the lower blood flow in tendons, ligaments, joints, and bones (passive musculoskeletal system) takes much longer than in the well-perfused skeletal muscles [21].

27.3.2.2 Off-Road Running

As most ski athletes live in or close to the mountains and being close to nature, off-road running enjoys great popularity. In alpine skiing, overload injuries are common, so one should avoid classic road running, in which overload injuries are caused by repeated impacts from hard concrete surfaces. Therefore, off-road running is a good alternative. The surface of the trails is softer, which means that knees, ankles, shins, and hips are better protected. The risk of injuries that are aggravated by impacts, such as knee pain or even shin splints, is thus reduced. Winding trails, varied surfaces, and the constant avoidance of roots and branches, rocks, and stones forces the trail runner to be focused, thereby improving neuromuscular coordination. This is an important prevention tool for recreational skiers who are also confronted with different challenges, such as different terrains, slope conditions, snow quality, and slope frequenting by other skiers. For the purpose of optimal injury and strain prevention, the athlete has to learn and consider signs of fatigue within the scope of risk management [17, 22].

In regard of competitive alpine skiing, the following factors of active prevention are essential [17, 22]:

- Optimization of mental and physical agility using coordination training: Trail runners must adapt quickly to the changing terrain. This means improved balance and stability as well as an improved feeling for the ground. Trail running increases the use of muscles, and it increases the awareness of the body's movement and position in space (i.e., proprioception).
- Torso-stabilizing training (i.e., core strengthening) to counteract the whiplash-like inertia torque during turn changes and the compression forces in the spine with neutralized S form: Running off-road trains the same muscles as road running but also the smaller,

stabilizing muscles that are required for lateral movement, balance, and proprioception. This is activating and training the inner and outer thighs (i.e., adductors and abductors) and is strengthening the feet and ankles. If the step height is taller, the abdominal muscles and the lower back are trained as well, and the body activates other stabilizing muscles for an improved balance. Off-road running supports stability, strengthens feet and ankles, and provides more mobility.

- Specific training of physical endurance and force: A longer trail run in an alpine terrain is a good alternative to strength training. Different inclines and hill lengths and surfaces demand different muscles more often, which activates more secondary muscle groups. Softer terrain also means less rebound. This means additional resistance, which leads to a muscle training of the quads, hip flexors, and glutes.

important. This is specifically trained by interval training and long-distance runs. Neuromuscular coordination and proprioception prevent injuries and are improved by off-road running.

27.4 Conclusion for Practice

In conclusion, a continuous, structured, and diversified running training should be part of alpine skier's all-year training program for improvement of coordination and endurance in order to prevent injuries. For competitive alpine skiing, it is also recommended to integrate aerobic running into the warm-up and cooldown program.

Training strategies in alpine ski racing need to meet multiple physical demands and include several different modalities. Aerobic and anaerobic endurance capacity is key for being able to perform on the highest level. Despite running being only a little aspect in the whole training schedule of elite alpine skiers, it is evident that when performed as short-duration HIIT session in the preparation season, it can significantly improve aerobic and anaerobic endurance capacity and is therefore a feasible option to improve the overall performance level.

Running also acts as a prevention tool. To practice alpine skiing at the lowest possible risk, a general endurance and strength training is

References

1. van Poppel D, van der Worp M, Slabbekoorn A, van den Heuvel SSP, van Middelkoop M, Koes BW, et al. Risk factors for overuse injuries in short- and long-distance running: a systematic review. J Sport Health Sci. 2021;10(1):14–28.
2. Knechtle B, Nikolaidis PT. Physiology and pathophysiology in ultra-Marathon running. Front Physiol. 2018;9:634.
3. Hespanhol Junior LC, van Mechelen W, Verhagen E. Health and economic burden of running-related injuries in Dutch Trailrunners: a prospective cohort study. Sports Med. 2017;47(2):367–77.
4. Bontemps B, Vercruyssen F, Gruet M, Louis J. Downhill running: what are the effects and how can we adapt? A narrative review. Sports Med. 2020;50(12):2083–110.
5. Bere T, Flørenes TW, Nordsletten L, Bahr R. Sex differences in the risk of injury in world cup alpine skiers: a 6-year cohort study. Br J Sports Med. 2014;48(1):36–40.
6. Posch M, Schranz A, Lener M, Tecklenburg K, Burtscher M, Ruedl G. In recreational alpine skiing, the ACL is predominantly injured in all knee injuries needing hospitalisation. Knee Surg Sports Traumatol Arthrosc. 2021;29(6):1790–6.
7. Bere T, Flørenes TW, Krosshaug T, Haugen P, Svandal I, Nordsletten L, et al. A systematic video analysis of 69 injury cases in world cup alpine skiing. Scand J Med Sci Sports. 2014;24(4):667–77.
8. Csapo R, Juras V, Heinzle B, Trattnig S, Fink C. Compositional MRI of the anterior cruciate ligament of professional alpine ski racers: preliminary report on seasonal changes and load sensitivity. Eur Radiol Exp. 2020;4(1):64.
9. Andersen RE, Montgomery DL. Physiology of alpine skiing. Sports Med. 1988;6(4):210–21.
10. Gilgien M, Reid R, Raschner C, Supej M, Holmberg H-C. The training of olympic alpine ski racers. Front Physiol. 2018;9:1772.
11. Stoggl T, Kroll J, Helmberger R, Cudrigh M, Muller E. Acute effects of an ergometer-based dryland alpine skiing specific high intensity interval training. Front Physiol. 2018;9:1485.
12. Gibala MJ, Little JP, Macdonald MJ, Hawley JA. Physiological adaptations to low-volume, high-intensity interval training in health and disease. J Physiol. 2012;590(5):1077–84.
13. Breil FA, Weber SN, Koller S, Hoppeler H, Vogt M. Block training periodization in alpine skiing:

effects of 11-day HIT on VO2max and performance. Eur J Appl Physiol. 2010;109(6):1077–86.

14. Gross M, Hemund K, Vogt M. High intensity training and energy production during 90-second box jump in junior alpine skiers. J Strength Cond Res. 2014;28(6):1581–7.

15. Mayr HO. Prophylaxe von Verletzungen. Sportärztezeitung; 2018.

16. und SSiSU. https://www.stiftung.ski/fileadmin/user_upload/ASU_Analyse_2017_2018.

17. Mayr HO, Zaffagnini S. Prevention of injuries and overuse in sports: Directory for physicians, physiotherapists, sport scientists and coaches: Springer; 2015:137–54.

18. Natri A, Beynnon BD, Ettlinger CF, Johnson RJ, Shealy JE. Alpine ski bindings and injuries. Curr Find Sports Med. 1999;28(1):35–48.

19. Bere T, Florenes TW, Krosshaug T, Koga H, Nordsletten L, Irving C, et al. Mechanisms of anterior cruciate ligament injury in world cup alpine skiing: a systematic video analysis of 20 cases. Am J Sports Med. 2011;39(7):1421–9.

20. Schmid P. Dauermethode versus Intervalltraining in der kardiologischen Rehabilitation. Österreichisches J Sportmed. 1999;29:20–7.

21. Gibala MJ. High-intensity interval training: a time-efficient strategy for health promotion? Curr Sports Med Rep. 2007;6(4):211–3.

22. Easthope CS, Hausswirth C, Louis J, Lepers R, Vercruyssen F, Brisswalter J. Effects of a trail running competition on muscular performance and efficiency in well-trained young and master athletes. Eur J Appl Physiol. 2010;110(6):1107–16.

Running in Kiteboarding

28

Francesco Feletti, Mirco Babini,
and Michele Felisatti

28.1 Introduction

The use of kites as a propulsion system dates back to the thirteenth century in Asia, but their introduction in modern outdoor sports goes back to the beginning of the 1980s [1]. In kiteboarding, a board allows gliding over the water's surface using the power of the wind.

This sport has tremendously increased in popularity, and the number of participants today exceeds 2.8 million worldwide [2]. Most participants are male, and the age ranges between 10 and 75, with a peak between 18 and 40 [3]. Since the equipment can be easily transported, exotic locations have become very popular kite tourism destinations. Kiteboarding received important official recognition in 2008 when it was adopted as an international sailing class by the World Sailing under the IKA, the International Kiteboarding Association (www.kiteclasses.org) [2].

Kiteboarding was included for the first time in the Olympic programme for the Youth Olympic Games in Buenos Aires in 2018 and in the Summer Olympic Games for Paris 2024, for the debut of the Formula Kite Olympic format (www.formulakite.org).

The acrobatic nature of this sport makes essential adequate physical preparation, and, although differences between the various disciplines may be significant, aerobic and endurance trainings are crucial for kiters at any level.

Amateur participants can benefit from running as a tool to develop general resistance and ensure functional reserve in unexpected situations, resulting in improved comfort and safety.

Instead, competitive athletes generally include running in their training programs because it can directly affect performance.

Indeed, running may allow kiters to meet their sport's physiological and cardiovascular requirements and maintain a high capacity for reaction even in unforeseen situations frequently caused by the unpredictability of atmospheric conditions.

F. Feletti (✉)
Dipartimento Diagnostica per Immagini - Ausl della Romagna, U.O. Radiologia - Ospedale S. Maria delle Croci, Ravenna, Italy

Dipartimento di Medicina Traslazionale e per la Romagna, Università degli Studi di Ferrara, Ferrara, Italy
e-mail: feletti@extremesportmed.org

M. Babini
International Kiteboarding Association (IKA), Montreaux, Switzerland
e-mail: president@kiteclasses.org

M. Felisatti
Esercizio Vita Medical Fitness, Ferrara, Italy

28.2 The Sport and the Equipment

28.2.1 Equipment

Power kites are specifically developed to transfer wind energy to the athlete. The inflatable kites commonly used in kiteboarding are constructed using a single layer of ultra-lightweight, water-repellent and airtight canvas applied to a supporting structure inflated with a pump or a compressor before use. Most kites used in kiteboarding have this structure and measure between 5 and 20 m^2.

Lines have a length between 20 and 30 m, a thickness of 1–3 mm and a load capacity of approximately 200 kg each [4]. The two lines attached to the two sides of the leading edge are known as front lines and transfer the kite's power directly to the kiter through the chicken loop, a loop device hooked into the kiteboarding harness. The chicken loop allows a rapid release for safety reasons. The lines attached to the kite's two trailing edges (backlines) are connected directly to the handlebar and steer the kite. The bar is a tubular structure in carbon fibre measuring between 45 and 60 cm, depending on the kite sizes.

A new generation of high-performance kites has been rapidly developed over the last 5 years due to the needs required by the racing disciplines. They are foil kites, built with a ram air structure of communicating cells inflated by the wind and allowing high-performance sailing starting from five knots. A different line measure is adopted with the foil kite than inflatables; the average length is between 11 m and 15 m; the diameters range is 1.5/1.8 mm for the front and 1.1/1.3 mm for the back lines.

Different harness models are available, including waist and seat types, chosen depending on the discipline practiced and individual preferences.

The traditional twin-tip boards have a symmetrical shape, allowing the rider to change the direction of travel without changing stance on the board, and they are particularly suited to freestyle, free-ride and wake style.

On the contrary, directional boards travel only in a direction and require a manoeuvre (tack or jibe) to change direction. They are preferred in for wave riding, where their higher volume and rideability allow the kiter to better interact with the hydrodynamic forces of the waves, and are commonly used in course racing because they allow to achieve optimal upwind and downwind angles [5].

A recent innovation in course racing kiteboard manufacturing has seen the introduction of hydrofoil technology which allows a reduced area of contact with water, resulting in even greater speeds and keep the board gliding fast even if the wind drops.

The feet are usually fixed to the board by sandal-type foot straps, similar to those used in windsurfing. Some expert riders alternatively use shoe-type bindings, used mainly in wake style.

Finally, some young riders do not adopt any feet fixation.

28.2.2 Styles and Disciplines

Kiteboarding includes some amateur styles, such as free riding and wave riding, and competitive disciplines.

Free riding is the amateur style par excellence and includes simple cruising, improvised stunts and manoeuvres of various types extemporaneously chosen by the rider.

Another popular style is *wave riding*, which combines kiteboarding with surfing; it is practised using strapless surfboards or specific directional kiteboards.

The most popular competitive disciplines are course racing and freestyle.

Course racing is similar to yacht racing and requires both speed and tactics [6].

The racing discipline tremendously developed in recent years, and the introduction of hydrofoil boards allows speeds of over 40 knots and entails a significantly increased physical demand.

Freestyle consists of a rider performing jumps, manoeuvres and evolutions of varying technical complexity. Being the 'mother' discipline of kiteboarding, it is constantly evolving and includes some subdisciplines. For example, *big air* mainly consists of 'hooked' evolutions, characterized by high jumps with long flight phases. The *wake style* includes tricks and aerials borrowed from wakeboarding, generally performed with the kite unhooked and full power. Finally, the *strap-less freestyle* is the younger style; it involves tricks and jumps with a directional board, without any feet fixation, allowing riders to move and flip the board freely in the air.

28.3 Role and Characteristics of Running in Kiteboarding

28.3.1 Physiological Demands of Kiteboarding

Knowledge of the physiological demands of kiteboarding is crucial to develop proper training methods, especially with the evolution of competitive disciplines such as freestyle and course racing.

28.3.1.1 Course Racing

During racing events, the rider stays hooked to the kite while standing on the board and with the back slightly leaned back to balance the traction of the kite, while the upper part of the trunk faces in the direction of travel. The knees maintain an angle between 135° and 150°, while the elbow remains bent at about 90° to exert the traction on the bar necessary to maintain adequate tension on the back lines. In this position, the kite's pulling force on the waist keeps the lumbar spine in hyperextension, and abdominal muscles exert isometric work to counteract the lordosis.

Instead, the thigh muscles' strain to balance the water pressure against the board is semi-isometric. Specifically, the isometry is interrupted by continuous small flexion-extension movements of the hips and knees necessary to follow the waves and adjust the course in response to changes in the wind direction and intensity.

Vercruyssen et al. [7] assessed the physiological demands of a crossing trial, adopting the heart rate (HR) as an estimation of the oxygen uptake (VO2) and measuring the blood lactate concentration. Overall, in light wind conditions, ranging from 12 to 15 knots (6.2–7.7 m s-1), they described a prevalently aerobic effort of moderate intensity, characterized by significant HR and blood pressure increases but low increments in VO2 [7].

Specifically, an average HR of 72–85% of the maximum HR was found, while the mean blood lactate concentration (La(b)) ranged between 1.2 and 4.4 mmol l-1 and the estimated % VO2 max remained between 65.5 and 74.3 [7]. The high levels of HR may be a consequence of the isometric contraction of large muscle masses of the lower and upper limbs, during which the muscle perfusion is impaired by mechanical compression of blood vessels from the contracting skeletal muscles.

Instead, the low La(b) values observed during kiteboarding crossing might depend on the isometric effort's discontinuous nature owing to relief intervals (in the range of 10–20 s) caused by leg movements to adjust the direction and adapt the direction gliding to the waves [7].

These rest intervals would partially restore the muscles' oxygen accessibility, thus promoting a more oxidative degradation of glycogen and a low lactate concentration [8, 9].

28.3.1.2 Freestyle

Freestyle consists of performing jumps, manoeuvres and tricks with different styles and in different combinations.

Jumps which exploit the lift of the kite are specific to kiteboarding. The jump is prepared by maximizing the lines' tension by directing the gliding board into the wind. In this phase, the isometric effort performed by the kiter increases considerably. Slight flexion of the hips is also required to prepare the slight push needed to help the rider take off from the water's surface [10]. The jump begins when the kiters suddenly invert the direction of the kite's flight, abruptly discharging the energy accumulated in the previous phase towards the zenith, which lifts them into the air.

The jump can lift the kiter to a height of several meters, and the flight phases can last several seconds. Various tricks can then be performed during the hang time, such as horizontal and vertical rotations or grabs with particular body positions.

The sequences of movements executed during this phase are similar to those of gymnastics. They involve extending the limbs and bust, crouches and flexion of the limbs and torso and grasping the board, all involving different physical effort types. The descent phase is slow, and the kiter exploits the kite's parachute effect by modifying the tension on the lines. During landing, the legs bend and absorb part of the impact.

The effort of the postural muscles, especially those of the back and the abdomen, allows the body to take up the correct position for landing, while the lower limb muscles contract during landing.

Overall, freestyle is an intense activity involving both aerobic and anaerobic metabolisms.

During a freestyle event in mid-wind conditions (ranging from 15 to 22 knots), the average HR was quantified 85.4 ± 3.0% of the maximum HR; VO2 values were estimated as 80.0 ± 4.5% of the maximum oxygen uptake, while mid-values for La(b) of 5.2 ± 0.8% mmol/L were observed at the end of a freestyle trial [11, 12].

28.3.2 Main Objectives of Athletic Training

Notwithstanding kiteboarding's physiology and physical needs depending on the specific discipline practised, the performance model of kiteboarding has general elements that are always valid, starting from the learning phase to amateur and competitive practice.

Kiteboarding is a situational sport because environmental-related parameters change continuously and are often unpredictably.

For example, the wind and wave characteristics may influence athletic performance by requiring fine-tuning of the kite and board handling or affecting the choice of the equipment and the manoeuvre speed.

For these reasons, training must prepare kiters for various motor patterns and develop a strong ability to adapt.

Training must aim at improving both short- and long-term athletic performances.

The physical qualities that are most important to be empowered are general endurance, aerobic capacity and mixed aerobic-anaerobic work. The enhancement of resistant strength and maximal strength is also necessary [7].

In terms of energy metabolism, the kitesurfer must develop the efficiency of both the aerobic and anaerobic systems and their combined and alternating use.

It is also crucial to frequently change workout schemes both in qualitative and quantitative terms.

Specific kiteboarding preparation training must also include exercises to optimise specific kiteboarding athlete gestures, correct movement automation, correct distribution of effort and good coordination.

28.3.3 The Role of Running in Training for Kiteboarding

General resistance, which expresses the ability to maintain the effectiveness of the athletic gesture over time [13], is a crucial quality for practising kiteboarding.

Fatigue may increase the risk of incidents, while the general resistance guarantees a high reaction capacity and reduces technical errors, consequently reducing the small and large traumatology linked to kiteboarding [4, 10, 14].

Improving the endurance of the kiter is essential for amateur athletes, but it is even more necessary for agonists.

In fact, in course-racing event participation without adequate preparation, fatigue can compromise performance capacity after only 30′ [9], especially in sessions with significant waves and currents.

In addition, the general resistance allows improving the performance in racing by increasing the physical capacity, improving recovery and increasing the psychic load capability, thus reducing tactical errors related to fatigue.

It is therefore essential to enhance general endurance and aerobic capacity with activities such as running.

The aerobic activities that can be integrated into specific training for kiteboarding may include running, cycling and rowing.

Running, however, is generally preferred for at least two reasons [7].

Firstly, running involves the same kinetic chains involved during kiteboarding, and running protocols seem to be the most relevant for describing aerobic fitness levels in kitesurfers [7].

Even if running induces types of muscle contractions different from those observed in kiteboarding, it involves the same muscle groups: the legs and the postural muscles of the trunk.

Secondly, it is simple to administer and does not require specific equipment or dedicated spaces. Running is, in general, a simple training tool to implement even for those kitesurfers who are often travelling, both for competitions and the search for the best weather and sea conditions for training.

28.4 Training Strategies

28.4.1 Training Strategies for Beginners and Amateurs

In the beginner, previously sedentary or coming from non-endurance sports, it is advisable to introduce running with a slow and long type of work, starting from 20 to 30-min sessions, with HR between 130 and 150 bpm and lactate percentages below 2 mmol/L. This approach can improve cardiorespiratory and circulatory efficiency with measurable potential effects in heart rate reduction and cardiac output improvement and increase capillarization and mitochondrial function [13]. Furthermore, this approach can improve thermoregulation, particularly important in this sport involving highly thermo-dispersive situations such as immersion or semi-immersion and the convective effect by the wind.

Table 28.1 Example of use of the training run for an amateur free-rider

Warming up
Slow running (time: 5′)[a]
Repeated running sprints (time: 30″–60″)[b]
Slow running (time: 5′)[a]
Repeated running sprints (time: 30″–60″)[b]

[a]This exercise can be alternate/replaced in the interval training scheme reported above with/by swimming or stand-up paddleboarding
[b]This exercise can be alternate/replaced in the interval training scheme reported above with/by workout circuits, jumping rope or boxing

Amateurs may also progress to interval training (IT), an appropriate method for the typically intermittent type of kiteboarding effort [7]. The alternation of high-energy demands with recovery phases enables more prolonged and intense training even among non-professionals.

As part of interval training, running can effectively train the respiratory and cardiovascular systems, improving the anaerobic threshold and maximum oxygen consumption (VO2 max) [15].

More specifically, in the basic programs intended for amateur free-riders, running is used very simply in repeated sprints lasting between 30 seconds and 60 seconds (Table 28.1).

As reported in Table 28.1, running is combined or alternated with high-energy explosive power exercises and activities requiring long and slow-energy production such as jogging, swimming and stand-up paddleboarding.

28.4.2 Training Strategies for Agonists

Agonist kiters always have at least one running session per week [7], and training with running is crucial in agonists to increase the VO2 max, a critical factor for performance. For example, in a crossing event, it has been shown that kiters with higher VO2 max values cover greater distances in the unit of time (Fig. 28.1).

In general, kiteboarding requires low-intensity and long-duration energy production associated with periods of explosive energy use, and it is

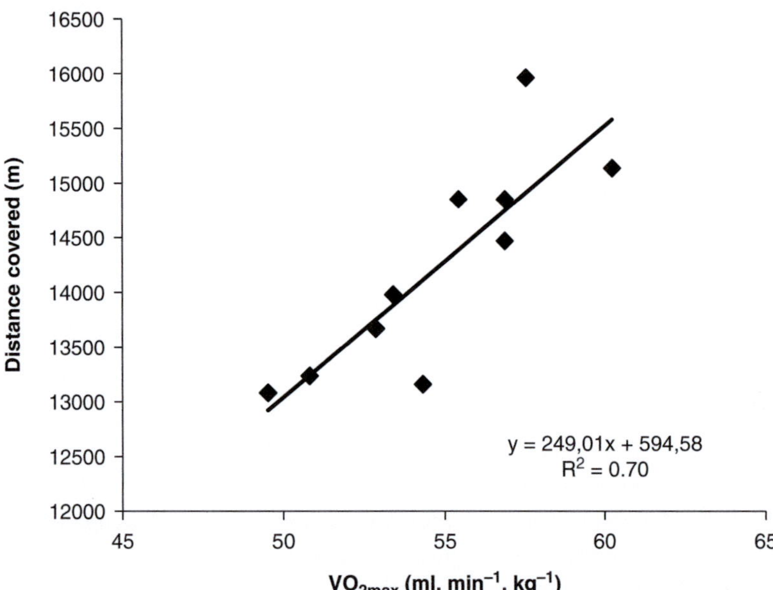

Fig. 28.1 Linear regression between the distances covered during a crossing trial and the maximal oxygen uptake (VO2 max). Reprinted with permission from Vercruyssen F et al. [7]

essential to work on both. Therefore trainers often adopt interval training and high-intensity interval training (HIIT), where running is interspersed with weight training and exercises focused on the expression of strength or resistant force to improve muscle recruitment and synchronisation, central and peripheral nervous activation, hypertrophy, lactic acid potency and lactic acid tolerance [15]. Similarly to other water sports, anaerobic endurance is necessary to withstand long durations of gliding, while power training should emphasise maximal force production for optimal counteraction to the water pressure against the board during gusts [16]. The prescription for anaerobic power should detail the loading parameters required by the movement patterns involved in the specific kiteboarding discipline, such as direction changes in crossing competitions or the specific practised tricks in freestyle.

For example, strength training focused on upper limbs may be crucial for freestyle competitors engaged in unhooked manoeuvres, which place the shoulders and elbows under significant strain [4].

In the initial stages of agonists preparation or during recovery after an injury, running should be introduced as a slow work, with the purpose to build baseline aerobic endurance.

After the initial training phase, interval training adapted to the specific kiteboarding discipline practised can be introduced. Trainers should custom the IT acting on parameters such as the type of test, the number, duration and intensity of series and repetitions and the duration and type (i.e., active or passive rest) of breaks between sets [13].

A training program intended for agonists running work shall target the intensity and maximum oxygen consumption required in the practised kiteboarding discipline. For example, in course racing, it may be adequate using minimum speeds corresponding to 90% of VO2 max, leading to lactate concentrations of 3–5 mmol/L [7, 13, 17].

In a typical training program for agonists and elite athletes, running is modulated and administered in several weekly sessions. Uphill running of 50 and 100 m and repeated sprints are coupled with running rhythmic exercises and elastic-explosive training (Tables 28.2 and 28.3).

Running rhythmic exercises address the specific needs of kiters in terms of speed and balance.

Speed is a fundamental aspect in the movement required by freestyle manoeuvres and course racing changes of direction. To improve speed, exercises to run with a wide or narrow stride width can be used alternatively by exploit-

Table 28.2 A training program using running for the aerobic component dedicated to advanced and competitive athletes involves strength and elastic-explosive training (I), running rhythmic exercises (II), uphill running (III) and fractional aerobic power (IV)

I. Strength and elastic-explosive training	II. Running rhythmic exercises[a]	III. Uphill running	IV. Fractional aerobic power[b]
A. Continuous squat-jump[c] (sets: 3; reps.: 5)	*E.* *E1. Wide-stride-width running* (distances: 50 m; number: 3) *E2. Narrow-stride-width running* (distances: 50 m; number: 3)	*H1. Sprints* (distances: 50 m; number: 2–5); recovery period: 3–8′) *H2. Sprints* (distances: 100 m; number: 3–5; recovery period: 6–8′)	*I1. Sprints* (distances: 100 m; number: 6–8; speed: Distance-related, short distances/high speeds; long distances/medium speeds) *I2. Stationary sprints* (time: 1.5′; reps.: 2–3) *I3. Accelerated sprints[d]* (distances: 300 m; reps: 2–3; recovery period: 3′)
B. Continuous depth jump (sets: 3; reps.: 6–8)[e]	*F. Skip* (150–200 ground touches; number of series: 2/3; recovery period: 30″)		*L. Sprints* (distances: 300 m; number: 10; recovery period: 3′)
C. Half squat jumps (continuous) *C1.* (sets: 2; reps.: 6–8; load: 50% bodyweight) *C2.* (sets: 2; reps.: 5 load: Natural)[e] *D. Barbell skip with limited knee rise (45°)* (sets: 3–4; ground touches: 50–60; load: 50% bodyweight)[e]	*G1. Alternating leaps Wide-stride-width* 1 × 50 m (up to 4 × 50 m); 1 × 100 m (up to 5 × 100 m) *G2. Alternating leaps Wide-stride-width* 1 × 50 m (up to 4 × 50 m); 1 × 100 m (up to 5 × 100 m)	*Possible alternatives/ integrations* *Sprints* (distances: 50 m; number: 4; recovery period: 3–8′) *Sprints* (distances: 80 m; number: 4; recovery period: 4–8′) *Sprints* (distances: 100 m; number: 2; recovery period: 6–8′) *Acceleration sprints* (distances: 80 m; number: 4; recovery period: 5–10′) *Acceleration sprints* (distances: 100 m; number: 4; recovery period: 6–8′)	*M. Sprints* (distances: 300–600 m; total distance: 3000 m; recovery period: 3–4′) For example: 5 × 600 m 6 × 500 m 3 × 600 m + 3 × 400 m 3 × 600 m + 4 × 300 m

[a] Exercises to be performed continuously in sequence

[b] To be divided with appropriate breaks for incomplete recovery

[c] The exercise should be performed with legs shoulder-width apart (knees flexed); hands on hips, making sure that the spine is as erect as possible; and buttocks push back as if to sit on a chair. When maximum flexion is achieved, the knees must be precisely above the ankles to not be subjected to harmful stress. Keeping the head high, the athlete must jump to the ceiling, pushing the elevation off the ground, landing with a spring on the knees to reduce impact

[d] Acceleration, sprint and progressive exercises should be performed immediately after the climbs

[e] For exercises a, b and c, it is necessary to decrease the times and then accelerate, increasing the explosive thrust power. The recoveries must be about 3′, and at the end of each exercise, some pre-athletic activities must be included (e.g. jump, side run, speed ladder)

Table 28.3 An example of weekly training microcycle

Weekly training microcycle					
Monday	Tuesday	Wednesday	Thursday	Friday	Saturday
Strength and elastic-explosive strength training (I)	Alternating leaps (G1)	Skip (F)	Running rhythmic exercises (II)	Alternating leaps (G2)	Fractioned aerobic power (IV)
Running rhythmic exercises (II)	Uphill running (H1) Acceleration sprints (I1, I3)	Fractioned aerobic power (IV)	Continuous endurance training	Uphill running (H2) Acceleration sprints (I1, I3)	

ing their complementarity. On the one hand, the narrow stride width allows for modulating the frequency of the steps, and the rapid alternation of many movements per second implies rapid alternation of muscular excitations and inhibitions and improves the athlete's ability to relax. On the other hand, the wide-stride-width running promotes the athlete's ability to quickly express high-strength peaks (explosive-elastic-reactive). Therefore, an athlete's speed optimisation may be achieved by the best mixture or sequence of wide- or narrow-stride-width running.

To improve balance, running rhythmic exercises are combined with specific exercises.

It is possible to work on general balance with proprioceptive gymnastics exercises that use proprioceptive or skimmy tablets. Instead, the specific balance can benefit from walking on a slackline or standing on a fit ball or from exercises borrowed from artistic gymnastics like the handstand.

Elastic-explosive training includes bouncing, which involves exaggerated run with extended vertical and horizontal displacement. This exercise emphasises lifting rather than advancing the body and is helpful to develop and increase strength in its explosive and reactive elastic expression.

28.5 Prevention Strategies

When introducing running into kiteboarding training, it should be considered that many kiters have no experience with running. Starting with slow running workouts can minimize the adverse effects of any errors in running technique and represent an opportunity to correct them.

Kiters must use suitable footwear for their running technique and possibly orthotics capable of introducing adequate compensation whenever necessary. Uphill running should not exceed the 5–8% slope to avoid modifying the individual running mechanics too deep [13]. A heart rate monitor is essential to know in which physiological quality the athletes are training.

The functional load of running on the lower limbs is added to kiteboarding with the potential risk of overuse injuries. The knee is the joint that is most often affected by overuse injuries related to running [18, 19]. At the same time, in a prospective study on 43 kiters engaged in 3 different disciplines (i.e. freestyle, course racing and wave riding), knee overuse injuries emerged as an important cause of reduced participation, damaged performance and discomfort. Specifically, the person-days of reduced participation were 8%; damaged quality of the performance was affected in 16% of person-days, while knee pain was reported in 30% of person-days [20]. Among overuse injuries reported in kiters were patellofemoral pain syndrome, patellar tendinopathy and iliotibial band syndrome, common overuse injuries in running [19]. It is, therefore, necessary to consider the risk of a cumulative, synergistic effect of absorbed vibrations, repetitive micro-trauma and overload [4, 7, 8, 20]. Therefore, running training must be modulated based on suitable predictive models of joint stress to limit exposure and accompanied by constant monitoring aimed at detecting any early onset of overuse symptoms.

28.6 Conclusions

Due to the demand for aerobic endurance and high VO2 max, running training plays a crucial role in kiteboarding for amateurs and agonists.

Since kiteboarding is a situation sport, running is generally included in interval training schemes alongside exercises that aim to achieve other physical qualities.

Given the lack of literature regarding running in the preparation dedicated to kiteboarding, training schemes are not codified, and much is left to empiricism and approaches borrowed from other sports.

Further studies should be conducted to define the best training model in each kiteboarding discipline, the better way to administer running and the dose to avoid adverse effects related to the additional overload effect potentially resulting in overuse injuries.

Contribution Statement The three authors cooperated in the formulation of the main conceptual ideas. In particular, MB and FF co-worked on the section relative to the sport and the equipment; FF formulated the part about the role and characteristics of running in kiteboarding. MF mainly provided the data about the training strategies and examples of a training program.

Altogether the authors discussed the data and contributed to the final manuscript.

References

1. Feletti F, Brymer E. Injury in kite buggying: the role of the 'out-of-buggy experience'. J Orthop Surg Res. 2018;13(1):104. Published 2018 May 2. https://doi.org/10.1186/s13018-018-0818-x.
2. ISAF Kiteboarding Format Trials. Technical report. Events committee – May 2012 item: 8(b) Santander. p. 1–33.
3. Kristen K, Syré S, Humenberger M. Kitesurfen – sportmedizinische Aspekte, Risikofaktoren und Verletzungen. OUP 2014;6:306–311. doi: 10.3238/oup.2014.0306–0311.
4. Feletti F. Kitesports medicine. Kiteboarding and snowkiting. In: Feletti F, editor. Extreme sports medicine. Chapter 15. Springer; 2017. p. 177–95.
5. International Kiteboarding Association - Formula Kite. Class Rules. 2021. https://www.sailing.org/tools/documents/FK2021CR20210201-[26950].pdf. Accessed June 2021.
6. Kiteboarding styles. In: Kitesurfing handbook. http://kitesurfi ng handbook.peterskiteboarding.com/kitesurfi ng-styles. Accessed Jan 2014.
7. Vercruyssen F, Blin N, L'Huillier D. Assessment of the physiological demand in kitesurfing. Eur J Appl Physiol. 2009;105:103–9.
8. Spurway NC. Hiking physiology and the "quasiisometric" concept. J Sports Sci. 2007;25:1081–93.
9. Vogiatzis I, Tzineris D, Athanasopoulos D, Georgiadou O, Geladas N. Quadriceps oxygenation during isometric exercise in sailing. Int J Sports Med. 2008;29:11–5. https://doi.org/10.1055/s-2007-964993.
10. Lundgren L, Brorsson S, Hilliges M, Osvalder A. Sport performance and perceived musculoskeletal stress, pain and discomfort in kitesurfing. Int J Perform Anal Sport. 2011;11(1):142–58.
11. Bourgois JG, Boone J, Callewaert M, Tipton MJ, Tallir IB. Biomechanical and physiological demands of kitesurfing and epidemiology of injury among kitesurfers. Sports Med. 2014;44(1):55–66.
12. Camps A, Vercruyssen F, Brisswalter J. Variation in heart rate and blood lactate concentration in freestyle kytesurfing. J Sports Med Phys Fitness. 2011;51(2):313–21.
13. La Torre A. Lo sviluppo della resistenza. In: La Torre A, editor. Allenare per Vincere – Metodologia dell'allenamento. Chapter 6. Roma: Scuola dello Sport Coni; 2018. p. 217–59.
14. Berneira O, Domingues R, Medeiros A, Vaghetti O. Incidência e característiscas de lesões em practicantes de kitesurf. Rev Bras Cineantropom Desempenho Hum. 2011;13(3):195–201.
15. Lundby C, Montero D, Joyner M. Biology of VO_2 max: looking under the physiology lamp. Acta Physiol (Oxf). 2017;220(2):218–28. https://doi.org/10.1111/apha.12827. Epub 2016 Nov 25. PMID: 27888580
16. Farley ORL, Harris NK, Kilding AE. Physiological demands of competitive surfing. J Strength Cond Res. 2012;26(7):1887–96.
17. Buchheit M, Laursen PB. High-intensity interval training, solutions to the programming puzzle. Part II: anaerobic energy, neuromuscular load and practical applications. Sports Med. 2013;43(10):927–54. https://doi.org/10.1007/s40279-013-0066-5. PMID: 23832851
18. Messier SP, Martin DF, Mihalko SL, Ip E, DeVita P, Cannon DW, Love M, Beringer D, Saldana S, Fellin RE, Seay JF. A 2-Year prospective cohort study of overuse running injuries: the runners and injury longitudinal Study (TRAILS). Am J Sports Med. 2018;46(9):2211–21. https://doi.org/10.1177/0363546518773755. Epub 2018 May 23. Erratum in: Am J Sports Med. 2021 Feb;49(2):NP13. PMID: 29791183
19. Francis P, Whatman C, Sheerin K, Hume P, Johnson MI. The proportion of lower limb running injuries by gender, anatomical location and specific pathology: a systematic review. J Sports Sci Med. 2019;18(1):21–31. Published 2019 Feb 11
20. Paiano R, Feletti F, Tarabini BP. Use of a prospective survey method to capture a picture of overuse injuries in Kiteboarding. Muscle Ligam Tendons J. 2020;10(165) https://doi.org/10.32098/mltj.02.2020.02.

Running in Sailing

Francesco Feletti and Andrea Madaffari

29

29.1 Role and Characteristics of Running in Sailing

29.1.1 Evolution of Physical Performance in Sailing

Traditionally, the pursuit of performance in sailing was based on the setting up of the boats and improved technical management skills rather than on physical performance.

Over the past two decades, however, the physical demands on sailors have increased dramatically as a consequence of three factors:

- Thc introduction of highly spectacular boats.
- A different management of more traditional boats.
- New regatta formats [1].

F. Feletti (✉)
Dipartimento Diagnostica per Immagini - Ausl Romagna, U.O. Radiologia - Ospedale S. Maria delle Croci, Ravenna, Italy

Dipartimento di Medicina Traslazionale e per la Romagna, Università degli Studi di Ferrara, Ferrara, Italy
e-mail: feletti@extremesportmed.org

A. Madaffari
Club Nautico Marina di Carrara, Carrara, Italy

Compagnia della Vela Venezia, Venezia, Italy
e-mail: madafit@tim.it

Among high-performance boats are planing hull boats, the so-called skiffs, such as the 49er and 29er, and hydro-foiling sailing classes. The latter includes a wide variety of boats spanning from single-handed dinghies (e.g. Moth class); the racing yacht used in the 2021 America's Cup (i.e. AC75), up to the hydro-foiling windsurfing (i.e. iQFoil class), which will be included in the 2024 Olympics.

Both skiffs and hydro-foiling boats involve significantly higher physical performance than more traditional boats in terms of accuracy of execution, speed and endurance.

Also, the management of more traditional boats, like the Laser or the 470, dramatically evolved in the last two decades. Exploiting every slight variation in wind and waves resulted in a significant increase in physical effort.

Finally, the modern regatta formats involve shorter heats and increase the sport's pace dramatically, transforming sailing from a traditionally intermittent activity to a continuous athletic performance.

29.1.2 The Role of Running in Sailing

A day of regatta generally includes several heats, with durations between 30′ and 1 h. This implies several hours of constant activity, making resistance a critical aspect to be improved in sailors.

Resistance in sailing can be defined as the ability to produce the motor activities required to steer the boat while maintaining the effectiveness of the technical gesture unchanged and without a decline in work intensity and mental concentration. Resistance also allows the capacity for action and reaction to be kept constantly high, reduces technical and tactical errors and decreases the risk of trauma due to fatigue [2].

Due to the crucial role of endurance, running has been included since the first systematic experiences of physical preparation in the training programs dedicated to sailing [3]. Although some authors identify rowing as the aerobic activity par excellence preparatory to sailing, running offers undoubted advantages; it can be carried out anywhere, without the need for specific equipment, and it is very quickly modulable and adaptable to specific needs [4–9].

In modern sailing, running can be crucial for health and well-being in all types of boats and all classes, including amateur sailors and cruising enthusiasts. At the same time, it can variably contribute to improving the performance of competitive athletes.

The present work aims to outline the role of running for sailing, focusing mainly on dinghies and sailboards. Besides, the role of racing in keelboats will also be briefly discussed, from the America's Cup to offshore sailing.

29.1.3 Running for Beginners and Amateurs

Among beginner and amateur sailors, running is essential to improve endurance for preventive purposes. The majority of amateur participants and beginners, especially the younger ones, sail in dinghies and sailboards.

Dinghies demand the maintenance of forced body stances, such as leaning out (hiking) to move the crew's body weight upwind to decrease the boat heeling caused by the wind pressure on the sail.

In conditions of general fatigue, these actions may be executed incorrectly, with potential damage to the musculoskeletal system [1].

For example, children who sail with Optimist or Open-Bic dinghies, often assume wrong postural attitudes in strong wind conditions, including lumbar hyperlordosis resulting in the

vicarious contraction of the iliopsoas following the fatigue of the abdominal muscles.

Endurance training through running is an excellent preventive tool when accompanied by specific work in postural education and compensatory and preventative exercises. Similarly, on sailboards, running is usefully associated with specific exercises for the back muscles, arms and trunk [10]. Especially in strong wind conditions, rollover or errors in manoeuvrers can lead beginners to efforts exceeding the VO2 max threshold. The consequent rapid exhaustion can be dangerous, especially for individuals who are not adequately trained and in cold waters. Having developed an adequate general endurance with running allows young participants and beginners to have a sufficient safety margin in unexpected situations.

29.1.4 Olympic Classes

The classes for which medals for sailing are contemplated at the Tokyo Olympic Games are shown in Table 29.1. In all Olympic and youth classes, cardiovascular and aerobic efforts are significant (see Figs. 29.1, 29.2, 29.3, and 29.4), and running is often used as a training tool. However, running training in Olympic sailing and the cadet classes must be tailored case by case, starting from a systematic identification of needs, the level of the sailor and the objectives pursued.

Table 29.1 Olympic classes at the Tokyo Olympic Games [11]

Type of boat	Class	Crew
Sailboard	RSX (sailboard)	SH; F
	RSX (sailboard)	SH; M
Traditional dinghy	Laser	SH; M
	Finn	SH; M
	Laser radial	SH; F
	470	DH; M
	470	DH; F
Skiff	49er	DH; M
	49er FX	DH; F
Hydrofoil catamaran	NACRA 17	DH; M/F (mixed)

SH single handed, *DH* double handed, *M* male, *F* female

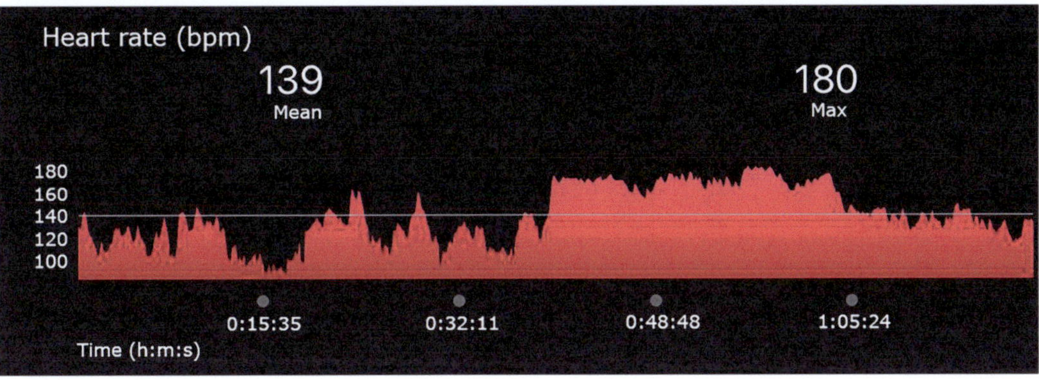

Fig. 29.1 Example of the electrocardiogram track of an RSX class athlete, in the top ten of the ranking, during a final medal race

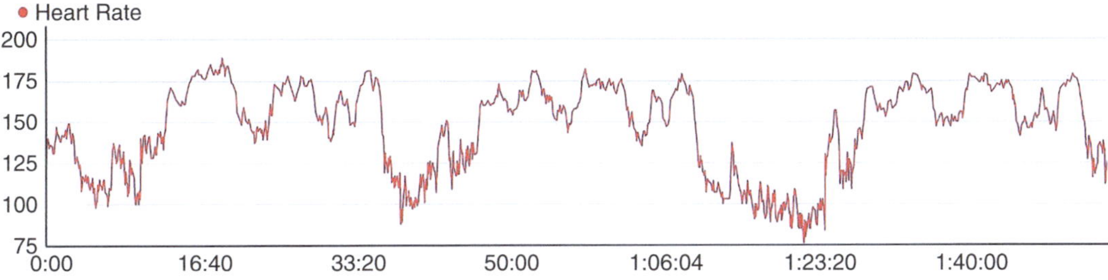

Fig. 29.2 Example of the electrocardiogram track of a Laser Radial class athlete, during three races of 1 day of the regatta

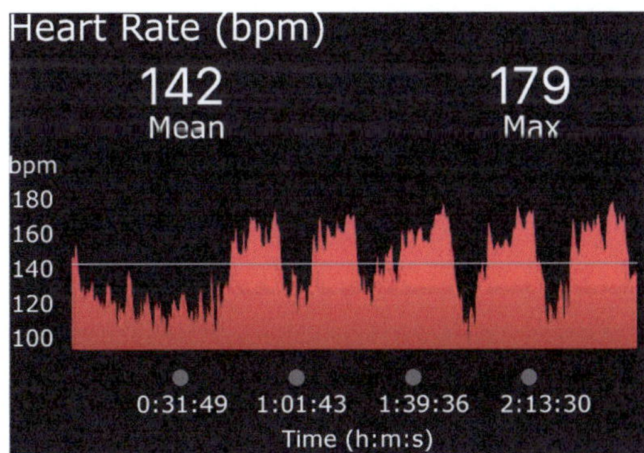

Fig. 29.3 Example of the electrocardiogram track of an RSX class female athlete during the four races of 1 day of the regatta

29.2 Training Strategies

29.2.1 Periodisation

In competitors, running training should be scheduled, taking into account loading and unloading periods and distinguishing the preparation phases from maintenance and implementation according to the scheduled objectives.

Out of season, a long and slow running training may be adequate to maintain the habit of fatigue and the aerobic capacity and to reach the optimal BMI whenever necessary.

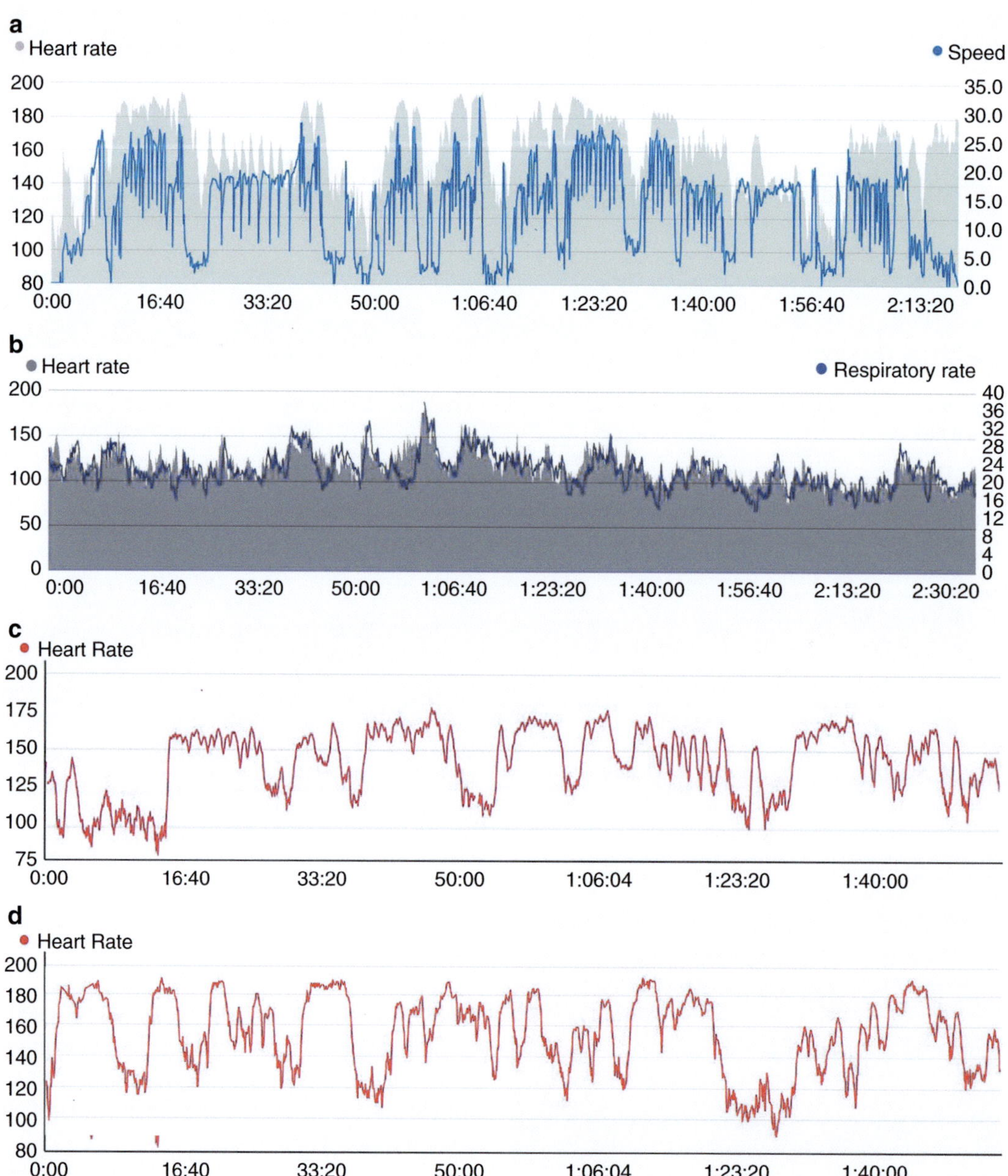

Fig. 29.4 The graphs show the heart rate trend of different crew roles in Olympic classes (**a**). 49er bowman during training (**b**). 49er helmsman during training (**c**). RSX athlete during training in weak wind conditions (mean wind speed <10 knots) (**d**). Laser Radial athlete during training with moderate wind (mean speed between 10 and 20 knots)

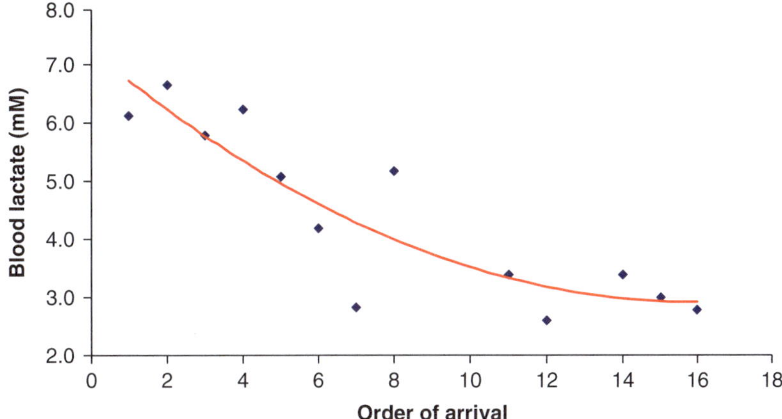

Fig. 29.5 Unpublished data collected and processed by Gianfelici A, Ghione P, De Pedrini L, Gallozzi C and Faina M as part of a scientific project aiming to define the physiological functional model of windsurfing. The study was conducted during national RSX sailboard competitions. It was a cooperation between the Italian Sailing Federation (FIV) and the Italian National Olympic Committee (CONI). Figure courtesy of Antonio Gianfelici

In the initial stages of training, the aim is to increase the aerobic capacity and medium-paced running workouts with an intensity higher than 80% of the VO2 max, and a speed about 10% slower than that of the anaerobic threshold can be introduced. Sessions of between 30 and 90 min with this speed accustom the athlete to a medium-high degree of fatigue [2].

Since a correlation has been described between the order of arrival in a regatta and the lactatemia (see Fig. 29.5), after the initial conditioning phase, running sessions at 97–100% of the anaerobic threshold lasting between 15 and 50 min should be introduced. Uphill running training may also be adequate, with the foresight, however, to avoid slopes that exceed the inclination of 8% so as not to perturb the mechanics of running in athletes not specialized in this activity [2].

Interval training can produce adaptations on the aerobic side by training in a purely anaerobic regime and may be an effective tool in a phase where the sailor must concentrate on work onboard.

When athletes need to limit changes in body weight during the season, shorter runs or repeated sprints interspersed with a full-recovery workout are preferred. Both of these strategies make it possible to minimize the impact on BMI, even with high volumes and intensities of work and significant lactic acid processes.

Finally, slow running (130–150 bpm; % lactate <2 mmol) outdoors is encouraged by sailing coaches as a physical warm-up on competition day because it allows the athlete to start monitoring the evolution of weather conditions and the intensity and direction of the wind, some hours before the competition.

Examples of running programs for participants involved in the Olympic class without specific BMI control needs are reported in Tables 29.2, 29.3, and 29.4.

Table 29.2 Running schemes

Running training	Description
Long (L)	50–60′
Medium (M)	40′
Repeated sprints (R)	8 × 400 m Rest until 130 bpm
Short-high intensity (S)	1 × 800 m 2 × 400 m 3 × 200 m 4 × 100 m Rest until 130 bpm

All training schemes involve 10′ of warming up

Table 29.3 Periodization

Classes	Number and type of sessions/week		
	October–November	December–February	March–July
Sailboards	3 M	2 M + 1R	1 M +2R +1S
Skiffs	1 L + 2 M	1 M + 3R	1 M + 2 S
Dinghies	3 M	1 M + 2R	1 M + 2R
Offshore	2 L + 1 M	1 L + 1 M + 1R	2 M + 1R

Table 29.4 Specific recommendations

1	Adoption of suitable clothing
2	Careful choice of footwear based on weight experience
3	Where the temperature permits, conclude the workout with stretching
4	Do not run R and S in two consecutive sessions

29.2.2 Training Strategies in Olympic Classes

The use of running must be adapted to the characteristics of each sailing class, the sailor's role in the crew and the regatta formats that affect the performance duration.

29.2.2.1 49er Class

In the 49er, the boat's acrobatic nature requiring a high degree of coordination between the two crew members makes it particularly likely for a handling error to result in a capsize.

Due to the vast sail area (up to 59.2 m²), capsizing may lead to supramaximal efforts for recovery that can quickly lead to exhaustion and can compromise the crew's performance for the rest of the racing day (Fig. 29.6) [11]. Therefore, preparatory running work aimed at raising the VO2max is recommended to ensure the maintenance of concentration and rapid physical recovery even after errors and unexpected situations.

Furthermore, in the 49er, both the helmsmen and the bowmen benefit from running to improve the reactive capacity of the lower limbs, particularly concerning the ankle joint. When sailors are standing on the edges of the terraces, they are called on to carry out small and continuous movements to adapt to the wave motion balance (Fig. 29.7). Therefore, in the 49er, sailors can take advantage of activities and gaits frequently used in running training such as leap run, back kick

Fig. 29.6 Explanation in the text. Photo captured during training before the Miami World Cup 2016. Team Bildstein/Hussl. Courtesy of Ivan Bulaja, Olympic team coach at the Austrian Sailing Federation

Fig. 29.7 Correct position of the 49er crew on terraces during navigation. Photo captured during the World Cup regatta, Palma de Mallorca 2013. Courtesy of Ivan Bulaja, Olympic team coach at the Austrian Sailing Federation

run, high skip, low skip, short uphill-repeated sprints and rhythmic runs and hops, to improve the reactivity and stability of the ankle.

29.2.2.2 RSX Sailboards

In RSX sailboards, pumping, an athletic gesture characterized by the manual grip on the boom and cyclical flexion and extension movements with the upper limbs to generate propulsion exploiting muscle energy, is allowed. Consequently, despite races generally not exceeding 40′, it becomes crucial to improve the aerobic capacity and increase the anaerobic threshold ([lactate-] = 4 mmol/L). These results can be achieved with running through medium and short/fast workouts.

The rowing machine is often used as a more specific exercise than running at muscular level; however, in the gesture of rowing, the contribution of the lactic acid mechanism is less than in running because of the different type of muscle contraction and the smaller muscle masses involved compared to running [2]. Therefore running is generally used in association with the rowing machine.

29.2.2.3 Laser, Finn and 470

For the helmsmen of Laser, Finn and 470, hiking in strong wind conditions represents the most significant physical effort [1]. It is a semi-isometric activity because the more properly isometric effort required by maintaining the position is interrupted by the small movements required to counteract wave motion [12]. Different training systems have been implemented to increase performance in hiking: running, cross-training, circuit training, cycling and lower limb strengthening [13, 14]. Starting from the London Olympic Games, some Laser sailors introduced cycling as a form of training in particular periods of the year, as it is more specific for the quadriceps muscle than running. However, targeted quadriceps workouts with squats, isometric leg extensions and simulators are probably much more efficient than cycling in training for the quadricep isometric effort. At the same time, running is undoubtedly a faster and more practical way to improve general endurance.

Even though there is still no consensus on which training activity is optimal for increasing hiking performance [13, 14], it is necessary to emphasize that economizing each manoeuvre's energy expenditure has become a fundamental prerequisite for achieving top-level performance in contemporary sailing.

Therefore, running endurance training for sailing must always be accompanied by work aiming at the economy of specific athletic gestures [15]. Specifically, one of the authors (AM) recently developed a simulation test using patented software and a simulator that measures the boat's righting capacity in the form of average data over 2 min [16]. The data obtained with this systematic approach will better clarify the role of running and other activities in specific training for hiking.

29.2.3 Training Strategies to Achieve the Ideal BMI

Running is also an easy and rapid tool, alongside the diet, in achieving the ideal BMI.

In all the Olympic classes, the boat's weight may be less than that of the crew, and the sailor's weight must satisfy a refined compromise between the boat's righting movement requirements and the need for lightness. For example, in the Laser Standard, the hull weight is 58.97 kg and with a sail area of 7.06 m^2, while the weight of the performing sailor is around 82–83 kg.

Similarly, in double crews, running is often used to calibrate each crew member's optimal body weight.

Specifically, to promote the oxidation of fatty acids, training with slow running over long distances must be introduced. The sessions last a minimum of 20/30 min with heart rates of 130–150 bpm and the maintenance of lactates below 2 millimoles [2].

29.2.4 Training Strategies in Rehabilitation

Running has a role in recovery after sailors' injuries.

For example, Hunt et al. reported how running was used to rehabilitate a 21-year-old Finn Class sailor after a conservatively treated MCL knee injury 2 months before a world-class and national ranking event [13]. Running was initially included to promote reconditioning; a progression treadmill up to 30′ was followed by sprint 75 m × 3 sessions up to 100% of maximum speed. After the third week, running training aiming to improve resistance was reintroduced.

In the final phases of rehabilitation, the inclusion of accelerations, lateral run and side shuffling to improve power, agility and knee stability is also effective [13].

29.2.5 Training Strategies in the America's Cup

The America's Cup is the most significant professional competition on the sailing scene regarding economic investments and human and technological resources [1].

Today it is disputed with AC75 boats equipped with hydrofoil technology, with a mast length of 26.50 m and managed by a crew of 11 sailors, who glide on the water, reaching speeds close to 50 knots [17]. Technological exasperation inevitably leads to the search for maximum physical performance, from which the role of running derives.

The America's Cup is a competition that requires extraordinary endurance from all the crew members, including helmspersons and tacticians, since sailors are engaged for 2–4 years in land-based activities and water sailing for 9–13 h a day [1, 18].

However, the grinders require a more significant aerobic performance, which was the object of sports-medical and physiological studies as it was considered among the critical success factors for the New Zealand crew during the 2017 edition [19, 20]. New engineering solutions adopted in the 2021 edition for the INEOS team UK require a constant number of turns/time unit at winches and lead to aerobic effort at a constant rate without searching for power peaks. Such a novelty led to the design of workouts for the grinders, which aimed at improving VO2max and functional threshold power (FTP), i.e. the power in watts that the subject can maintain for 1 h [19, 21].

As a result, specific tests and training protocols are required, including incremental tests and specific adaptations of the Coggan test at the upper limb cycle ergometer simulators that repro-

duce the grinders' natural position. Medium- and high-intensity training will be privileged, working in proximity to the FTP. Therefore, the grinders combine running with work on particular simulators that are programmed based on appropriate FTP and power range; training tests and protocols are codified.

29.2.6 Training Strategies in Offshore Sailing

Over recent decades, physical effort has significantly increased also in ocean racing.

While in the past, sailors managed variations in the wind or waves by varying the route and keeping the sails fixed; nowadays, the regulation of sails is constantly modified to take advantage of each slight environmental modification. The need for the crew's real-time intervention significantly increases the sailors' physical effort, and a high level of resistance is required both to optimize performance and warrant safety.

Specific training requires an initial period focused on aerobic endurance exercise. This phase aims to obtain an excellent VO2 max level; running can be introduced in a long and slow formula, with 1-h sessions.

Since the muscular effort required for manoeuvres can also be very intense—for operating the winches, for example – training at intensities closer to the anaerobic threshold must subsequently be included.

Besides running and specific muscle exercise, stress management plays a fundamental role when preparing for ocean regattas. Therefore, endurance training should be integrated with the preparatory works on circadian rhythms, nutrition and sleep, which are generally started weeks before the competition [1].

29.3 Prevention Strategies

Running is unlikely to cause sailors' injuries if it is conducted with good periodization and attention to individual characteristics. However, many sailors have never played more popular sports such as football, volleyball, basketball or athletics; therefore, many do not have a correct running technique. In the initial stages of training, the proper running cadence, foot strike and posture must be carefully considered and, if necessary, corrected.

Specifically, running may be contraindicated in the presence of pre-existing knee or ankle disorders that may have occurred due to a previous sailing activity not sustained by proper athletic training.

In particular, among dinghy sailors, lesions of the knee cartilage, even of such severity as to require surgical treatment, have been attributed to prolonged hiking sessions [1, 22]. Therefore, functional overload injuries to the knees and lower back must be prevented by improving posture and technique during water practice and modulating load progression in running training [1, 23].

Despite offering many advantages, running was reported as a cause of injuries also among the sailors participating in the America's Cup.

In a prospective study involving 35 crew members, during the 74 weeks of sailing and training for the preparation and participation in the America's Cup, interval running was the third cause of injury.

Overall, running caused 26 (12%) injuries resulting in 1.8 days of absence from sailing due to injury [24]. The monitoring of total workloads may be crucial when prescribing running to America's Cup sailors to manage and implement appropriate recovery strategies [24].

29.4 Conclusions

Sailors do not traditionally have a habit of running, running is not among the motor patterns of sailing, and training with running is not codified in this sport. However, sailing has changed profoundly in recent years, making it essential for participants to have adequate physical preparation with good general endurance and aerobic and lactacid capacity. Running has therefore entered the training programs of sailors at all levels, although special considerations may apply to individual sailing classes. Consequently, future research is needed to validate the effectiveness of running as part of training for sailing and to define how to administer running to the participants in each sailing classes and to the different phases of preparation. Finally, it may be crucial to obtain evidence-based data about the best way to integrate running with the other land-based and onboard training components and adopt appropriate recovery strategies to prevent overuse injuries.

Acknowledgements The photos (Figs. 29.6 and 29.7) used for the technical topics of this chapter are courtesy of Ivan Bulaja, coach of the Olympic team of the Austrian Sailing Federation.

Figure 29.5 and related data are courtesy of Dr. Antonio Gianfelici, Institute of Sports Medicine and Science of the Italian National Olympic Committee (CONI).

Contribution Statement FF conceived the paper, gathered data from literature and experts, received information from the co-author by interviewing him and wrote the first draft of the paper.

AM formulated the main conceptual ideas discussed in the paper based on his know-how resulting from his unique experience as an athletic trainer and sailor spanning over four decades (including four America's Cup campaigns, two World Championships and one Olympic campaign, Atlanta 1996, as a trainer in charge of all the sailing classes of the Italian national team).

Both authors discussed the results and contributed to the final manuscript.

References

1. Feletti F, Aliverti A. Extreme sailing medicine. In: Feletti F, editor. Extreme sports medicine, injuries and illnesses. Chapter 22. Springer; 2017. p. 275–87. https://doi.org/10.1007/978-3-319-28265-7_22.
2. La Torre A. Lo sviluppo della resistenza. In: La Torre A, editor. Allenare per Vincere – Metodologia dell'allenamento. Chapter 6. Scuola dello Sport Coni; Roma. 2018. p. 217–59.
3. Wright G, Clarke J, Niinimaa V, Shephard RJ. Some reactions to a dry-land training programme for dinghy sailors. Br J Sports Med. 1976;10(1):4–10. https://doi.org/10.1136/bjsm.10.1.4.

4. Allen JB, De Jong MR. Sailing and sports medicine: a literature review. Br J Sports Med. 2006;40(7):587–93. Published online 2006 Mar 17. https://doi.org/10.1136/bjsm.2002.001669.

5. Crafer S. Taking the strain. Yachts and yachting Feb 1995: 4–7.

6. Cunningham P. Olympic fit. Yachts and Yachting Oct 1999: 51–56.

7. Legg SJ, Mackie HW, Slyfield DA. Changes in physical characteristics and performance of elite sailors following introduction of a sport science programme prior to the 1996 Olympic games. Appl Human Sci. 1999;18:211–7.

8. Cunningham P. Fighting fit. Yachts and Yachting Oct 1999: 40–43.

9. Cunningham P. Let's get physical. RYA News Winter 1997: 56–57.

10. Dyson R, Buchanan M, Hale T. Incidence of sports injuries in elite competitive and recreational windsurfers. Br J Sports Med. 2006;40:346–50. https://doi.org/10.1136/bjsm.2005.023077.

11. World Sailing. Classes & Equipment. www.Sailing.org/An. Official Website of World Sailing. Accessed on March 2021, https://www.sailing.org/classesandequipment/olympic/index.php.

12. Vogiatzis I. Employment of near-infrared spectroscopy to assess the physiological determinants of hiking performance in single-handed dinghy sailors. Chapter 29. In: Feletti F, editor. Extreme sports medicine. Springer; 2017. p. 383–7.

13. Hunt SE, Herrera C, Cicerale S, Moses K, Smiley P. Rehabilitation of an elite olympic class sailor with MCL injury. N Am J Sports Phys Ther. 2009;4(3): 123–31.

14. Tan B, Aziz AR, Spurway NC, et al. Indicators of maximal hiking performance in laser sailors. Eur J Appl Physiol. 2006;98:169–76.

15. Santos-Concejero J, Oliván J, Maté-Muñoz JL, Muniesa C, Montil M, Tucker R, Lucia A. Gait-cycle characteristics and running economy in elite Eritrean and European runners. Int J Sports Physiol Perform. 2015;10(3):381–7. https://doi.org/10.1123/ijspp.2014-0179. Epub 2014 Oct 13

16. Harken Italy S.P.A., Merello A, Madaffari A. A training equipment for sailing athletes. 1998; Patent No.:WO/1998/015324 (A2).

17. America's Cup Event Limited. The official website of the America's Cup, AC75. 168 Beaumont Street, Auckland, (1010) New Zealand https://www.americascup.com/en/ac75.

18. Hadala M, Barrios C. Sports injuries in an America's cup yachting crew, a 4 -year epidemiological study covering the 2007 challenge. J Sports Sci. 2009;27(7):711–7.

19. Tognozzi M. America's Cup: a proposito dei grinder di BRITAnnia, il contributo di Andrea Madaffari. Fare Vela Jan 21, 2021. https://farevela.net/2021/01/19/americas-cup-a-proposito-dei-grinder-di-britannia-il-contributo-di-andrea-madaffari/

20. Bernardi M, Quattrini FM, Rodio A, Fontana G, Madaffari A, Brugnoli M, Marchetti M. Physiological characteristics of America's cup sailors. J Sports Sci. 2007;25(10):1141–52. https://doi.org/10.1080/02640410701287172.

21. Tognozzi M. America's Cup: il big match Luna Rossa vs Britannia, la conferenza stampa con Bruni, Spithill, Ainslie e Scott. Fare Vela Jan 21, 2021. https://farevela.net/2021/01/21/americas-cup-la-diretta-della-conferenza-stampa-ore-22-con-bruni-spithill-ainslie-e-scott/

22. Nathanson AT, Baird J, Mello M. Sailing injury and illness: results of an online survey. Wild Env Med. 2010;21:291–7.

23. Shephard RJ. Biology and medicine of sailing. An update. Sports Med. 1997;23:350–6.

24. Neville VJ, Molloy J, Brooks JHM, Speedy DB, Atkinson G. Epidemiology of injuries and illnesses in America's Cup yacht racing. Br J Sports Med. 2006;40(4):304–12. https://doi.org/10.1136/bjsm.2005.021477.

Running in Rowing

30

Antonio Spataro

30.1 Role and Characteristics of Running in Rowers

We used to say that running was not useful to the rower.

If we consider running just for its peripheral effects (e.g., on the lower limbs), this statement *could* also be justifiable, but if we look at its central effects, especially the cardiac ones, then we must affirm that running is very important in rowers' training.

Rowing is an endurance sport in which both aerobic and anaerobic (lactic acid) metabolisms are used, resulting in a high oxygen consumption ($VO2$) of about 7–8 L/min. A high VO2 max depends on the function of three different systems:

- The respiratory system capable of extracting from the inhaled air an high concentration of oxygen.
- The cardiac system capable of pumping high quantities of blood (and O_2) at each cardiac output.
- The muscle system capable of extracting oxygen from the peripheral circulation.

In this specific discipline, the main component is represented by the cardiovascular system.

More specifically, about 75% of the ability to transport oxygen to the muscles is given by the maximum cardiac output.

Cardiac output is determined by both stroke volume (SV) and heart rate (HR). In endurance sports, bradycardia is one of the main adaptations, so only the stroke volume can play a fundamental role. In fact, during the workloads, the SV increases from values of about 90–100 ml to 140–150 ml. Consequently, the cardiac output progressively rises from 5–6 L/min to over 18–25 L/min [1].

The consequence of these phenomena, determined by the continuous workload that rowers carry out every day, is represented by adaptations on the heart morphology.

In athletes, the heart morphology varies depending on the sport practiced. It can be thick or thin-walled, with small or large cavities and having different combinations of these characteristics.

The heart of athletes who practice aerobic sports (such as cycling, cross-country running, or cross-country skiing) has mild hypertrophic walls and large cavities (eccentric hypertrophy) (Fig. 30.1).

On the other hand, the heart of a rower, who does not exercise with running, has hypertrophic walls and relatively small cavities (concentric hypertrophy, type 1) (Fig. 30.2).

This is an intermediate entity between those athletes who practice running (phasic and cyclic

A. Spataro (✉)
Italian Rowing Federation, Rome, Italy
e-mail: antonio.spataro@coni.it

Fig. 30.1 Result of repeated cardiac adjustments and adaptations over time in endurance elite athletes characterized from mild hypertrophy walls and large cavities (eccentric hypertrophy)

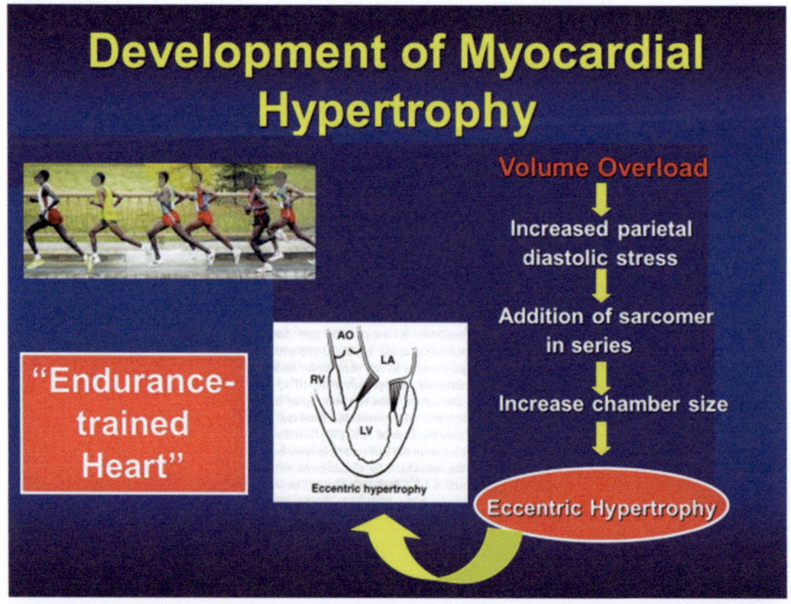

Fig. 30.2 Result of repeated cardiac adjustments and adaptations over time in strength elite athletes characterized from hypertrophy walls and small cavities (concentric hypertrophy)

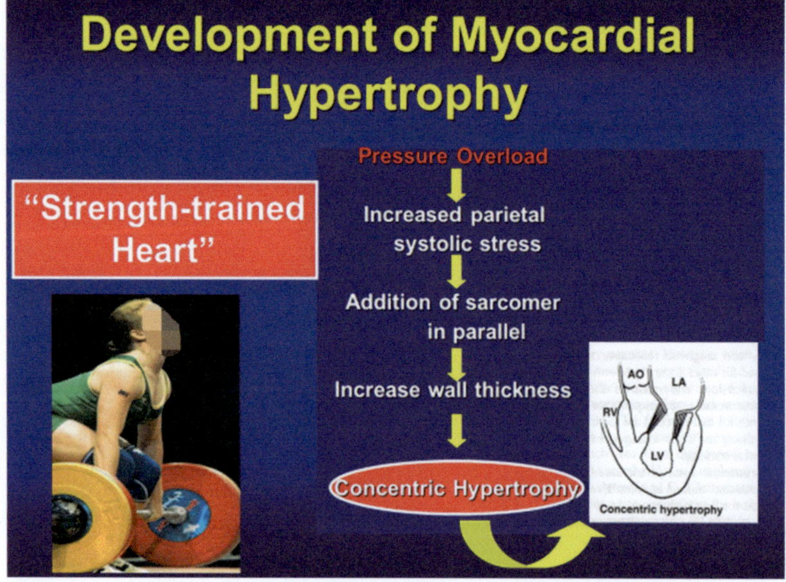

muscle action) and those athletes that practice isometric sports (such as powerlifting).

30.2 Training Strategies

These changes occur mainly during pubertal development, even if we can further modify the situation over time. Therefore, it is very important to have an optimal training program that takes care of heart development too (during 14–18 years).

The somatotropic hormone (STH or GH, growth hormone) acts by promoting the growth of both the skeletal system and internal organs. Therefore, since its secretion is maximum during puberty and immediately thereafter, appropriate stimuli multiply the effect on target organs.

A wrong training during prepubertal, pubertal, and postpubertal age (approximately 4 years) has repercussions on the entire career of the rower.

An interesting Italian study has reported increased LV wall thicknesses (13–15 mm) on elite male rowers undergoing a systematic pre-participation medical and physiological screening (that included an echocardiography) before their participation in Olympic events, associated with distinct cavity enlargement (end-diastolic transverse dimension, 55–65 mm) [2, 3]; each athlete had also normal LV systolic function (ejection fraction >50%) and diastolic filling pattern (Fig. 30.3).

The female rowing athletes showed larger LV dimension and greater wall thickness and mass than the control women [4] (Fig. 30.4).

The LV dimension ranged from 40 to 66 mm, exceeding the upper limit of normal range (54 mm) in 8% of the cases. In addiction, a marked dilatation (>60 mm) was found in 1% of the cases.

The LV wall thickness ranged from 6 to 12 mm, thus remaining within the normal limits in all the athletes studied.

The lesser hypertrophic response to training of women's heart is probably due to different factors, including both smaller body size and lean body mass and both lower blood pressure response to exercise and lower level of natural androgenic anabolic hormones.

One of the most important characteristics of physiological hypertrophy, confirming its adaptive nature to the athletic training, is the regression as a consequence of deconditioning.

For example, athletes with LV hypertrophy may show reduction in wall thickness (of about 2–5 mm) within 3 months of detraining (Fig. 30.5) [5].

Morphological heart adaptations seem to be a necessary element to reach high sport performance. Numerous observations show a positive correlation between heart size and aerobic power in elite athletes.

In a long-term study, a positive correlation of both ventricular size and wall hypertrophy versus max aerobic power has been found [6]. The idea was that max aerobic power is determined for about 70–80% by heart size. The limiting factor may be the capability of the heart to deliver oxygenated blood to the exercising muscles. Indirect support to this idea are the results gained by athletes during long periods of training.

In Olympic rowers the maximum performance corresponds to the peak of aerobic capacity and maximal heart adaptation.

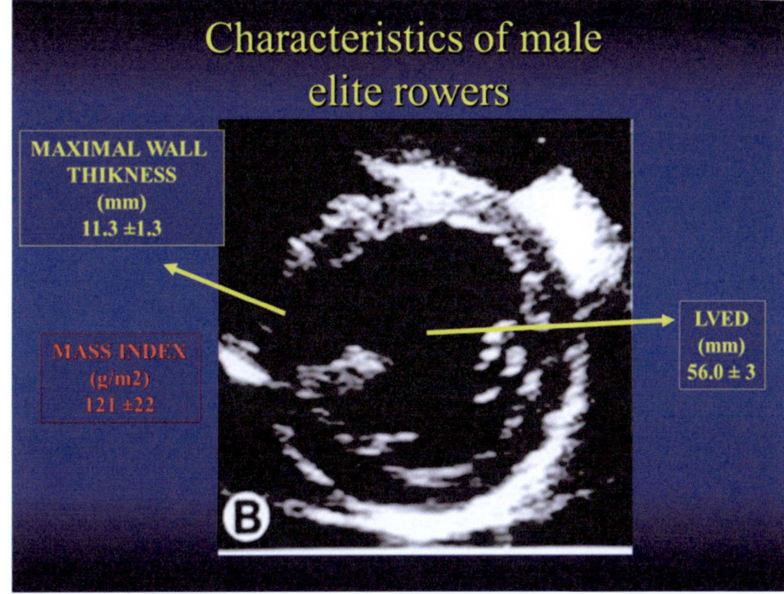

Fig. 30.3 Stop-frame two-dimensional echocardiograms obtained during diastole from a 28-year-old male elite rower. In the short axis view, it is visible a mild thickening of the posterior ventricular septum (maximal wall thickness), an enlarged left ventricular cavity (LVED), and high LV mass index

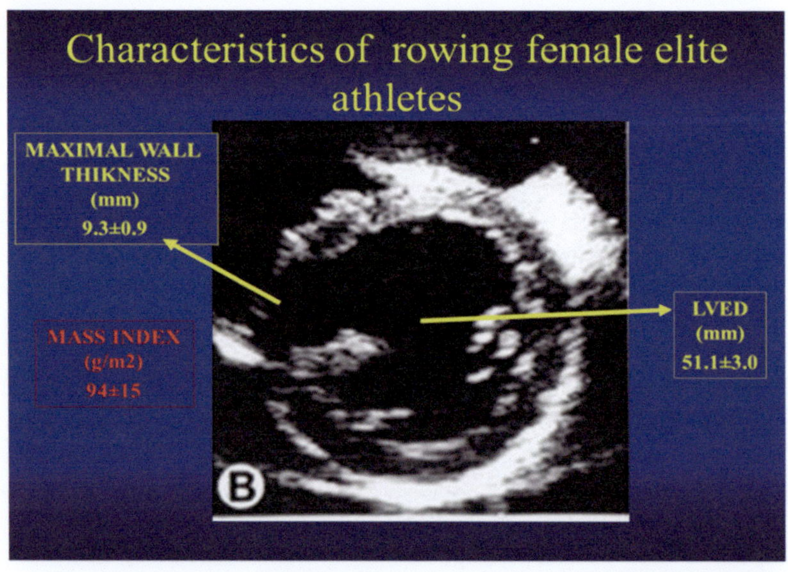

Fig. 30.4 Stop-frame two-dimensional echocardiogram obtained during diastole from a 24-year-old female elite rower. In a short axis view at the papillary muscle level, a mild enlargement of the left ventricular cavity (LVED) is evident. Although, no increase of the thickness can be seen on the left ventricular wall

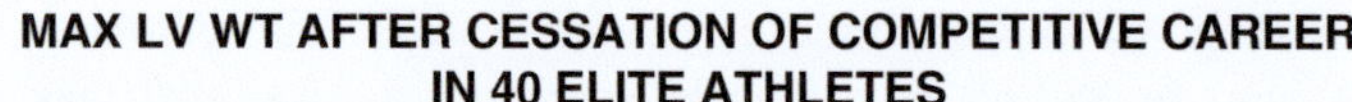

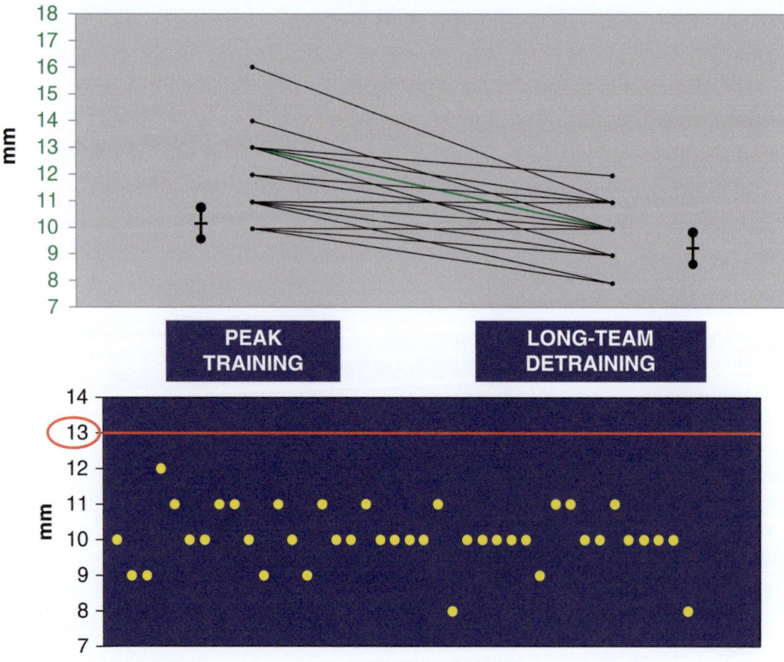

Fig. 30.5 Maximum LV wall thicknesses at peak training and after long-term detraining shown individually in the 40 elite athletes that had been formerly engaged in rowing (n_29)

30.3 Prevention Strategies

An adequate training and technical running skills are important.

The most common musculoskeletal pathology in rowing athletes is low back pain. This disease is caused by the functional overload of the muscular, tendon, and bone structures of the vertebral column, due to the specific technical gesture of the athletes, capable of vigorously soliciting an hyperextension (of the column) associated with an important stretching and contraction of the sacroiliac muscles [7].

Less common is the Achilles tendinopathy, an overuse pathology that afflicts usually amateur runners, but it can also affect other rowers who include running in their athletic training program. Its pathogenesis is caused by repetitive microtraumas to this structure, which is mechanically stressed by the repetitive athletic gesture, in a context of insufficient tissue repair capacity. Intrinsic risk factors such as a high BMI, vascularization, age, abnormal tendon structure, and excessive pronation of the foot and extrinsic factors, such as the use of unsuitable shoes for running on slopes, can facilitate the disease onset.

Ankle sprain is another common running trauma. Acute sprain is most often due to marked plantar flexion with an excessive foot inversion, an event that can occur on rough and running surfaces, favored by adverse weather conditions. Other risk factors are ankle instability, history of previous sprains, high BMI, and running on unsuitable surfaces. The most frequently damaged anatomic structure are the anterior talofibular ligament (Fig. 30.6) and the calcaneofibular ligament.

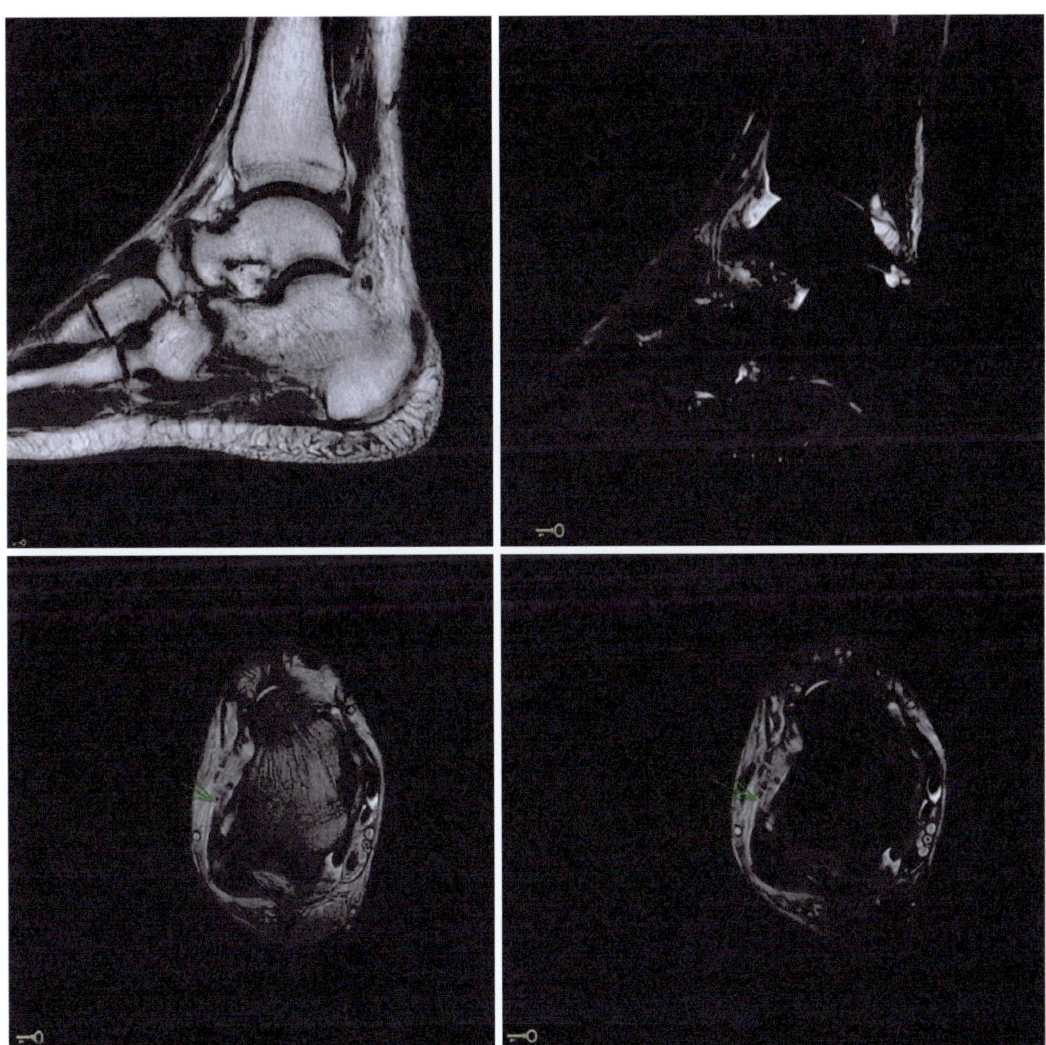

Fig. 30.6 RM 1.5 Tesla right ankle of a 26-year-old elite rower with sprain trauma while running. Laceration of the antero-external joint capsule with high degree (III degree) of distraction of the anterior talofibular ligament (arrows) due to the recent spraining trauma in inversion. It is associated with abundant ankle joint effusion and a widespread edematous-hemorrhagic imbibition of the ankle subcutaneous fat, especially in the external peri-malleolar area and of the adipose triangle of Kager

In conclusion, the correction of predisposing factors in the rower who adds a running training to his workout may reduce the risk of running injuries. Furthermore, a careful physical and instrumental examination (X-ray, MRI) is essential to reach an accurate diagnosis in order to begin a prompt rehabilitation treatment.

In addition to normal physical and pharmacological therapies, a multidisciplinary approach such as osteopathic treatment, manual therapy, stretching, postural reeducation, ozone therapy, and the use of customized orthoses are important preventive measures.

References

1. Rosiello RA, Mahler DA, Ward JL. Cardiovascular responses to rowing. Med Sci Sports Exerc. 1987;19(3):239–45.
2. Pelliccia A, Di Paolo FM, Corrado D, Buccolieri C, Quattrini FM, Pisicchio C, Spataro A, et al. Evidence for efficacy of the Italian national pre-participation screening programme for identification of hypertrophic cardiomyopathy in competitive athletes. Eur Heart J. 2006;27(18):2196–200. https://doi.org/10.1093/eurheartj/ehl137.
3. Pelliccia A, Maron BJ, Spataro A, Proschan MA, Spirito P. The upper limit of physiologic cardiac hypertrophy in highly trained elite athletes. N Engl J Med. 1991;324:295–301.
4. Pelliccia A, Maron BJ, Culasso F, Spataro A, Caselli G. Athlet's heart in women Jama. 1996;276:211–5.
5. Pelliccia A, Maron BJ, De Luca R, Di Paolo FM, Spataro A, Culasso F. Remodeling of left ventricular hypertrophy in elite athletes after long-term deconditioning. Circulation. 2002;105(8):944–9. https://doi.org/10.1161/hc0802.104534.
6. Wieling W, Borgholds EAM, Hollander AR, Danner SA, Dunning AJ. Echocardiographic dimension and maximal uptake in oarsmen during training. Br Heart J. 1981;46(2):190–5. https://doi.org/10.1136/hrt.46.2.190.
7. Hickey GJ, Mc FPA, Donald WA. Injuries to elite rowers over a 10-yr period. Med Sci Sports Exerc. 1997;29(12):1567–72. https://doi.org/10.1097/00005768-199712000-00004.

Andrey Korolev, Nina Magnitskaya,
Mikhail Ryazantsev, Alexey Logvinov,
Zhanna Pilipson, and Dmitriy Ilyin

31.1 Role and Characteristics of Running in Space

It is 60 years now since the beginning of the space era, starting with the historical flight of Yuri Gagarin in 1960 [1]. Space industry is developing rapidly; every year the number of cosmonauts and space tourists is increasing. Private spaceflights were established, launching astronauts to the International Space Station (ISS); personal spaceflights around the Moon and Mission to Mars are being planned. Despite the considerable knowledge gained in space physiology, many questions regarding the response of human body to microgravity still need to be answered.

On Earth gravity creates constant longitudinal mechanical load to musculoskeletal system. The shape and structure of the earthly animal bones and muscles are supported on micro- and macro levels (cells, minerals, etc.) to resist the gravity. During the spaceflight, these mechanisms interrupt. It has been shown that a person's stay in microgravity is associated with reductions in bone mass, muscle volume and strength, and cardiovascular capacity and changes to blood pressure regulation and vestibular and sensorimotor function [2–5].

Maintaining the integrity of the musculoskeletal system during spaceflight is essential to astronaut health and successful mission accomplishment.

Living in weightless conditions leads to high level of bone resorption and low level of bone formation due to reduced mechanical loading, altered calcium homeostasis and metabolism According to Smith, during the long-term flights, calcium absorption decreases up to 50%, urinary calcium excretion increases up to 50%, and bone resorption increases over 50% [6]. Mineral loss wouldn't be critical if it was uniform—the so-called phenomenon of skeleton mineral redistribution [7]. This phenomenon leads to the formation of critical points of bone mass loss (pelvis, lumbar spine, hip) and increase of the mineral amount in the upper body (Fig. 31.1) [8]. There is an assumption that mineral transfer to the upper body occurs due to the "phenomenon of cranial body fluids redistribution" because of interrupted hemodynamics mechanisms, which evolutionarily formed for 1 g gravity [7].

A. Korolev (✉) · D. Ilyin
European Clinic of Sports Traumatology and Orthopaedics (ECSTO), European Medical Center, Moscow, Russia

Department of Traumatology and Orthopedics, Medical Institution, Peoples Friendship University of Russia (PFUR), Moscow, Russia
e-mail: akorolev@emcmos.ru

N. Magnitskaya · M. Ryazantsev · A. Logvinov
Z. Pilipson
European Clinic of Sports Traumatology and Orthopaedics (ECSTO), European Medical Center, Moscow, Russia

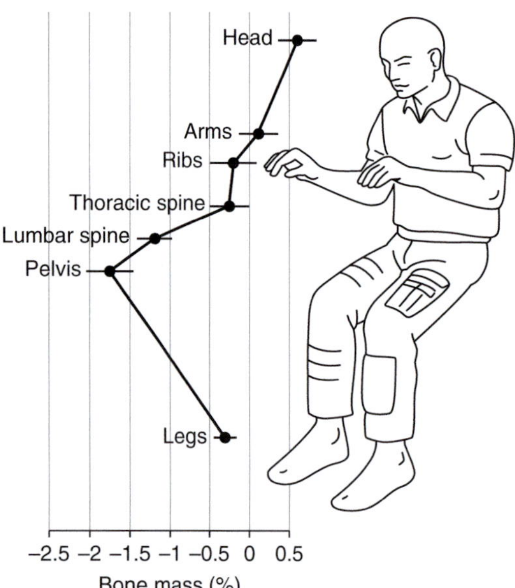

Fig. 31.1 Bone mass changing during long-term flight (Clément G, Hamilton D, Davenport L, Comet B (2010) Medical survey of European astronauts during MIR missions. Advances in Space Research 46: 831–839)

According to 11 years of research carried out at the "MIR" orbital station, it was proved that there is no large loss of calcium—100 mg per day or 0.25–0.3% per month from its total amount in the human body [1] during the 6-month flight. However, its much faster than on the Earth. Interestingly, according to the Stavnichuk meta-analysis, long-duration missions reported less bone loss than intermediate missions, suggesting that microgravity-induced bone loss may gradually decrease (or decrease in course of time) [9]. Lang reported a difference in trabecular and cortical bone loss; according to his work, hip mass mineral loss over 90% was from the cortical bone with a rate of 1.5–1.7% per month [10]. Mineral transfer and as a consequence bone loss lead to the increase in calcium concentration resulting in renal stone formation and due to bone marrow changes—decreasing blood cells [11].

The loss of skeletal muscle mass during the spaceflight has been a medical concern since the early Gemini, Soyuz, and Skylab missions [12]. Postural muscles (core stability muscles) that support standing and walking are more affected than non-postural muscles [13–15]. Significant atrophy was shown as soon as 5 days of spaceflight;

Fig. 31.2 Cosmonaut Oleg Kotov, Treadmill BD-2, in-flight program

extensors are generally more affected than flexors. Muscle loss varied significantly between cosmonauts, but the maximum loss reached up to 10%. The decrease of quadriceps and triceps surae by 6% was shown after 8 days of spaceflight, increasing up to 15.9% after 9–16 days [16, 17]. Studies of MIR cosmonauts after 6 months of microgravity have indicated declines in the calf plantar flexors varying from 6 to 20% [18]. Not only loss of fiber mass but also biomechanical and structural changes have been confirmed with muscle biopsies collected after spaceflights [19, 20].

Taking into account all the abovementioned risks, specific physical training plan should be drawn as a countermeasure to musculoskeletal and cardiovascular deconditioning in zero-gravity environment. The program includes pre-flight (ground) training, in-flight program, and postflight recovery (Fig. 31.2) [21].

Training program includes aerobic exercise, strengthening, stretching, and neuromuscular electric stimulation. Before starting to run on the treadmill, crewmembers must strap themselves with bungee cords to replicate the moment of landing and to keep from floating off. NASA astronaut Sunita Williams was the first astronaut who ran a Boston Marathon in space on a treadmill in 2007.

31.2 Training Strategies

Currently, crewmembers are prescribed to train 2.5 h a day 6–7 days a week with one strengthening session and one aerobic session a day, suggested to be separated [22]. The current training strategies during human spaceflight involve concurrent strength and aerobic exercises, with strength training recommended to comprise 54% of the total training volume for European Space Agency (ESA) crews [23]. Aerobic exercises are required 6 days a week for about 60 min each day and can be performed on either CEVIS (cycle ergometer with vibration isolation system) (Fig. 31.3) or T2 (treadmill) [24].

Fig. 31.3 Cosmonaut Anatoly Ivanishin exercises on CEVIS

T2 or a new generation of treadmill, the so-called COLBERT, was installed in 2010 on ISS to replace a previous model. It is connected to a vibration isolation system as well as CEVIS.

Exercise program including T2, ARED (advanced resistive exercise device), CEVIS, and training is part of a fundamental long-duration spaceflight strategy to decrease bone mineral density loss [25].

31.3 Prevention Strategies

The guidelines of ground and space training have not been defined in detail because of individual astronauts' source data and plan. However, Astronaut Strength, Conditioning, and Rehabilitation (ASCR) specialists prescribe individual crew members' daily schedule based on their previous microgravity training experience and personal fitness parameters (ground ARED training sessions and peak cycle test results performed 2–3 months before the flight). However, the majority of preflight exercises are performed on traditional on-earth equipment, because ISS training devices are designed for microgravity conditions [24].

Running in space is currently the main way to reduce the effects of microgravity on the human body and maintain the musculoskeletal system at a safe level. Future research is needed for a complete and detailed understanding of all the changes and solving the issue of preventing during longer and more distant spaceflights.

References

1. Ostroumov G. An interview with Yuri Gagarin. Sov Rev. 1961;2(5):47–52.
2. Tanaka K, Nishimura N, Kawai Y. Adaptation to microgravity, deconditioning, and countermeasures. J Physiol Sci. 2017;67(2):271–81.
3. Schneider V, Oganov V, LeBlanc A, Rakmonov A, Taggart L, Bakulin A, и др. Bone and body mass changes during space flight. Acta Astronaut. 1995;36(8–12):463–6.
4. Narici MV, De Boer MD. Disuse of the musculoskeletal system in space and on earth. Eur J Appl Physiol. 2011;111(3):403–20.

5. Iwase S, Nishimura N, Tanaka K, Mano T. Effects of microgravity on human physiology. B: Beyond LEO-human health issues for deep space exploration. IntechOpen; 2020.

6. Smith SM, Wastney ME, Morukov BV, Larina IM, Nyquist LE, Abrams SA, и др. Calcium metabolism before, during, and after a 3-mo spaceflight: kinetic and biochemical changes. Am J Physiol-Regul Integr Comp Physiol. 1999;277(1):R1–10.

7. Oganov VS, Bakulin AV, Novikov VE, Murashko LM, Kabitskaia OE, Morgun VV, и др. Reactions of the human bone system in space flight: phenomenology. Aviakosmicheskaia Ekol Meditsina Aerosp. Environ Med. 2005;39(6):3–9.

8. Clément G. Fundamentals of space medicine. T. 23. New York: Springer; 2011.

9. Mariya S, Nicholas M, Tatsuya C, Morris M, Komarova SV. A systematic review and meta-analysis of bone loss in space travelers. NPJ Microgravity. 2020;6(1).

10. Lang T, LeBlanc A, Evans H, Lu Y, Genant H, Yu A. Cortical and trabecular bone mineral loss from the spine and hip in long-duration spaceflight. J Bone Miner Res. 2004;19(6):1006–12.

11. Smith SM, Zwart SR, Heer M, Hudson EK, Shackelford L, Morgan JL. Men and women in space: bone loss and kidney stone risk after long-duration spaceflight. J Bone Miner Res. 2014;29(7):1639–45.

12. Gazenko OG, Schulzhenko EB, Grigoriev AI, Atkov OY, Egorov AD. Review of basic medical results of the Salyut-7-Soyuz-T 8-month manned flight. Acta Astronaut. 1988;17(2):155–60.

13. Gardetto PR, Schluter JM, Fitts RH. Contractile function of single muscle fibers after hindlimb suspension. J Appl Physiol. 1989;66(6):2739–49.

14. Ohira Y, Jiang B, Roy RR, Oganov V, Ilyina-Kakueva E, Marini JF, и др. Rat soleus muscle fiber responses to 14 days of spaceflight and hindlimb suspension. J Appl Physiol. 1992;73(2):S51–7.

15. Roy RR, Bello MA, Bouissou P, Edgerton VR. Size and metabolic properties of fibers in rat fast-twitch muscles after hindlimb suspension. J Appl Physiol. 1987;62(6):2348–57.

16. LeBlanc A, Rowe R, Schneider V, Evans H, Hedrick T. Regional muscle loss after short duration spaceflight. Aviat Space Environ Med. 1995;66(12):1151–4.

17. Akima H, Kawakami Y, Kubo K, Sekiguchi C, Ohshima H, Miyamoto A, и др. Effect of short-duration spaceflight on thigh and leg muscle volume. Med Sci Sports Exerc. 2000;32(10):1743–7.

18. Zange J, Müller K, Schuber M, Wackerhage H, Hoffmann U, Günther RW, и др. Changes in calf muscle performance, energy metabolism, and muscle volume caused by long-term stay on space station MIR. Int J Sports Med. 1997;18:S308–9.

19. Fitts RH, Trappe SW, Costill DL, Gallagher PM, Creer AC, Colloton PA, и др. Prolonged space flight-induced alterations in the structure and function of human skeletal muscle fibres. J Physiol. 2010;588(18):3567–92.

20. Trappe S, Costill D, Gallagher P, Creer A, Peters JR, Evans H, и др. Exercise in space: human skeletal muscle after 6 months aboard the international Space Station. J Appl Physiol. 2009;106(4):1159–68.

21. Kozlovskaya IB, Yarmanova EN, Yegorov AD, Stepantsov VI, Fomina EV, Tomilovaskaya ES. Russian countermeasure systems for adverse effects of microgravity on long-duration ISS flights. Aerosp Med Hum Perform. 2015;86(12):A24–31.

22. Loehr JA, Guilliams ME, Petersen N, Hirsch N, Kawashima S, Ohshima H. Physical training for long-duration spaceflight. Aerosp Med Hum Perform. 2015;86(12):A14–23.

23. Petersen N, Jaekel P, Rosenberger A, Weber T, Scott J, Castrucci F, и др. Exercise in space: the European Space Agency approach to in-flight exercise countermeasures for long-duration missions on ISS. Extreme Physiol Med. 2016;5(1):1–13.

24. Hackney KJ, Scott JM, Hanson AM, English KL, Downs ME, Ploutz-Snyder LL. The astronaut-athlete: optimizing human performance in space. J Strength Cond Res. 2015;29(12):3531–45.

25. Swaffield TP, Neviaser AS, Lehnhardt K. Fracture risk in spaceflight and potential treatment options. Aerosp Med Hum Perform. 2018;89(12):1060–7.